WITHDRAWN

D1313858

Residual Deposits:
Surface Related Weathering Processes and Materials

551,3
R311w

Residual Deposits:
Surface Related Weathering
Processes and Materials

edited by

R. C. L. Wilson

Department of Earth Sciences, The Open University,
Milton Keynes

1983

Published for
The Geological Society of London,
by Blackwell Scientific Publications
Oxford London Edinburgh
Boston Melbourne

Published by

Blackwell Scientific Publications
Osney Mead, Oxford OX2 0EL
8 John Street, London WCIN 2ES
9 Forrest Road, Edinburgh EH1 2QH
52 Beacon Street, Boston, Massachusetts 02108, USA
99 Barry Street, Carlton, Victoria 3053, Australia

First published 1983

© 1983 The Geological Society. Authorization to
photocopy items for internal or personal use, or the
internal or personal use of specific clients, is granted by
The Geological Society for libraries and other users
registered with the Copyright Clearance Center (CCC)
Transactional Reporting Service, provided that a base
fee of $02.00 per copy is paid directly to CCC, 21
Congress Street, Salem, MA 01970, U.S.A.
0305-8719/83 $02.00.

DISTRIBUTORS

USA
  Blackwell-Mosby Book Distributors
  11830 Westline Industrial Drive
  St Louis, Missouri 63141
Canada
  Blackwell-Mosby Book Distributors
  120 Melford Drive, Scarborough
  Ontario M1B 2X4
Australia
  Blackwell Scientific Book Distributors
  31 Advantage Road, Highett
  Victoria 3190

British Library Cataloguing in Publication Data

Wilson, R.C.L.
  Residual deposits: surface related weathering
  processes and materials.—(Geological Society
  special publication, ISSN 0305-8719; no. 11)
  1. Weathering
  I. Title   II. Series
  551.3'02   QE570
  ISBN 0-632-01072-X

Printed in Great Britain by
Butler & Tanner Ltd, Frome and London

Contents

CAT May 13 '85

84-6247

Preface

This volume contains papers presented at the Geological Society on March 25th and 26th 1981, plus three additional contributions by researchers who were unable to be present at the meeting. The meeting brought together earth scientists with interests in geomorphology, geochemistry, pedology, sedimentology and applied geology. The multidisciplinary approach to the study of residual deposits is reflected in the 25 chapters of this book, which are arranged in four main groups:

Weathering processes (chapters 1–3);
Kaolinites, laterites and bauxites
   (chapters 4–11);
Red beds (chapters 12–14);
Duricrusts: calcretes, silcretes and gypcretes
   (chapters 15–25).

The last two chapters of the book deal with karst related fluorite–baryte deposits, and Cenozoic pedogenesis and landform development in south-east England.

Richard Crockett, Andrew Goudie and Don Highley provided invaluable suggestions during the planning of the meeting that led to this book.

R. C. L. WILSON, Department of Earth Sciences,
   The Open University, Walton Hall, Milton Keynes MK7 6AA.

WEATHERING PROCESSES

Lichen weathering of minerals: implications for pedogenesis

M. J. Wilson & D. Jones

SUMMARY: The weathering of primary rock-forming minerals beneath crustose lichens is due principally to the activity of the fungal component in excreting extracellular organic acids, particularly oxalic acid. This results in extensive surface etching of the minerals, the conversion of some minerals to siliceous relics, the formation of poorly ordered weathering products and the crystallization of oxalates at the rock/lichen interface and within the lichen thallus. Interpretation of these features is facilitated by the fact that they can be simulated experimentally. The relevance of lichen weathering to the pedogenic weathering predominant in soils under a cool temperature climate is discussed.

Pedogenic weathering of rock-forming minerals can occur in two fundamentally different ways (Duchaufour 1979). The first type of weathering takes place under the influence of water more or less charged with carbon dioxide and in the virtual absence of organic matter. Such weathering has been described as 'geochemical' and is characterized by the rapid crystallization of clay minerals like kaolinite and montmorillonite and of simple oxides and hydroxides like hematite and gibbsite. Geochemical weathering is best developed in the soils and weathered rocks of humid tropical regions and both decomposition of primary minerals and crystallization of clay minerals can be represented readily in terms of equilibrium reactions and stability diagrams. The second type of weathering is strongly influenced by the presence of organic matter and living organisms and is termed 'biochemical' weathering. Here, crystallization processes are extremely slow and the characteristic weathering products occur in the form of non-crystalline or poorly ordered materials and organo-mineral complexes. Biochemical weathering predominates in the soils of cool temperature climates and cannot be represented readily by calculated stability diagrams, because mineral decomposition is largely controlled by the action of organic acids and may involve the effects of straight-forward acidity, chelation or a combination of both. The effectiveness of organic acids in the dissolution of rock forming minerals has been illustrated experimentally by Huang & Keller (1970) and Huang & Kiang (1972).

In soils, organic acids can be produced either directly by living organisms or indirectly following the decomposition of organic matter. It is probable that the latter process predominates, although it is, in fact, very difficult to make any kind of assessment of the effects of direct biological weathering on primary minerals in soil. This is because the soil may involve many different types of minerals and species of micro-organisms leading to an extremely complex and heterogeneous environment. In addition, this environment may have undergone radical changes with time leading to a situation where it is impossible to assess the influence of a single type of organism on the weathering of a particular kind of mineral. However, such a direct relationship can be easily observed in lichen-encrusted rocks, which thus provide a convenient natural model for the investigation of biological mineral weathering.

Lichens as weathering agents

The lichen symbiosis usually involves only two organisms—the alga and the fungus—the fungal component (mycobiont) being dominant. In crustose lichens, it is generally found that the fungus is in close and intimate contact with the underlying substrate and that the algal cells are buried in the upper layers of the lichen thallus. The ability of lichens to weather rocks and minerals is, therefore, essentially due to the activity of the mycobiont and in particular to the excretion of organic acids during growth. Many crustose lichens generate oxalic acid in abundance as well as a variety of weak phenolic acids (known as lichen acids) which can react with minerals to form metal complexes. The weathering ability of lichens has not always been accepted unreservedly, but can no longer be doubted since scanning electron microscopy and electron microprobe analysis have made it possible to study the rock/lichen interface in much more detail than could be done with the optical microscope and ordinary chemical techniques. Weathering effects observed include marked etching of the rock-forming minerals, the conversion of some minerals to siliceous relics, the precipitation of poorly ordered aluminous and ferruginous material on the rock surface beneath the lichen and the occurrence of crystalline organic salts in the lichen thallus.

The interpretation of these features as weathering effects is facilitated by the fact that they can be readily simulated experimentally (Jones *et al.* 1980; Wilson *et al.* 1981). In this paper, examples of the weathering effects of lichens on various primary minerals are given and the implications for the characterization of 'biochemical' weathering processes in soil formation are discussed.

Mineral etching

This is the most spectacular manifestation of weathering at the rock/lichen interface. Etch marks form through the preferential dissolution of the mineral surface at particular sites and it is widely agreed that these sites mark the emergence of some kind of structural dislocation. Generally such dislocations are considered to be areas of high strain energy characterized by a distorted lattice. Therefore they tend to be attacked much more easily during the weathering process. The exact nature of the dislocations is a specialized study in its own right but certain features, frequently observed in mineral grains beneath encrusting lichens, can be interpreted readily. In the feldspar minerals, for example, the etching out of various kinds of lamellar intergrowths is often observed. These intergrowths usually have a chemical composition and structure somewhat different from that of the adjoining phase and, at the interface of the lamellae, this can result in lattice distortion which is accommodated by dislocations. Hydrothermal etching of matching cleavage faces of a microcline perthite showed the reality of these dislocations (Wilson & McHardy 1980) leading to the perthitic lamellae being selectively 'weathered' out. Fig. 1(a) shows the occurrence of etched out perthitic lamellae in the surface of a potassium feldspar directly beneath a growing lichen. Plagioclase feldspars frequently reveal similar features, the etched out lamellae probably representing Boggild and Huttenlocher intergrowths in feldspars of labradorite and bytownite/anorthite compositions, respectively (Fig. 1b and c). Lichen weathering can also emphasize the lamellar microstructure of other minerals. Fig. 1(d) shows a clinopyroxene grain (from the surface of a lichen-encrusted gabbro)

in which two intersecting sets of lamellae have been etched out into a most distinctive grid iron pattern. These lamellae probably represent exsolved intergrowths of orthopyroxene and pigeonite which are common in augite, although, again, a special study would be required to prove that this was the case.

It might be anticipated that etching would also be found along twin planes. Although this is sometimes observed it does not seem to occur very frequently with albite-type twins in plagioclase feldspar. For example, Fig. 1(e) shows complete etching out of intergrowths in a weathered anorthitic plagioclase feldspar but little or no etching along the albite twin planes. This is consistent with the observation (Smith 1974) that albite twinning involves no major structural misfit in the aluminosilicate framework, but merely a change in bond angles. However, this is not necessarily the case with pericline twinning.

Mineral surfaces beneath growing lichens are sometimes covered with geometrically shaped etch pits which may or may not be related to some obvious structural feature. Thus, square etch pits are observed in plagioclase feldspar, both randomly (Fig. 1f) and linearly distributed (Fig. 2a). Heavy pitting of potassium feldspar can also occur (Fig. 2b) eventually leading to an extremely rugged surface (Fig. 2c). Deep corrosion is particularly characteristic of olivine (Fig. 2d), the rounded edges and corners giving an impression of easy solubility, in keeping with the known susceptibility of this mineral to weathering.

Clearly, therefore, the early stages of the weathering of primary minerals by lichens essentially involve solution at particularly vulnerable sites from which the constituent elements are completely removed. Usually, there is no indication of the formation of a residual layer in keeping with recent evidence that silicate dissolution is controlled through surface reactions rather than diffusion through a leached layer (Berner 1978; Dibble & Tiller 1981). Continued etching may result in the thorough honeycombing of mineral grains, rendering them extremely fragile and ultimately leading to their disintegration into finer particles. Some minerals, however, become almost

---

Fig. 1. Features observed by SEM beneath crustose lichens. (a) Etched out perthitic lamellae in K-feldspar in granite, Bennachie. (b) Etched out lamellar (Boggild ?) intergrowths on labradorite on gabbro, Pitcaple. (c) Etched out lamellar (Huttenlocher ?) intergrowths in anorthitic plagioclase feldspar in gabbro boulder, Portsoy. (d) Etched out cross-cutting lamellar intergrowth in clinopyroxene in gabbro boulder, Portsoy. (e) Lamellar intergrowths in anorthitic plagioclase feldspar in gabbro boulder, Portsoy. Note relatively unetched albite twin plane running NNE–SSW across micrograph. (f) Etch pits in anorthitic plagioclase feldspar in gabbro boulder, Portsoy.

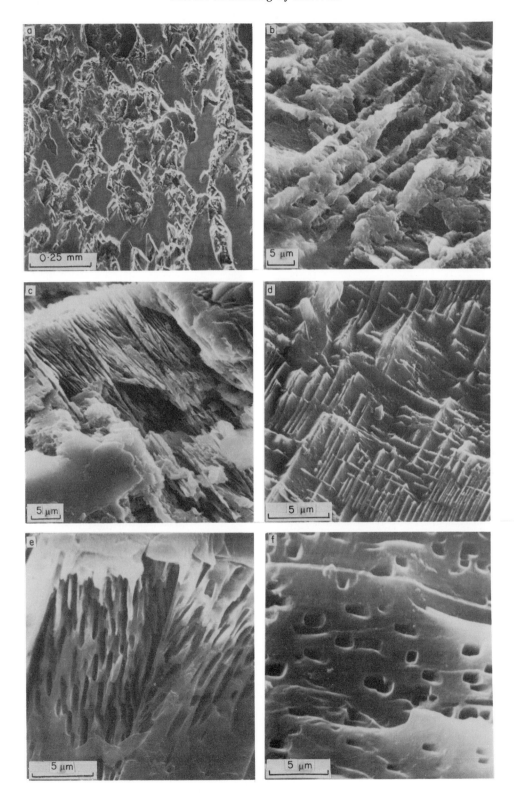

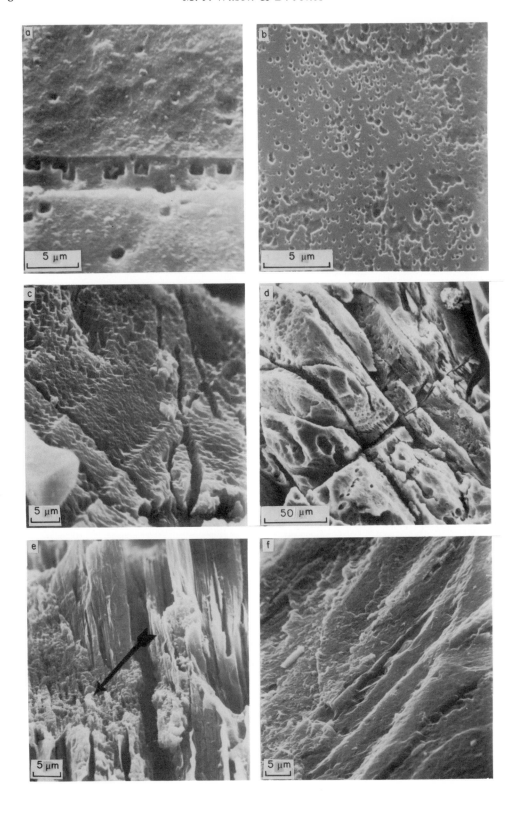

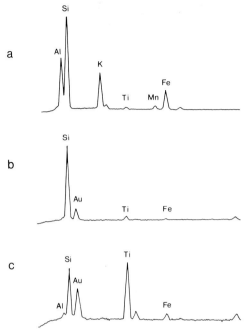

FIG. 3. Electron microprobe spectra of tri-octahedral mica (zinnwaldite ?) from Bennachie granite (a) fresh (b) directly beneath crustose lichen and (c), area arrowed in Fig. 2(e).

totally depleted of all constituents other than silica. This was found in a study of lichen weathering of serpentinite, where the dominant chrysotile mineral was entirely converted to fibrous silica following the removal of structural magnesium (Wilson *et al.* 1981). In this case a residual layer of silica was detected. Similarly, the weathering of trioctahedral mica (probably zinnwaldite) in a lichenized granite was found to involve the depletion of magnesium, iron, potassium and aluminium from the mineral, leaving a siliceous relic with concentrations of titanium (Figs 2e and 3). Moreover, lichen weathering of anorthitic plagioclase feldspar can yield an isotropic siliceous pseudomorph from which all calcium and aluminium have been removed (Figs 2f and 4). The apparent ease with which elements like iron and aluminium are removed from these minerals is undoubtedly more consistent with attack by

chelating acids rather than with normal hydrolytic decomposition.

Secondary weathering products

Typically, lichen weathering results in poorly ordered secondary products and although these may become better crystallized with time, this has not so far been observed. It seems likely that silica gel, which is often found, could derive from the eventual disintegration and dispersion of the siliceous relics described above, as has been indicated, for example, in the weathering

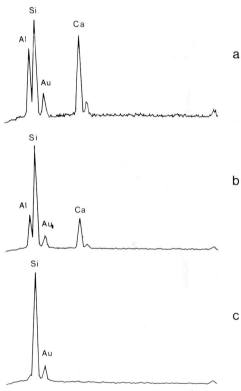

FIG. 4. Electron microprobe spectra of anorthitic plagioclase feldspar from gabbro near Portsoy (a), Fresh, (b) after partial alteration by *A. niger* and (c) directly beneath crustose lichen. Spectra similar to (c) were also recorded from grains leached with oxalic acid and grains incorporated into a culture with *A. niger*.

FIG. 2. Features observed by SEM beneath crustose lichen. (a) Etch pits associated with narrow linear feature in anorthitic plagioclase feldspar in gabbro boulder, Portsoy. (b) Etched surface of K-feldspar in granite, Bennachie. (c) Etched surface of albite in Aberchirder granite. (d) Etched and rounded olivine in Pitcaple gabbro. (e) Trioctahedral mica converted to siliceous relic in Bennachie granite. Fine-grained titanium-rich material, is arrowed. (f) Surface of siliceous relic of anorthitic plagioclase feldspar in gabbro boulder, Portsoy.

of chrysotile (Jones *et al.* 1981). Aluminous and ferruginous products are also found. Thus, in the weathering crust of lichenized basalt, aggregates of amorphous material, with a little silicon, occur in close association with ferrihydrite, a poorly ordered ferric oxide (Jones *et al.* 1980). Therefore it appears that the aluminium and iron were chelated by oxalic acid and translocated over small distances before being oxidized (possibly microbially) and precipitated on the rock surface. Poorly crystalline aluminous goethite is also associated with a lichen growing on biotite-chlorite schist (Jones *et al.* 1981).

At the rock/lichen interface and in the lichen thallus itself, there are often secondary crystalline oxalates, originating from the reaction between the oxalic acid produced by the mycobiont and the primary rock-forming minerals. Calcium oxalate is by far the most common form and, because of its insoluble nature, accumulates even in lichens colonizing rocks with only small amounts of calcium. However, where the substrate rock contains virtually no calcium, other oxalates may be found. For example, lichen weathering of serpentinite was found to involve the formation of crystalline magnesium oxalate—the mineral glushinskite (Wilson *et al.* 1980). It has also been found that in a coarse-grained lichenized allivalite from Rhum, magnesium and calcium oxalate are directly associated with olivine and anorthite, respectively (Fig. 5a and b). Moreover, there is evidence that lichen-encrusted iron and manganese ores from the Lecht Mines near Tomintoul contain the oxalates of these metals. These observations confirm the close relationship that can exist between a crustose lichen and its substrate rock.

Experimental weathering

Many of the features previously described were simulated experimentally (Jones *et al.* 1980; Wilson *et al.* 1981), either by direct treatment of the minerals with oxalic acid or by incubation with an oxalic acid-producing fungus like *Aspergillus niger*. (The use of mycobionts

for weathering experiments is impracticable because they grow very slowly and are difficult to culture.) Within four weeks both treatments had produced the distinctive etching patterns found, for example, in plagioclase feldspar and pyroxene (Fig. 5c and d) and eventually converted biotite, chrysotile and anorthite (Fig. 5e) to siliceous relics. The close association between the crystalline oxalates and the primary minerals following experimental weathering is another feature that can be readily observed (Fig. 5f).

Implications for soil processes

In general, the weathering effects from crustose lichens are similar to those observed in soils formed under cool temperature climates where biochemical weathering is predominant. In both instances there is active breakdown of primary minerals, combined with a virtual absence of well-ordered weathering products. For example, SEM observations show that the sand fractions of Scottish soils are often characterized by intensive etching of primary minerals (Wilson 1975). Furthermore, truly pedogenic weathering products—as opposed to the fine-grained material that might have been inherited from interglacial weathering episodes—are indeed of a poorly ordered nature. Thus, the form of iron in the thin iron pans of Scottish podzols is largely microcrystalline goethite (Goodman & Berrow 1976) and recently it has been found that podzolic B_2 and B_3 horizons contain an X-ray amorphous alumino-silicate complex called proto-imogolite allophane, which may well be the predominant form of mobile aluminium in podzols (Farmer *et al.* 1980). One possible mechanism for the formation of proto-imogolite is that aluminium is complexed by a simple chelating acid, like oxalic acid, and, following biological degradation, is liberated in a reactive form to combine with silica (Farmer 1979).

Although the role of biological weathering in soils remains to be more completely assessed, there are at least indications that oxalic acid

FIG. 5. Features observed by SEM beneath crustose lichens (a, b), after experimental weathering of minerals in a culture of *Aspergillus niger* (c, d) and after treatment with oxalic acid (e, f). (a) Magnesium oxalate crystals associated with olivine in allivalite, Rhum. (b) Calcium oxalate crystals associated with anorthite in allivalite, Rhum. (c) Etched out lamellar intergrowths in anorthitic plagioclase feldspar separated from gabbro boulder, Portsoy. Note the albite twin planes running ENE–WSW across the micrograph. (d) Etched out lamellar intergrowths in clinopyroxene separated from gabbro boulder, Portsoy. These features are poorly developed in control grains but become much more pronounced after alteration. (e) Surface of siliceous relic of anorthitic plagioclase feldspar separated from gabbro boulder, Portsoy. Control experiment showed no evidence of such siliceous particles. (f) Calcium oxalate crystals associated with etched anorthitic plagioclase feldspar separated from gabbro boulder, Portsoy. These crystals were absent from the control experiment.

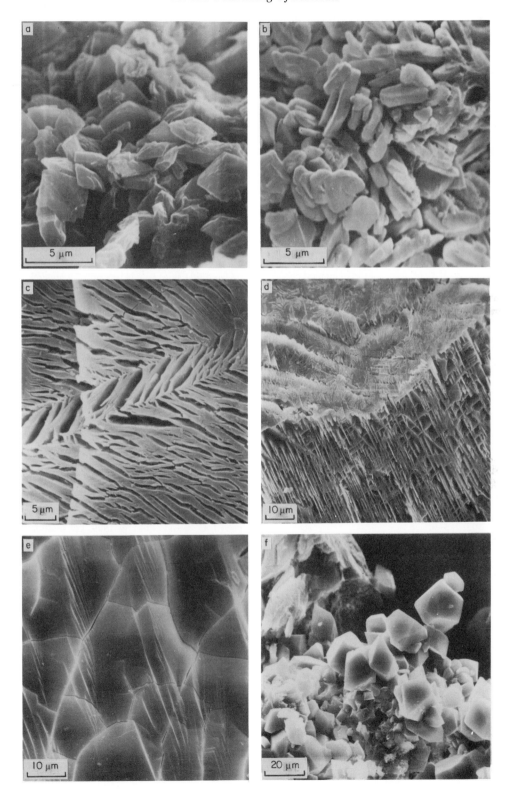

could be an important weathering agent, whether produced directly or indirectly. The general efficacy of this acid as an agent of mineral breakdown and cation release has been shown by the experiments of Robert *et al.* (1979). In fact, oxalic acid, because of its combined chelating and acidic properties, is much more active than either mineral acids or most of the common organic acids. Moreover, there is evidence that oxalic acid is one of the most abundant acids in the soil solution, particularly in capillary waters where strong concentrations are found (Vedy & Bruckert 1979). Some investigations have shown weathering effects directly attributable to oxalic acid-producing fungi. Graustein *et al.* (1977) noted the occurr-

ence of calcium oxalate crystals in the litter layer of some forest soils, where they adhere to the outer surfaces of fungal hyphae. thus indicating oxalate to be a major metabolic product of such fungi. Subsequently, Cromack *et al.* (1979) showed that there was intense chemical weathering in the immediate vicinity of the hyphae, involving the etching of primary minerals and the alteration of chlorite-like clay minerals.

In conclusion, it seems probable that studies of the decomposition of minerals beneath crustose lichens could provide information of some relevance to biological weathering in soils, especially with regard to the effects of growing fungi.

References

BERNER, R. A. 1978. Rate control of mineral dissolution under earth surface conditions. *Am. J. Sci.* **278**, 1235–52.

CROMACK, K., SOLLINS, P., GRAUSTEIN, W. C., SPEIDEL, K., TODD, A. W., SPYCHER, G., CHING, Y. L. & TODD, R. L. 1979. Calcium oxalate accumulation and soil weathering in mats of the hypogeous fungus *Hysterangium crassum*. *Soil Biol. Biochem.* **11**, 463–8.

DIBBLE, W. E. & TILLER, W. A. 1981. Non-equilibrium water/rock reactions—I. Model for interface controlled reactions. *Geochim. cosmochim. Acta*, **45**, 79–92.

DUCHAUFOUR, P. 1979. (Introduction générale.) Alteration des roches cristallines en milieu superficiel. *Sci. Sol*, 87–9.

FARMER, V. C. 1979. Possible roles of a mobile hydroxyaluminium orthosilicate complex (proto-imogolite) in podzolization. *In*: *Migrations organo-minérales dans les sols tempérés*. International Colloquim of C.N.R.S., Nancy No. 303, 275–9.

FARMER, V. C., RUSSELL, J. D. & BERROW, M. L. 1980. Imogolite and proto-imogolite allophane in spodic horizons. Evidence for a mobile aluminium silicate complex in podzol formation. *J. Soil Sci.* **31**, 673–84.

GOODMAN, B. A. & BERROW, M. L. 1976. The characterization by Mössbauer spectroscopy of the secondary iron in pans formed on Scottish podzol soils. *Suppl. J. Phys.* **37**, C6, 849–55.

GRAUSTEIN, W. C., CROMACK, K. & SOLLINS, P. 1977. Calcium oxalate: occurrence in soils and effect on nutrient and geochemical cycles. *Science*, **798**, 1252–4.

HUANG, W. H. & KELLER, W. D. 1970. Dissolution of rock forming minerals in organic acids. Simu-

lated first stage weathering of fresh mineral surfaces. *Am. Mineral.* **55**, 2076–94.

HUANG, W. H. & KIANG, W. C. 1972. Laboratory dissolution of plagioclase feldspar in water and organic acids at room temperature. *Am. Mineral.* **57**, 1849–59.

JONES, D., WILSON, M. J. & McHARDY, W. J. 1981. Lichen weathering of rock-forming minerals: application of scanning electron microscopy and microprobe analysis. *J. Microsc.* **124**, 95–104.

JONES, D., WILSON, M. J. & TAIT, J. M. 1980. Weathering of a basalt by *Pertusaria corallina*. *Lichenologist*, **12**, 277–89.

ROBERT, M., RAZZAGHE, M. K., VINCENTE, M. A. & VENEAU, G. 1979. Role du facteur biochimique dans l'altération des mineraux silicates in Altération des roches cristallines en nulieu superficiel. *Sci. Sol*, 153–74.

SMITH, J. V. 1974. *Feldspar Minerals. Vol. 2. Chemical and Textural Properties*. Springer-Verlag, Berlin.

VEDY, J. C. & BRUCKERT, S. 1979. Les solutions du sol. Composition et signification pedogenetique. *In*: BONNEAU, M. & SOUCHIER, M. (eds) 161–86. Masson, Paris.

WILSON, M. J. 1975. Chemical weathering of some primary rock forming minerals. *Soil Sci.* **119**, 349–55.

WILSON, M. J., JONES, D. & RUSSELL, J. D. 1980. Glushinskite, a naturally occurring magnesium oxalate. *Mineralog. Mag. London*, **43**, 837–40.

WILSON, M. J., JONES, D. & McHARDY, W. J. 1981. The weathering of serpentinite by *Lecanora atra*. *Lichenologist* **13**, 167–76.

WILSON, M. J. & McHARDY, W. J. 1980. Experimental etching of a microcline perthite and implications regarding natural weathering. *J. Microsc.* **120**, 291–302.

M. J. WILSON, Department of Mineral Soils, The Macaulay Institute for Soil Research, Craigiebuckler, Aberdeen AB9 2QJ, Scotland.

D. JONES, Department of Microbiology, The Macaulay Institute for Soil Research, Craigiebuckler, Aberdeen AB9 2QJ, Scotland.

Porewater reactions in the unsaturated zone with special reference to groundwater quality in England

D. A. Spears

SUMMARY: Porewater composition is an important control on the reactions leading to the formation of residual deposits. Although the mobility of porewater means it is not retained in ancient residual deposits, porewater movement through the unsaturated zone is a key factor in aquifer recharge. Porewater composition provides a sensitive means of monitoring reactions, and if infiltration has been operative over sufficient period of time whole rock changes may also be detected. Reactions involving different sedimentary fractions (resistate, hydrolysate and precipitate) are considered for infiltration into the Permo-Triassic sandstones and the Chalk of England. Externally derived elements may be identified and their behaviour in the unsaturated zone described. The examples considered are rainwater Na and Cl, fertilizer NO_3 and sewage effluent Cu, Zn and Ni.

Residual deposits result from weathering reactions in which water plays a key part. Although only the products and the residues from these reactions are preserved in the geological column, a full understanding of the reactions must involve a consideration of mineral stabilities and porewater compositions. The latter is controlled by the minerals which are either dissolving or precipitating, the length of time available for reaction and climatically controlled factors such as temperature and the volume of water flowing into and out of the system. Loss to the atmosphere by evaporation/transpiration leads to an increase in the ionic strength of the porewaters with the possibility of exceeding equilibrium activities and thus of causing precipitation, whereas downwards loss of porewater removes the ions enabling dissolution reactions to continue. On the one hand there is the formation of calcretes, gypcretes and related deposits (precipitates) and on the other, laterites, bauxites and kaolinites (resistates and hydrolysates). In the study of groundwater it is the downwards movement through the unsaturated zone which is of interest both from the point of groundwater quantity (recharge) and quality. In this field of study it is the water composition which is of prime interest, but again this is only part of the system, and a full understanding must involve an examination of the minerals present. The contemporary climate in the United Kingdom favours dissolution reactions, but a notable exception occurs in the case of the anthropogenic input of some toxic elements. In this paper I will consider some of the work done in conjunction with the Water Research Centre. The two most important aquifers in England are the Chalk and the Permo-Triassic sandstones. Reaction therefore involves sedimentary minerals and it is logical to consider separately the precipitate minerals from the resistate and hydrolysate fractions.

Precipitate minerals—carbonate and pyrite

In a number of borehole profiles through the unsaturated zone in calcareous sands and sandstones, loss of carbonate in the near-surface samples is a common feature. In the Vale of York loss of carbonate (calcite and dolomite) was established by X-ray diffraction and by whole rock analyses (Spears & Reeves 1975a). The presence of a weathering front at depth points to the rapid establishment of equilibrium with the infiltrating water and no further reaction at greater depths. The rapid rate of reactions was confirmed in a laboratory study (Spears 1976). Equilibrium with respect to the carbonate, or lack of it, may be confirmed from the analyses of the extracted porewaters. Porewaters were extracted using a pressure vessel (Manheim 1966) and in later work by the more convenient centrifuge method (Edmunds & Bath 1976), but which has the disadvantage of being open to the atmosphere. Porewater analyses, in conjunction with the whole rock analyses, may be used to calculate a rate of infiltration. It is possible to determine how much carbonate has been removed given the depth of weathering and, from the porewater concentrations below this zone, the volume of water required to remove the carbonate. In the Vale of York the age of the superficial deposits is known and therefore a rate of infiltration may be found, which corresponds with independently determined rates (Spears & Reeves 1975b). Alternatively, given the rate of infiltration the length of time required to establish the

13

zone of weathering may be calculated. This calculation applied to the Bunter Sandstone in Nottinghamshire (Gleadthorpe, N.G.R. SK 589707) suggests the profile developed through the whole of the Flandrian. A similar leached zone has been described by Reardon, Mozeto & Fritz (1980) in calcareous sandy soils from Ontario. The dissolution of calcite is rapid and equilibrium is quickly established at the 'carbonate leached/unleached zone interface'. However, their data suggest the dissolution of dolomite is slower leading to a more diffuse interface. Carbonate dissolution is responsible for an increase in pH, which leads to the precipitation of iron and manganese oxyhydroxide phases. Thus the dissolution of the carbonate controls other reactions.

In limestones the percentage non-carbonate fraction is small and therefore the rate at which this insoluble residue accumulates during weathering is probably too slow to survive erosion except in favoured circumstances. One such example has been described by Piggott (1965) from Derbyshire where superficial deposits overlying the limestone have protected the insoluble residue. A different form of protection is provided by glacial erratics in the northern Pennines which are now perched on limestone pillars. Calculation of the rate of limestone removal (Penny 1974), agrees with the rate determined from the amount of Ca removal in solution by streams and rivers (Sweeting 1965). The dissolution of carbonates is therefore an important source of Ca, Mg and HCO_3^- in groundwaters. In the Chalk the Ca present in porewaters has been observed to decrease with depth in the unsaturated zone, and equilibrium with respect to calcite is retained by changes in the HCO_3^- activity and the pCO_2 (Spears 1979). The Ca is believed to originate from within the system, and certainly the abundance of calcite and the rapid rate of reaction favours this view. However, the possibility that Ca was introduced into the system at activities equal to or greater than equilibrium values has to be eliminated. Evidence for internal derivation of Ca is provided first by similar depth profiles for Ca in both polluted and unpolluted Chalk boreholes and secondly by the absence of comparable profiles and concentrations in carbonate-free Permo-Triassic sandstones. The nitrate profile in one borehole examined in detail is comparable with the Ca, but the former is of external origin and the latter internal, and therefore the nitrate decrease is attributed to solute movement and not to change in the internal input with time.

The linear relationship between the Ca and the NO_3 (Spears 1979, fig. 4) has a small intercept value which is not significantly different to zero and thus loss of nitrate by denitrification is not important in the profile examined. The behaviour of an internal element in the unsaturated zone can therefore provide information on the anthropogenetic, fertilizer input. A comprehensive survey of nitrate in groundwater has been made by Young & Gray (1978).

Fortunately pyrite is more restricted in distribution and abundance than carbonates. It occurs primarily as a diagenetic mineral and is most common in marine black shales. The low Eh diagenetic environment differs markedly to the normal high Eh weathering environment. Breakdown of pyrite was observed in the Vale of York sediments in addition to the dissolution of carbonates (Spears & Reeves 1975a). The resulting sulphate passes into the groundwater whereas the iron remains in the unsaturated zone. In this system the low pH resulting from the pyrite is buffered by the carbonates. In colliery spoil heaps there are usually only small amounts of carbonate compared with pyrite and the resulting drainage is acid and a pollution problem. There is also enhanced alteration of the hydrolysate fraction under these conditions. The oxidation of pyrite involves a free energy change comparable to hydrocarbons and much greater than that resulting from structural modifications of silicates during weathering (Curtis 1976). The heat evolved is a significant factor in the spontaneous combustion of spoil heaps.

Resistate and hydrolysate minerals—quartz, feldspars and clay minerals

These minerals have been exposed in previous weathering cycles and therefore rates of reaction will be slow. Exchange reactions, particularly of the clay minerals, are an exception. In most mudrocks the clay minerals are dominantly detrital with diagenetic overprinting. In sandstones and Chalk diagenetic formation may be more important but nevertheless the assemblage is one stable at low temperatures. In the unsaturated zone of the sandstones it is difficult to detect whole rock mineralogical and chemical changes attributable to infiltration. Unlike the precipitate minerals these minerals react slowly, if at all. In such cases changes in the porewater composition are a more sensitive means of detecting reactions, but variations due to external input and solute movement must also be considered. In the Vale of York (Spears & Reeves 1975a) Na and K increase in the porewaters due to reaction. There is also a small, but significant change in the whole rock K_2O, which is probably due to loss of K from the interlayer site in the 10 Å clay minerals, rather than to loss from feldspars. The whole rock Na_2O does not change significantly and experimentally the release of Na differs to that of K (Spears 1976). Furthermore, the porewater Na is the only element to correlate with Cl, all of which suggests an external source for at least some of the Na. Comparable porewater concentrations of Na and Cl were subsequently described in Chalk porewaters (Spears 1979) confirming the external (sea-derived) input. Although the dissolution of feldspars, and silicates in general, has attracted widespread attention, with ideas on the rate determining step evolving (see Dibble & Tiller 1981), the present author has been unable to demonstrate a significant variation in feldspar abundance due to infiltration. In a borehole from the Middle Chalk (Deep Dean, N.G.R. TQ 539023 (there is a significant depth variation in the microline and albite contents (determined by XRD) of the acid insoluble residues, but it is the near surface samples which contain the most feldspar with none detected below 14.5 m. The distribution of kaolinite is similar and the quartz content is also higher near surface. The normal clay assemblage for the Middle and Upper Chalk (Perrin 1957; Young 1965) is present in all samples and only in the soil is there evidence of degradation. Although the kaolinite is associated with the feldspars, this probably represents a common eluvial origin rather than an *in situ* weathering product. An eluvial origin was proposed by Hodgson *et al*. (1967) for the kaolinite in the lower horizon of the Clay-with-Flint, and they also noted that there was no major mineralogical changes in the clay derived from the dissolution of the Chalk. In addition to the possibility of eluviation there is of course the original stratigraphic variation in rock composition. A variation in clay mineralogy with depth in the unsaturated zone of the Bunter Sandstone (near Stourbridge, N.G.R. SO 868819) which was possibly due to infiltration, was subsequently found in reverse sequence with respect to depth and was thus an original stratigraphic variation. The background variation should be established in investigations of the movement, and possible fixation, of an anthropogenetic input through the unsaturated zone. It is in this role of filtration medium that the resistate and hydrolysate fractions have the greatest influence on groundwater quality, rather than a source of elements in solution, unlike the precipitate minerals.

External input—natural and anthropogenic

An external source for an element present in the porewaters would be suspected in the absence of a suitable mineralogical control, i.e. a source within the system, and if the variation in the concentrations of the element in question differed to the elements known to be internally derived. This is the case for the Cl and part of the Na present in the Vale of York porewaters referred to earlier for which a rainwater source was postulated (Spears & Reeves 1975a). On the other hand it may be known from the history of the land utilization that there is likely to have been an external input into the system and the problem is one of potential aquifer contamination. The NO_3^- referred to earlier is in this category as fertilizers are known to be responsible for enhanced porewater concentrations.

The spreading of sewage effluent is another example where it is known that there are additions to the natural system. In an investigation of an effluent site near Stourbridge (N.G.R. SO 868819) the potential contamination of the groundwater in the Permo-Triassic sandstone was investigated by means of water and whole rock samples from boreholes drilled in polluted and unpolluted areas. There is a mineralogical variation due to grain size and as a result there are a number of correlations between whole rock elements. Fig. 1 shows the plot of Al_2O_3 v. Ni for samples covering the depth range 1–29 m from two boreholes, one of which is polluted. The plots for the polluted and unpolluted samples are not significantly different and furthermore there is not a significant variation with depth in the polluted borehole. Identification of the mineralogical control enables individual samples to be compared with the deviation about the regression line rather than the total sample deviation. In the case of Ni this means that contamination of a few parts per million could be detected in the whole rock. Zn and Cu are two elements which do not correlate with any major elements. Either these elements are present in several fractions of the sediment which do not vary in sympathetic manner, or the natural variation is overridden by the effect of an external addition. Positive identification of the latter is possible if the suspect sample (or samples) has a trace metal concentration which is significantly different to the normal distribution, that is the natural background variation, and if this sample comes from a depth where contamination could be expected. In the present case this would be nearness to the surface, that is to the authropogenetic input, but a situa-

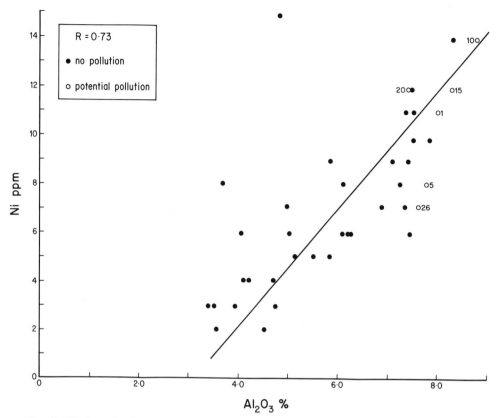

FIG. 1. Whole rock Ni ppm and Al_2O_3% concentrations in borehole samples of Permo-Triassic Sandstones, Stourbridge (N.G.R. 868819). The correlation is highly significant and demonstrates the clay association of both elements. The line of the reduced major axis regression equation is shown. Potentially polluted samples are shown as open circles with the depth in metres. These samples lie within the natural variation.

tion could be envisaged where contamination of the whole rock took place at depth in zones of maximum permeability. In only one sample, from a depth of 1 m in the polluted borehole, do Zn and Cu concentrations differ significantly to the normal distribution. The Zn value is 96 $\mu g\ g^{-1}$ compared with a mean value of 25.1 ± 8.6 $\mu g\ g^{-1}$ ($n = 30$) and the Cu value is 20 $\mu g\ g^{-1}$ compared with a mean value of 8.2 ± 2.1 $\mu g\ g^{-1}$ ($n = 30$). Exclusion of this contaminated sample from the correlation analysis (which is also justified on the grounds of non-normal distribution) does not reveal any specific mineralogical host for Zn and Cu. Therefore these elements are present in more than one fraction of the sediment but the external input is sufficiently great to be identified.

The importance of the near-surface sediments in removing trace metals was subsequently established by analysing the soils (Fig. 2). The external input is sufficiently great to

swamp the background variation and whole rock concentrations decrease exponentially with depth. In the case of Zn concentrations exceed the background level at a depth approaching 1 m, but not in the case of Ni, which thus confirms the analysis of the borehole samples. A similar study of pollutant migration in the unsaturated has been undertaken using lysimeters and synthetic leachates in the Lower Greensand (Ross 1980). Uptake of heavy metals by sesquioxides and clay minerals was thought to be particularly important. The unsaturated zone is therefore an important buffer which is capable of retaining the anthropogenetic input. In this case the role of the hydrolysate fraction, and to a lesser extent the resistate fraction, is removal of elements from the porewater, thus protecting the aquifer. The direct contribution of these fractions to the groundwater composition is of lesser importance because of the slow reaction rates. This

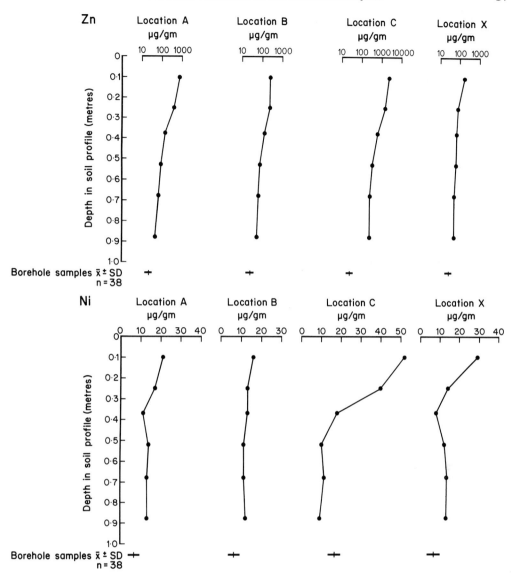

FIG. 2. Zn and Ni whole rock concentrations in four polluted soil profiles on Permo-Triassic Sandstones, Stourbridge (N.G.R. 868819). The mean values and standard deviations in the underlying sandstones are also shown.

does not, however, apply to the precipitate minerals, which not only contribute to the groundwater chemistry, but also exert a considerable influence on the chemical environment in the porewaters and thus on second stage reactions.

ACKNOWLEDGMENTS: The author would like to thank the Water Research Centre and the Natural Environmental Research Council for funding projects dealing with porewater reactions in the unsaturated zone. The co-ordinating efforts of Mr K. J. Edworthy, Dr K. Whitelaw and Dr L. Clarke have been much appreciated and the assistance and co-operation of many other members of the Centre is gratefully acknowledged. The author is also extremely grateful to officers of the Yorkshire Water Authority, the Severn-Trent Water Authority, and the Southern Water Authority for their willingness to provide borehole samples and supplementary information.

References

CURTIS, C. D. 1976. Stability of minerals in surface weathering environments: A general thermochemical approach. *Earth Surf. Process*, **1**, 63–70.

DIBBLE, W. E. & TILLER, W. A. 1981. Nonequilibrium water/rock interactions—1. Model for interface-controlled reactions. *Geochim. cosmochim. Acta*, **45**, 79–92.

EDMUNDS, W. M. & BATH, A. H. 1976. Centrifuge extraction and chemical analysis of interstitial waters. *Environ. Sci. Technol. 10*.

HODGSON, J. M., CATT, J. A. & WEIR, A. H. 1967. The origin and development of clay-with-flints and associated soil horizons in the South Downs. *J. Soil Sci.* **18**, 85–102.

MANHEIM, F. T. 1966. A hydraulic squeezer for obtaining interstitial water from consolidated and unconsolidated sediments. *U.S. geol. Surv. Prof. Pap.* **505-C**, 256–61.

PENNY, L. F. 1974. Ch. 9. Quaternary. *In:* RAYNER, D. H. & HEMINGWAY, J. E. (eds) *The Geology and Mineral Resources of Yorkshire*. Yorkshire Geological Society. 405 pp.

PERRIN, R. M. S. 1957. The clay mineralogy of some tills in the Cambridge district. *Clay Miner. Bull,* **3**, 193–205.

PIGGOTT, G. D. 1965. The structure of limestone surfaces in Derbyshire. *In*: Denudation in Limestone Region: a symposium. *Geogr. J. London,* **131**, 41–4.

REARDON, E. J., MOZETO, A. A. & FRITZ, P. 1980. Recharge in northern clime calcareous sandy soils: soil water chemical and carbon-15 evolution. *Geochim. cosmochim. Acta*, **44**, 1723–35.

ROSS, C. A. M. 1980. Experimental assessment of pollutant migration in the unsaturated zone of the Lower Greensand. *Q.J. eng. Geol. London,* **13**, 177–87.

SPEARS, D. A. 1976. Information on groundwater composition obtained from a laboratory study of sediment water interaction. *Q.J. eng. Geol. London*, **9**, 25–36.

SPEARS, D. A. 1979. Porewater composition in the unsaturated zone of the Chalk with particular reference to nitrate. *Q.J. eng. Geol. London*, **12**, 97–105.

SPEARS, D. A. & REEVES, M. J. 1975a. The influence of superficial deposits on groundwater quality in the Vale of York. *Q.J. eng. Geol. London,* **8**, 255–69.

SPEARS, D. A. & REEVES, M. J. 1975b. The infiltration rate into an aquifer determined from the dissolution of carbonate. *Geol. Mag.* **112**, 585–91.

SWEETING, M. M. 1965. Introduction in 'Denudation in Limestone Regions: A Symposium'. *Geogr. J. London,* **131**, 34–37.

YOUNG, B. R. 1965. X-ray examination of insoluble residues from the Chalk. Appendix D. *In:* GRAY, D. A. (ed.) *The Leatherhead (Fetcham Mill) Borehole. Bull. geol. Surv. G.B.* **23**, 110–14.

YOUNG, C. P. & GRAY, E. M. 1978. Nitrate in groundwater. *The distribution of nitrate in the Chalk and Triassic sandstone aquifers*. Technical Report TR69, Water Research Centre, Medmenham.

D. A. SPEARS, Department of Geology, University of Sheffield, Sheffield S1 3JD, England.

A review of experimental weathering of basic igneous rocks

David C. Cawsey & Paul Mellon

SUMMARY: Attempts to reproduce and accelerate in the laboratory the effects of weathering on minerals are reviewed, together with some observational evidence of actual weathering effects. Laboratory experiments confirm the importance of water as the main agent of chemical weathering and that the progress of weathering is controlled by drainage, oxidation and pH. Some practical applications of accelerated weathering to engineering problems are briefly described, including a short description of the authors' own work. The relative importance of chemical and physical weathering processes is considered briefly.

Weathering processes and products have been extensively reviewed by, for example, Reiche (1950), Keller (1957) and Ollier (1969). Weathering has been defined by Sanders & Fookes (1970) as 'the process of alteration of rocks occurring under the direct influence of the hydrosphere and atmosphere'. Such weathering may take place by physical disintegration and/or chemical decomposition. Peltier (1950) has related these types of weathering to mean annual rainfall and temperature, and stressed that any experimental investigation of rock weathering must reproduce as far as possible the natural conditions of weathering. However, as weathering is a process which is usually considered to have taken place over a relatively long period of time, experimental weathering usually involves an attempt to speed up the process.

Physical weathering

Blackwelder (1933) found that basalt and granite resisted fracturing when rapidly heated to 200–300°C. Griggs (1936) subjected rocks to repeated heating and cooling cycles involving sudden temperature changes of 110°C. Even after cycles considered equivalent to 244 years a polished granite surface was unaffected, but when cooled with tap water the equivalent of only $2\frac{1}{2}$ years of weathering led to loss of polish, surface cracking and incipient exfoliation. Griggs concluded that chemical weathering was more important than insolation weathering.

Partly as a result of the experiments by Blackwelder and Griggs, later work has tended to concentrate on chemical rather than physical weathering processes. However, examples of insolation weathering causing boulder cracking and blocky disintegration are described by Ollier (1969), who also drew attention to the limitations of Griggs' experiments. Physical processes, such as sheeting due to unloading, salt and ice crystal growth, and wetting and drying ('slaking weathering'), may be more important than previously thought. For example, specimens of basalt moisture saturated and then equilibrated at 65% relative humidity at 20°C showed a linear shrinkage of 0.015–0.020% (Nepper-Christensen 1965), and larger values have been reported by Edwards (1970).

Spheroidal weathering and corestones in many profiles show that tectonic and other discontinuities often determine how weathering proceeds. Nevertheless, spheroidal weathering is attributed primarily to chemical diffusion within the rock, producing concentric zones resembling Liesegang rings. Carl & Amstutz (1958) reproduced three-dimensional Liesegang 'rings' in an artificial quartz-gelatin 'rock', and compared these with the commonly observed bands of limonite staining in weathered rocks. Augustithis & Ottemann (1966) found that Ca and Fe are enriched in such brown zones, and Al, Si, K, Zr, Y and Rb are enriched in the leached zones between.

Chemical weathering

Chemical weathering of minerals

The resistance of rocks to chemical weathering depends on the susceptibility of the component minerals (Goldich 1938). Early experimental work on chemical weathering by Tamm (1924) demonstrated that hydrolysis is important in the breakdown of silicates, and also showed that different conditions of weathering give different alteration products. Using a filtration process, Correns (1961) examined the decomposition of various minerals by solutions of different pH, and found that high rates of filtration resulted in dissolution, leaving only

19

thin surface reaction layers. Closed systems with more ions in solution produced thicker crusts and pseudomorphs.

Several authors have suggested that chemical weathering of feldspars takes place by diffusion of ions through such a residual coating of hydrous aluminium silicate (Correns & von Engelhardt 1938; Wollast 1967; Helgeson 1971; Busenberg & Clemency 1976). Correns & von Engelhardt inferred that this layer maintained a constant thickness and influenced the rate of alteration. Helgeson postulated that reaction products such as gibbsite, kaolinite and mica formed zoned layers on the mineral surface.

Using a soxhlet extraction apparatus, Parham (1969) leached microcline and plagioclase for 140 days with distilled water at a temperature of approximately 78°C and pH of about 6.5. The conditions used were equivalent to a hot climate with high rainfall and good drainage. Weathering products examined with an electron microscope first appeared as bumps at random sites on cleavage surfaces and at exposed edges; these grew into tapered projections and then flame-shaped sheets which rolled to form tubes on the microcline. Sheets formed on the plagioclase feldspars developed into a platy mineral, probably boehmite. Parham noted similar weathering products penetrating feldspars exposed in a five year old road cut in Hong Kong. Other electron microscope studies (Tchoubar 1965; Wilson 1975; Berner & Holdren 1977) have revealed that dissolution during weathering of feldspars occurs mainly at sites of excess energy on the crystal surface, such as dislocations, and not by uniform attack over the entire mineral surface. Distinctive etch pits develop on the feldspar surface, characteristically on (001) with their long axes parallel to (010), and they may coalesce producing a very porous regular honeycomb structure with irregularly scattered clay mineral flakes (Dearman & Baynes 1979).

Weathering of feldspar may give clay minerals by poorly understood mechanisms (Wilson 1975), possibly involving intermediate amorphous products. In a confined environment mica may form, but in more open systems the potassium ions are removed and halloysite, vermiculite or montmorillonite form. In very leached profiles feldspars may be converted to gibbsite either directly or through an intermediate halloysite stage.

Ferromagnesian minerals are relatively easily weathered, mainly to Mg-rich mixed layer trioctahedral expansible minerals and chlorites (Wilson 1975). In poorly drained conditions olivine is altered to iddingsite, an orientated intergrowth of iron oxide and trioctahedral layer silicates (Brown & Stephen 1959). In freely drained profiles olivine dissolves rapidly, often showing deeply pitted surfaces, suggesting that, as in feldspars, dissolution occurs at sites of structural dislocations.

The pyroxene in some Australian basic volcanic rocks weathers directly to montmorillonite (Craig & Loughnan 1964), but in northeast Scotland Basham (1974) recorded the gradual alteration to vermiculite pseudomorphs, the c-axis of the vermiculite being normal to that of the pyroxene, indicating structural control by the primary mineral.

Amphiboles are usually considered to be more resistant to weathering than olivine or pyroxenes although the weathering products are often similar. Stephen (1952) and Kato (1965) showed that hornblende weathers to chlorite and then to interstratified chlorite-vermiculite, and Wilson & Farmer (1970) described hornblende containing lamellar intergrowths of a Fe-rich amphibole, which weathered preferentially to interstratified swelling chlorite-saponite. Wilson (1975) has suggested that oxidation of structural ferrous iron is important in the early weathering stages of amphiboles and other ferromagnesian minerals.

In the lower parts of weathering profiles examined by Craig & Loughnan (1964) the clay minerals were composed of elements derived essentially from parent minerals, such as Al-montmorillonite from plagioclase. As weathering proceeds, the more unstable minerals of this 'pseudomorphic stage' are destroyed, and some of the ions thereby released are incorporated

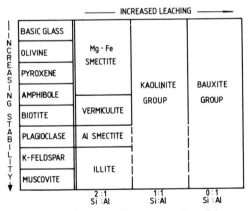

FIG. 1. Diagrammatic representation of the weathering sequence. Dashed vertical lines indicate that kaolinite and bauxite minerals may be formed directly from the primary mineral.

into other montmorillonite forming from primary minerals still present in the profile. The montmorillonite formed in many profiles is unstable in the conditions of near-surface horizons, and three possible degradation products result: poorly crystalline montmorillonite (with disordered stacking in the *c* crystallographic direction); halloysite; and kaolinite. Intense leaching may cause desilicification of kaolinite and the formation of bauxite minerals.

Fig. 1 shows diagrammatically the sequence of mineralogical changes as weathering proceeds, based on flow charts proposed by Fieldes & Swindale (1954) and Craig & Loughnan (1964). Previous work indicates that three types of mineral deterioration occur, namely:

(a) pseudomorphing of primary minerals by alteration products;
(b) etching of minerals and formation of residual alteration products;
(c) etching of minerals and removal of alteration products.

However, it is not suggested that deterioration is limited only to these three types. In any case, all three types may occur affecting minerals in the same small sample of rock. There is therefore no suggestion that they are sequential, or necessarily relate to particular weathering conditions, except for the microenvironment affecting the particular mineral concerned.

Chemical weathering of rocks

In previous experiments simulating the weathering of basic igneous rocks, drainage conditions which have been used are: (a) percolation; (b) immersion; and (c) capillarity. Experimental work by Pedro (1961), Pickering (1962) and White & Sarcia (1978) illustrates the effect of these different drainage conditions which also influence pH and oxidation.

Samples of basalt were experimentally weathered for more than two years by Pedro (1961) in a soxhlet apparatus with conditions simulating a hot (65°C), moist, tropical climate with abundant rainfall (1000 mm day^{-1}). The apparatus allowed observation of: (a) a moist upper aerobic zone, and (b) a periodically saturated lower zone corresponding to that of water table fluctuation. In the aerobic zone basalt fragments developed a 2–3 mm thick ochreous crust consisting of boehmite and limonite with small amounts of gibbsite and hematite. In the water table zone fragments only became greyish at the surface with gibbsite, and a white film of well-crystallized hydrated gibbsite and some goethite was deposited on the side of the apparatus. Afterwards the lessivate was

alkaline (pH 9–10), and its composition showed that silica, alkaline-earths and some aluminium had been leached out, whereas iron and titanium were not. The soft, friable, ochreous crust contained less silica, alkalis and alkaline earths but more iron and aluminium than the unweathered basalt. Comparable effects were found in naturally weathered basalt. Similar results were also obtained by Craig & Loughnan (1964) from weathering profiles on basic volcanic rocks in Australia.

Using stagnant solutions at several different pHs between 3.5 and 9.0, at 35°C and 0°C, and percolation tests using recycled solution at 23–25°C, Pickering (1962) experimentally leached three quartz-free rocks (a peridotite, a high glass 'potassium andesite' (latite) and a 'nepheline sodium andesite'). Similar results were obtained from both immersion and percolation tests with more silica being removed from the rocks than aluminium and iron, causing a residual enrichment of the two metals. In immersion tests larger amounts of silica were removed in slightly acid solutions than in neutral and slightly basic solutions. Rate of decomposition of rock is probably therefore dependent on pH since acid ground-water is more effective in breaking down silicates. Pickering suggested that this is due to 'the accelerating effect of the abundant hydrogen ions on the leaching of uni- and divalent cations from silicate lattices, thereby weakening their structures and making them more easily decomposed'. Frederickson (1951) also described such a process and suggested that ionic substitution causes the crystal lattice to expand contributing to eventual collapse of the crystal.

Although rate of dissolution increased at the higher temperature used, Pickering observed the same pattern of variation in dissolution rate with pH for 0°C and for 35°C. Chemical weathering should therefore be more rapid in tropical climates but the effects should be the same in cooler climates. Drainage is important since, although gibbsite can form as a result of prolonged weathering in temperate climates, when sufficient silica is present in groundwater due to slower leaching, Al-rich clay minerals, rather than gibbsite, are the stable minerals. Dissolved silica must be removed rapidly enough to prevent formation of clay minerals or resilication of previously formed gibbsite. Silica was dissolved from crushed latite at a higher rate than other rocks tested, probably due to its high content (25% by volume) of glass, which is leached more easily than crystalline material of the same composition. By contrast the lower silica content of leaching solutions from peridotite may reflect coarser grain size and reduced surface area, with fewer structurally incomplete but more easily dissolved, surfaces exposed to the liquid.

White & Sarcia (1978) experimentally weathered a fresh glass-bearing olivine-basalt by percolation tests using crushed rock and capillarity tests using sawn columns containing a vertical crack standing in fluid. In the percolation tests 400 ml of fluid circulated at the rate of 2600 ml day^{-1}. Capillarity tests were at a constant 30°C and 30% humidity for three weeks. Demineralized water dissolved small amounts of Si, Na and Ca from the basalt and was found to parallel natural weathering in a semi-arid climate. Similar results were obtained from both percolation and capillarity tests. Since water did not remove Mg or K which are depleted by weathering in a humid climate, use of a weak humic acid solution was suggested. In their tests, White & Sarcia found that dissolved silica was reprecipitated as a gel which in nature would probably dehydrate to form opal. Spherulites of chalcedony rimmed with opal have been observed on samples from weathering profiles developed on basalt (Craig & Loughnan 1964). These were considered to represent deposition of silica dissolved from above.

The separation of hydrothermal and other alteration products due to internal processes from those formed by weathering is often difficult since the products are frequently the same or similar. For example, Wilson (1975) comments that iddingsite may form from olivine by hydrothermal processes or by weathering in poorly drained conditions. A white reaction layer like that obtained in Pedro's soxhlet experiments was found by Morey & Chen (1955) using an open system at high temperatures. A layer of muscovite and boehmite developed around microcline and with albite the layer at 350°C consisted of muscovite with analcime and boehmite. At 200 and 100°C boehmite and kaolinite appeared as precipitates. At these temperatures the reaction product is crystalline with more Al than Si than at temperatures at which weathering reactions take place. Keene *et al.* (1976) have shown experimentally that montmorillonite may be formed from tholeiitic glass and artificial sea-water at temperatures of 175–300°C. As in naturally occurring submarine basaltic glass, analcime and phillipsite are also formed. The presence of zeolites in an igneous rock may be taken as an indicator that if has been affected by hydrothermal activity, and that not all observed changes are the result of weathering. The results of experimental chemical weath-

ering indicate that weathering may take place over a relatively short time-scale and this is supported by studies of recent volcanic materials. In a 45 year period considerable weathering occurred of pyroclastics derived from the 1883 Krakatoa eruption (van Baren 1931). Silica, potassium and sodium were leached from the surface where alumina and iron oxides accumulated. Hay (1960) described a clayey soil 2 m thick developed on a 4000 year old andesitic ash on St Vincent, West Indies. Volcanic glass altered to halloysite, allophane and hydrated iron oxides. Anorthite, hypersthene, and some augite and olivine crystals were etched, while labradorite, hornblende and magnetite were unaltered. The soil is calculated to have formed from ash at the rate of 0.5 m/1000 years, mainly by weathering of glass. This compares with an average rate of only 58 mm/1000 years for volcanic ash in Papua (Ruxton 1966).

Experimental investigations have confirmed that water is the most important agent of chemical weathering, acting through the processes of dissolution and hydrolysis, and that the progress of chemical weathering in a given rock body is influenced by three principal factors: oxidation; pH; and drainage.

Oxidizing conditions convert Fe^{++} to Fe^{+++}, affecting Fe^{++} within silicate minerals and aiding disruption of the crystal lattice, and leached-out Fe^{++} in solution. The insoluble Fe^{+++} hydroxides which tend to form remain in the zone of weathering, leading to a residual concentration of iron in the weathering profile. Under reducing conditions, the opposite effect occurs, enabling iron to be mobilized as soluble Fe^{++}, and leached from the profile.

Low pH promotes hydrolysis of silicate minerals by providing additional H^+ ions which enter the crystal lattice, displacing metal cations, and disrupting the silicate framework. The solubilities of several components common in the weathering system, have shown in the laboratory to be dependent on pH, but within the normal range of pH of natural waters (pH 4–9), these solubilities are relatively constant. Only when the pH is greater than 9 does the solubility of silica increase appreciably, and aluminium and iron hydroxides are soluble at pH less than 4.

The quantity of water passing through the zone of weathering influences the nature of the secondary minerals which form. Where there is a large amount of water entering the zone of weathering, and drainage is good, even poorly-soluble products of hydrolysis may be leached out, ultimately leaving an insoluble residue, largely of Fe^{+++} and Al hydroxides. Where drainage is impeded, or where there is insufficient water to leach out all the products of weathering, clays will be the stable secondary minerals.

Engineering applications

Accelerated weathering experiments have been applied to certain engineering problems, particularly in relation to aggregates, where knowledge of the weatherability of rock is highly desirable. Several cases of premature failure of aggregate have been documented (e.g. Wylde 1976).

Some engineering tests have been developed which attempt to assess resistance of rock to weathering, or to detect potentially unsound rock. The sodium or magnesium sulphate soundness test, where rock is alternately soaked in saturated sodium or magnesium sulphate solution and dried, is a standard aggregate test in both Australia and the U.S.A., and standard impact tests have been modified to include a soaking period (Hosking & Tubey 1969). The slake durability test has been developed to assess the mechanical resistance of rock to wetting and drying (Franklin & Chandra 1972). These tests usually attempt to simulate aspects of physical weathering.

Farjallat *et al.* (1974) subjected a variety of basalts and gneisses to artificial weathering by several processes: outdoor exposure; the sodium sulphate soundness test; wetting and drying; ethylene glycol saturation and drying; and continuous leaching in a soxhlet apparatus, all over a period of thirty days. A measurable deterioration in the mechanical strength of the rock occurred, the sodium sulphate soundness test having the greatest weakening effect.

The authors are at present engaged in an extensive programme of artificial weathering, as part of a study of weathered basalt gravels from Ethiopia, comprising wetting and drying cycles, at 60°C, to simulate the seasonal rainfall of the source area. Preliminary results suggest that physical disintegration is much more important than chemical decomposition. The weathered basalt contains interstitial expanding clay minerals and it is thought that alternate wetting and drying promotes cracking of the rock due to the expansion and contraction of the clay minerals. It is intended to publish full results at a later date.

Concluding remarks

Weathering is a process involving the interaction of many different factors and the resulting products are complex. Experimental investigations have confirmed that water is an important weathering agent and that the progress of weathering is influenced by oxidation, pH and drainage. Drainage conditions are particularly important in determining if alteration products are retained or leached out.

Many of the weathering experiments to date have been concerned with simulating these chemical effects rather than physical weathering processes. Physical disintegration caused by weathering, however, can be of considerable importance in affecting the engineering performance of rock. For example, the effects of wetting and drying of rocks containing expanding clay minerals has probably been underestimated.

The experimental evidence indicates that significant physical weathering may occur over relatively short periods of time, whereas chemical weathering processes require much longer periods before the effects are readily observed.

References

AUGUSTITHIS, S. S. & OTTEMAN, J. 1966. On diffusion rings and sphaeriodal weathering. *Chem. Geol.* **1**, 201–9.

BASHAM, I. R. 1974. Mineralogical changes associated with deep weathering of gabbro in Aberdeenshire. *Clay Miner. Oxford,* **10**, 189–202.

BERNER, R. A. & HOLDREN, G. R. 1977. Mechanism of feldspar weathering: Some observational evidence. *Geology,* **5**, 369–72.

BLACKWELDER, E. 1933. The insolation hypothesis of rock weathering. *Am. J. Sci.* **26**, 97–113.

BROWN, G. & STEPHEN, I. 1959. A structural study of iddingsite from New South Wales, Australia. *Am. Mineral.* **44**, 251–60.

BUSENBERG, E. & CLEMENCY, C. V. 1976. The dissolution kinetics of feldspars at 25°C and 1 atm CO_2 partial pressure. *Geochim. cosmochim. Acta,* **40**, 41–50.

CARL, J. D. & AMSTUTZ, G. C. 1958. Three-dimensional Liesegang rings by diffusion in a colloidal matrix, and their significance for the interpretation of geological phenomena. *Bull. geol. Soc. Am.* **69**, 1467–8.

CORRENS, C. W. 1961. The experimental weathering of silicates. *Clay Miner. Bull.* **4**, 249–65.

—— & VON ENGELHARDT, W. 1938. Neue Untersuchungen uber die Verwitterung der Kalifeldspates. *Chemie Erde,* **12**, 1–22.

CRAIG D. C. & LOUGHNAN, F. C. 1964. Chemical and mineralogical transformations accompanying the weathering of basic volcanic rocks from New South Wales. *Aust. J. Soil Res.* **2**, 218–34.

DEARMAN, W. R. & BAYNES, F. J. 1979. Etch-pit weathering of feldspars. *Proc. Ussher Soc.* **4**, 390–401.

EDWARDS, A. G. 1970. Scottish aggregates: their suitability for concrete with regard to rock constituents. *Building Research Station Current Pap. 28/70.*

FARJALLAT, J. E. S., TATAMIYA, C. T. & YODHIDA, R. 1974. An experimental evaluation of rock weatherability. *Proc. 2nd int. Congress Int. Ass. Engng Geol.* IV-30.1 to 9.

FIELDES, M. & SWINDALE, L. D. 1954. Chemical weathering of silicates in soil formation. *N. Z. J. Sci. Technol.* **36**, 140–54.

FRANKLIN, J. A. & CHANDRA, A. 1972. The slake-durability test. *Int. J. rock mech. Ming Sci. Geomech. Abstr.* **9**, 325–41.

FREDERICKSON, A. F. 1951. Mechanism of weathering. *Bull. geol. Soc. Am.* **62**, 221–32.

GOLDICH, S. S. 1938. A study in rock-weathering. *J. Geol. Chicago,* **46**, 17–58.

GRIGGS, D. T. 1936. The factor of fatigue in rock exfoliation. *J. Geol. Chicago,* **44**, 781–96.

HAY, 1960. Cited in Ollier (1969).

HELGESON, H. C. 1971. Kinetics of mass transfer among silicates and aqueous solutions. *Geochim. cosmochim. Acta,* **35**, 421–69.

HOSKING, J. R. & TUBEY, L. W. 1969. Research on low-grade and unsound aggregates. *Road Research Laboratory Rep. LR293,* Crowthorne, Berkshire.

KATO, Y. 1965. Mineralogical study of weathering products of granodiorite at Shinshiro City (II). Weathering of primary minerals. Stability of primary minerals. *Soil Sci. Plant Nutrition,* **10**, 34–9.

KEENE, J. B., CLAGUE, D. A. & NISHIMORI, R. K. 1976. Experimental hydrothermal alteration of tholeiitic basalt: Resultant mineralogy and textures. *J. sedim. Petrol.* **46**, 647–53.

KELLER, W. D. 1957. *The Principles of Chemical Weathering.* Lucas, Columbia, Miss.

MOREY, G. W. & CHEN, W. T. 1955. Cited in Correns (1961).

NEPPER-CHRISTENSEN, P. 1965. Shrinkage and swelling of rocks due to moisture movement. *Medd dansk geol. Foren,* **15**, 548–55.

OLLIER, C. D. 1969. 4th imp. 1979. *Weathering,* Longmans, London.

PARHAM, W. E. 1969. Formation of halloysite from feldspar: Low temperature, artificial weathering versus natural weathering. *Clays Clay Miner.* **17**, 13–22.

PEDRO, G. 1961. An experimental study of the geochemical weathering of crystalline rocks by water. *Clay Miner. Bull.* **4**, 266–81.

PELTIER, L. 1950. The geographic cycle in periglacial regions as it is related to climatic geomorphology. *Ann. Ass. Am. Geol.* **40**, 214–36.

PICKERING, R. J. 1962. Some leaching experiments on three quartz-free silicate rocks and their contribution to an understanding of laterization. *Econ. Geol.* **57**, 1185–206.

REICHE, P. 1950. A survey of weathering processes and products. *New Mexico Univ. Publ. Geol.* **3**.

RUXTON, B. P. 1966. The measurement of denudation rates. *Inst. Aust. Geo. 5th Meeting, Sydney.*

SANDERS, M. K. & FOOKES, P. G. 1970. A review of the relationship of rock weathering and climate and its significance to foundation engineering. *Engng Geol., Amsterdam,* **4**, 289–325.

STEPHEN, I. 1952. A study of rock weathering with reference to the soils of the Malvern Hills. Part 2. Weathering of appinite and Ivy Scar Rocks. *J. Soil Sci.* **3**, 219–37.

TAMM, O. 1924. Cited in Keller (1957).

TCHOUBAR, C. 1965. Formation de la kaolinite a partir d'albite altarée par l'eau a 200°C. Etude en microscope et diffraction electroniques. *Bull. Soc. Fr. Mineral. Cristall.* **88**, 483–518.

VAN BAREN, J. 1931. Properties and constitution of a volcanic soil built in 50 years in the East-Indian Archipelago. *Comm. Geol. Inst. Agr. Univ. Wageningen, Holland,* **17**.

WHITE, R. W. & SARCIA, C. 1978. Natural and artificial weathering of basalt, northwestern United States. *Bull. Bur. Rech. geol. min. Paris* (deuxieme série) Section II, **1**, 1–29.

WILSON, M. J. 1975. Chemical weathering of some primary rock-forming minerals. *Soil Sci.* **119**, 349–55.

—— & FARMER, V. C. 1970. A study of weathering in a soil derived from biotite-hornblende rock. II. Weathering of hornblende. *Clay Miner.* **8**, 435–44.

WOLLAST, R. 1967. Kinetics of the alteration of K-feldspar in buffered solutions at low temperature. *Geochim. cosmochim. Acta,* **31**, 635–48.

WYLDE, L. J. 1976. Literature review: Crushed rock and aggregate for road construction—some aspects of performance, test methods and research needs. *Aust. Road Res. Board Rep.* **43**.

DAVID C. CAWSEY & P. MELLON, School of Natural Sciences, The Hatfield Polytechnic, P.O. Box 109, College Lane, Hatfield, Herts AL10 9AB, England.

KAOLINITES, LATERITES
AND BAUXITES

Kaolinisation and the formation of silicified wood on late Jurassic Gondwana surfaces

H. Wopfner

SUMMARY: In some parts of Australia and Africa intense kaolinisation took place on late Jurassic land surfaces, transforming considerable thicknesses of bedrock into masses of pure kaolin and quartz. Overlying these surfaces occur fluviatile, kaolinitic sandstones of late Jurassic or earliest Cretaceous age. They contain an abundance of silicified wood, including whole trunks of trees. The formation of silicified wood has previously been attributed to hot and arid desert environments, but in the examples described here, silicification proceeded under constantly high groundwater levels, requiring humid and probably warm climatic conditions. Silica was liberated by the kaolinisation of silicate minerals. Humic acids, produced during the initial decay of the fossil timber provided the low pH microenvironment for the fixation of SiO_2.

During the Jurassic a considerable change in depositional conditions was experienced in many parts of Gondwana. This change was related primarily to movements initiating the final break-up of Gondwana. In the interior parts of Australia and Africa long periods of non-deposition were terminated by the development of extensive intracratonic basins, within which, initially, fluviatile deposits were laid down. At the same time pericratonic basins developed in response to rift movement and distension along the south eastern coasts of both Australia and Africa.

The intracratonic basins considered here are the Great Artesian Basin in Australia and the eastern Iullemmeden Basin in north-western Nigeria. In both these basins intensive kaolinisation took place, affecting not only basement rocks but also overlying sandstones. Locally, silicified wood is abundant in these rocks. Whilst the author has carried out extensive research into the Great Artesian Basin, his knowledge of the Iullemmeden and other African basins was derived from short field trips and literature studies (for localities, see Fig. 1).

Great Artesian Basin, Australia

Deposition within the central and eastern portion of the Great Artesian Basin commenced in the early Jurassic. Essentially a freshwater, dominantly fluviatile succession was laid down, amounting to about 900 m of sedimentary accumulation in the depositional centres. Along the western and southern margins, however, the thickness of the section is severely reduced and only the uppermost unit, the Algebuckina Sandstone, is present. Palynology indicates a late Jurassic age for that sequence (Wopfner *et al.* 1970). It is generally a white,

medium-grained sandstone with about 5% of kaolin matrix. Interbedded conglomerate lenses, consisting of well rounded quartz, also contain completely kaolinised pebbles and cobbles of Precambrian rhyolites and other basement rocks. The basal conglomerates are sometimes auriferous and have supported quite sizeable mining communities, like Milparinka in north-western New South Wales. Here, silicified wood is also most abundant; trunks of up to 0.60 m in diameter are scattered across the sandstone outcrop, or are still standing

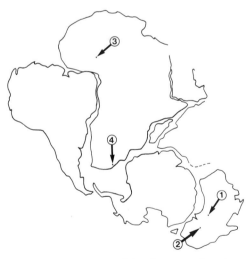

FIG. 1. Assembly of Gondwana continents in the Jurassic. Arrows indicate localities discussed in this paper. (1) Oodnadatta, western Great Artesian Basin. (2) Milparinka area, south-western Great Artesian Basin. (3) Libba, western Iullemmeden Basin, Nigeria. (4) Locality on Umfolosi River, South Africa.

27

upright in their growth positions, encased in sand. *Otozamites* is also abundant. Unfortunately, ruthless exploitation for semi-precious jewellery has ruined these magnificent exposures. The wood exhibits well developed growth rings, but its genetic affinities have never been investigated. Under the microscope wood cells are clearly visible, the cell lamina being filled with microcrystalline quartz and/or chalcedony. Radial cracks in the stems and other cavities are filled with white, blue and amber coloured agate.

The basement underlaying the late Jurassic sandstone consists of Precambrian micaschists, gneisses and phyllites and is usually kaolinised. That this kaolinisation is indeed a Jurassic feature and not due to some younger weathering process is demonstrated by kaolinised basement underneath an equivalent of the Algebuckina Sandstone. This was encountered at 1050 m below the surface in a stratigraphic borehole, drilled about 60 km north-west of Milparinka, across the border in South Australia (Wopfner & Cornish 1967). Both the altered basement and Algebuckina Sandstone contain abundant pyrite crystals.

Intensely kaolinised basement also underlies the Algebuckina Sandstone along the western basin margin, south of Oodnadatta (Wopfner *et al.* 1970). There, early Precambrian migmatites have been completely decomposed, all minerals but quartz having been altered to kaolinite. Fig.

FIG. 2. Contact between Precambrian migmatite and Algebuckina Sandstone, about 40 km south-east of Oodnadatta (South Australia). Migmatite is completely kaolinised, except for residual quartz of small pegmatite vein. White protruding bank is basal unit of Algebuckina Sandstone. Height of exposed basement is about 2 m.

FIG. 3. Totally kaolinised feldspar-mica-schist overlain by medium grained, kaolinitic Algebuckina Sandstone. Between sandstone and schists about 10 cm of kaolinitic siltstone is visible. Original schistosity, dipping at about 45° to the right is barely recognizable. Scale on hammer in inches.

2 shows an example of the intensely weathered basement, which is now only recognizable due to the presence of remnant quartzes of a small pegmatite. Very fine grained feldspar-mica-schists, also of Precambrian age have been changed to an almost featureless mass of kaolinite, shown in Fig. 3. The kaolinisation of the basement appears to be uniform throughout and does not show any differentiation or specific zonation.

In several localities along the margin of the western and southern outcrop quartzitic silicretes (Group I–1b in Wopfner's 1978 classification) occur about 20 m above the basement contact. They contain well preserved moulds of plant fossils, including *Cladophlebis australis, Hausmannia* and *Cycadites sp.* (Wopfner 1969; Wopfner *et al.* 1970) and some small stems of silicified wood.

Examples from Africa

In north-western Nigeria, along the eastern margin of the Iullemmeden Basin, similar kaolinised basement profiles, capped by kaolinitic sands and conglomerates, containing silicified wood, were observed. These fluviatile deposits, termed the Gundumi Formation (Kogbe 1976a) flank the north-western slope of the Nigerian platform. The Gundumi Formation is part of the 'continental intercalaire' of West Africa. Investigations of the fossil wood suggest an uppermost Jurassic to lowermost Cretaceous age (Kogbe 1976b); the formation is thus time-equivalent with the Algebuckina Sandstone in Australia. At Talata Mafala, east of Sokoto on the road to Gusau and at a road cut near Libba, about 40 km south of Jega, contacts between the Gundumi Formation and the underlaying basement are exposed. In both cases the basement is intensely kaolinised. At the exposure near Libba (Figs 4 & 5) the basement consists apparently of contorted schistose gneiss with quartz veins. The overlying gritty sandstone is also kaolinitic. Silicified wood occurs most abundantly about 4–5 m higher up in the section. It is fawn in colour and consists primarily of cryptocrystalline quartz. Most of it is probably of gymnosperm origin.

For the sake of completeness some lower Cretaceous–(?)uppermost Jurassic sands, containing silicified wood, resting on weathered gneiss or basalts should be mentioned. They occur along the South African coast north of Durban (e.g. near Mtubatuba on the Umfolosi River, see Fig. 1).

FIG. 4. Contact between kaolinised Precambrian basement and Gundumi Formation in road cut near Libba, about 100 km south of Sokoto, north-west Nigeria. Note progressive onlap of sandstone beds on to basement.

H. Wopfner

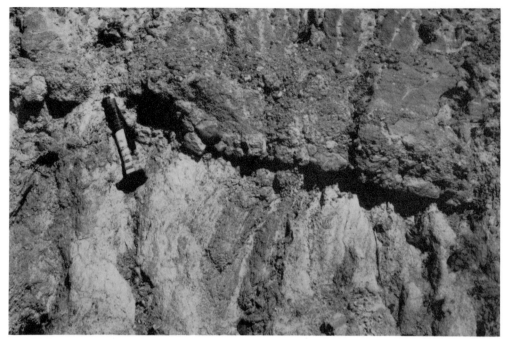

FIG. 5. Close up of contact shown in Fig. 4. The light coloured featureless areas are kaolinised schist, the darker areas quartz veins. Above contact kaolinitic grit of basal Gundumi Formation, north-west Nigeria.

Discussion and conclusions

The widespread, approximately time-equivalent occurrences of kaolinised basement, kaolinitic sandstones and silicified wood are difficult to accept as just mere coincidences. It seems likely that kaolinisation of basement, kaolinitic sandstone and the occurrence of silicified wood are linked in some way. Silicification of wood has been related to hot and arid desert conditions, but such conditions are incompatible with those likely to induce kaolinisation. What then were the conditions which led to the development of these profiles in such widely separated areas?

The intense kaolinisation of the basement could be interpreted as a deeply weathered soil-profile the development of which pre-ceeded the deposition of the sandstone. If this were the case, then there would be no genetic connection between kaolinisation of the base-ment and the kaolinitic sands, except that kaolin eroded from the former was incorpor-ated into the latter. Such an interpretation was proposed by Wopfner *et al.* (1970) for the Great Artesian Basin examples described ear-lier.

Assuming that the kaolinisation of the base-ment developed as part of a laterite profile, a characteristic zonation would have developed

and one would expect at least some of it to be preserved. Such well-zoned weathering profiles on gneiss have been described, for example, from the subsurface of northern Poland (Kabata-Pendias & Ryka 1968). However, no such zonation has been recognized within the kaolinised basement of the Great Artesian Basin, and no obvious differentiation was observed in the exposures of the Iullemmeden Basin.

It is difficult to envisage fluvial transport and depositional processes that would permit the combined deposition of coarse clastic grains and kaolin. It also appears doubtful that com-pletely kaolinised pebbles of regolith could have withstood mechanical pounding over long distances to become so well rounded. Thus the evidence appears to favour a common origin for the kaolinised basement rocks overlain by kaolinitic sands.

Fossilization of wood fibre, apart from coalification, is achieved by the substitution of organic substances by mineral matter. The main replacement minerals are either carbonate, pyrite or various forms of silica. Wood structure is preserved by all these minerals, but the best preservation is usually achieved by silica impreg-nation.

In the cases described in this paper, the fossil

timbers are generally undeformed. The substitution of the organic material must have taken place, therefore, with very little overburden pressure. Rapid sand accumulation is indicated by the trunks which were silicified in their growth positions.

It has been pointed out by Willstätter (1958) that infiltration of wood-fabric by silica can only take place in true solution. The preservation of minute details of the wood structure requires molecular exchange of matter, but also a minimum of decomposition. It would appear that all these requirements can only be realized by continued water saturation; in other words by continuing ground water flow through the porous sand encasing the trunks and logs. Kaolinisation of silicate minerals would have provided the necessary silica, whereas the wood acted as acceptor, possibly enhanced by a lowering of the pH.

In conclusion, the following depositional model is suggested. Progressive tilt established large flood plains of braided or low sinuousity streams. Humid climatic conditions provided prolific plant growth, and maintained continuously a high ground water level. Under these conditions low pH groundwaters promoted kaolinisation. The liberated silica was removed from the system by intergranular flow, except where it was 'trapped' by organic material or accumulated as silcrete in areas of extremely low pH environments (Wopfner 1978).

As mentioned above, the systems were primarily initiated by regional tilt. In the case of the Great Artesian Basin, this has been achieved by a northward tilt of the rift shoulders which developed along the present southern coast of Australia. A similar cause is suspected in the case of West Africa; but without exact knowledge of sediment intake and palaeo transport direction I prefer to leave this question open to further discussion.

References

KABATA-PENDIAS, A. & RYKA, W. 1968. Weathering of biotite gneiss in the Palaeozoic period. *9th int. Congr. Soil Science—Adelaide. Transact.* **IV**, 381–90.

KOGBE, C. A. 1976a. Outline of the geology of the Iullemmeden Basin in north-western Nigeria. *In:* KOGBE, C. A. (ed.) *Geology of Nigeria*, Elisabethan Publishing Co. Lagos. 331–8.

KOGBE, C. A. 1976b. The "Continental Intercalaire" in north-western Nigeria. *J. Min. Geol.* **13**, 45–50.

WILLSTATTER, R. 1958. Aus meinem Leben. *Verlag Chemie G.m.b.h. Weinheim-Bergstraße*, 356–60.

WOPFNER, H. 1969. The Mesozoic Era, *In:* PARKIN, L. W. (ed.) *Handbook of South Australian Geology*, pp. 133–71. Geological Survey of South Australia.

WOPFNER, H. 1978. Silcretes of northern South Australia and adjacent regions. *In:* LANGFORD-SMITH, T. (ed.) *Silcrete in Australia*, University of New England Press, Armidale. 93–141.

WOPFNER, H. & CORNISH, B. C. 1967. S.A.G. Portville No. 3 well completion report. *Geol. Surv. S. Aust. Rep. Invest.* No. 29, 62 pp.

WOPFNER, H., FREYTAG, I. B. & HEATH, G. R. 1970. Basal Jurassic—Cretaceous rocks of Western Great Artesian Basin, South Australia: Stratigraphy and environment. *Bull. Am. Ass. Petrol. Geol.* **54**, 383–416.

H. WOPFNER, Geologisches Institut der Universität zu Köln, 5000 Köln, Zülpicher Strasse 49, W. Germany.

Kaolinitic weathering profiles in Brittany: genesis and economic importance

J. Esteoule-Choux

SUMMARY: In Brittany, kaolinitic weathering profiles are developed on a variety of sedimentary, metamorphic and igneous rocks. In many places, laterite is associated with ironstone and/or silicified horizons. These profiles developed during the Cretaceous and Eocene, under hot wet climates: some of them are important commercial deposits.

In Brittany (Fig. 1), kaolinitic weathering profiles are developed on a variety of sedimentary, metamorphic and igneous rocks (Esteoule-Choux 1967). For some time, the kaolinitic alteration of granitic rocks was considered to result from hydrothermal activity, but recent studies show that the granites have been weathered in a manner similar to other rock types. Such weathered rocks are sometimes only covered by Quaternary deposits, but in some places, they are overlain by Ypresian sediments, Upper Eocene sedimentary kaolins or Miocene limestones. This paper describes four examples of weathering profiles developed on a granite, a metamorphic rock, an Ordovician sandstone and an Ordovician shale (see Fig. 1).

Description of the profiles

Ploëmeur area

The parent rock is a Westphalian leucogranite that has been tectonised and which contains quartz veins. The alteration is irregular: there are zones of unaltered granite between zones which are entirely kaolinised. The maximum depth of kaolinisation is about 60 m, but is more generally around 30 m. From the top to the bottom the profile shows the following units:

a zone of variable depth (2–45 m), where the granite is entirely kaolinised (feldspars have completely disappeared);
a zone where the altered rock still contains feldspars;
the fresh granite.

In most places, the rock is completely decomposed, although the original granitic texture can always be distinguished. The colour is generally white but sometimes ochre, greenish or light purple.

The clay fraction is composed of kaolinite whose crystallinity (Hinckley index 1963) varies between 0.625 and 1.225 (Bellion 1979).

Transmission electron microscopy shows a great variability in the shape and in the size of the particles: euhedral crystallites (0.1–2 μm), elongated in one direction and also particles without any geometrical shape. Some tubes and spheres of halloysite have been found in samples from two quarries. Scanning electron microscopy shows that the kaolin has a coarse texture composed of loosely packed books and a high porosity (Fig. 3a). This texture is characteristic of weathering environments (Keller 1976). Goethite has been identified only in a few samples.

Above kaolinised granite, some parts of the deposit contain large blocks of sandstones (called grès ladères or grès à Sabals) which correspond to climatically silicification (Fig. 2, Kergantic pit).

The Ploëmeur deposit has been considered to be a hydrothermal kaolin (Nicolas 1956; Charoy 1975) but a recent study (Bellion 1979) demonstrated the weathering origin of this kaolin; the texture seen under the SEM is additional support for this interpretation.

Profile of 'Le Bohu-Robien'

In this pit (Fig. 2), Precambrian schists have been weathered to a depth of 25 m. Numerous small vertical veins of quartz cut the schists and are also completely disaggregated. The schists, which appear to have vertical schistosity, are weathered to a very white kaolinitic clay which contains pieces of quartz, 2 or 3 cm long. Along the quartz veins the schists appear less weathered with only small greenish plates.

At the surface, there is a layer of ochreous loams (about 1–2 m in thickness) which contain ironstone pisoliths and gravels consisting of quartz (similar to those of the weathered schists), and ironstone. These loams are probably formed mainly of material eroded from the weathered mantle and from an ironstone crust.

The clay fraction of the weathered schist is

33

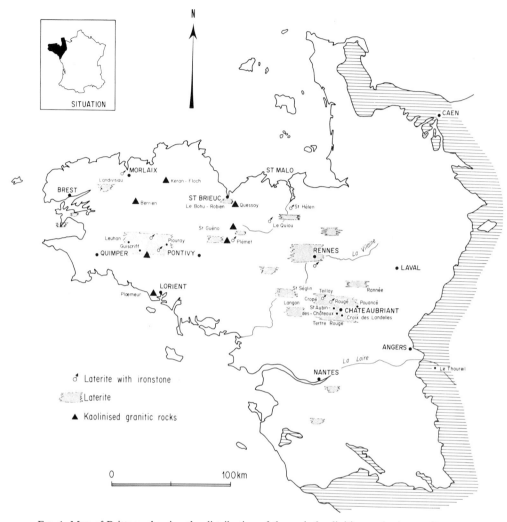

FIG. 1. Map of Brittany showing the distribution of the main kaolinitic weathering profiles.

composed of kaolinite. Under TEM large hexagonal plates can be seen (Fig. 3b) and under SEM the texture appears the same as that of the Ploëmeur kaolin.

Weathering profiles in Châteaubriant area

In an area about 40 km long and 18 km wide around Châteaubriant, sedimentary rocks have been extensively weathered, although the depth of kaolinisation is very irregular (more than 30 m in places, and only 1 or 2 m in others). The structure of the area consists of a series of synclines and anticlines formed of rocks of Ordovician and Silurian age, with fold axes orientated NW–SE. Within the 'Grès Armoricain' (Ordovician sandstones) and 'Schistes à Calymènes' (Ordovician shales) are numerous weathering profiles. The 'Grès Armoricain' is a quartzarenite which may contain a little muscovite and sometimes some feldspars. This formation is divided into 'Grès Armoricain inférieur', 'Schistes intermédiaires' and 'Grès Armoricain supérieur', but when weathered these divisions are not easily recognized.

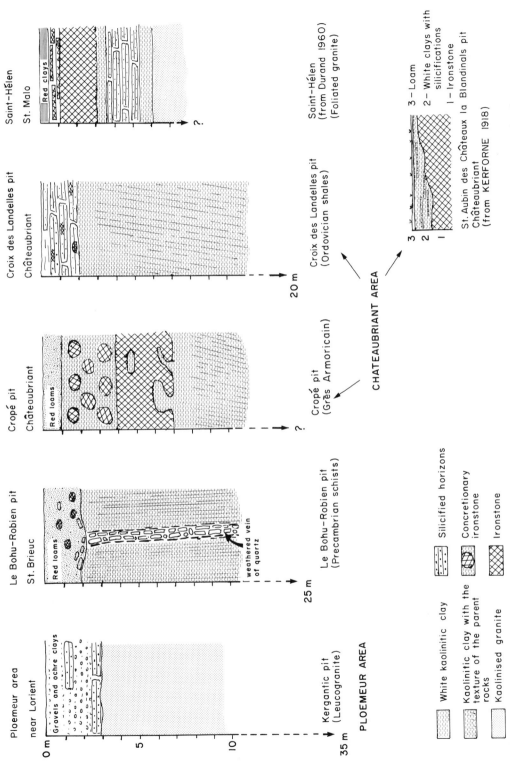

FIG. 2. Schematic sections showing the relationships between weathered rocks, ironstones and silicified horizons (for localities see Fig. 1).

Weathering of the 'Grès Armoricain'

A typical section in the 'Grès Armoricain' consists of the following units, from the top down.

  silicification, of variable development, forming huge blocks of sandstone ('Grès ladères') overlying white or ochre clays;
  ironstone crust;
  white or ochre, or mottled clays;
  clay with texture similar to the parent rock;
  parent rock.

This sequence is not always complete. Sometimes the ironstone crust is absent and its thickness is very variable; it is 15 m thick at Rougé, but more often 3–6 m (Cropé pit, Fig. 2). The silicification may not be developed (Cropé pit) or else may have been eroded because pieces of 'Grès ladères' may be found scattered on the soil surface.

The ironstone crust shows different facies: ironstone pisoliths in a matrix of ferruginous cement, massive, vesicular, or foliated. Mineralogically, these ironstones consist principally of goethite with sometimes a little hematite and a small amount of kaolinite.

Blocks of the 'Grès ladères' have a similar appearance in the field and polished surfaces. But they vary considerably when examined under the microscope, when quartzarenite sandstones, quartzarenites with a matrix composed of opal, chalcedony, small quartz grains or clay minerals can be seen. The quartz grains are not well sorted, and are angular or subangular, or sometimes subrounded. Sometimes these 'Grès ladères' show a very close association of silica with goethite, in which case they are reddish with white or colourless patches of silica.

The white, ochre or mottled clays are composed of kaolinite with a little micaceous clay. Quartz is present in variable, but minor amounts. Goethite appears in ochre clays in variable but significant quantities.

Under the TEM (Fig. 3c) the white kaolinite shows well developed hexagonal plates, but when the clay is ochreous the particles of kaolinite have jagged outlines and star-shaped twins of goethite are attached to them (Fig. 3d). Under the SEM these kaolinitic clays show a coarse texture composed of loosely packed books and a high porosity (Fig. 3e).

Weathering of the 'Schistes à Calymènes'

In Tertre Rouge pit, the following succession occurs under a bed of reddish sandy clay (1–2.5 m thick):

  dark grey clay, about 10–12 m thick, in which can be found the pygidia of trilobites more or less pyritized. Veinlets of weathered quartz are numerous, and iron oxides occur either as diffuse trails or as an iron oxides zones;
  dark grey clay in which the texture of the parent rock is well preserved (1.50 m thick);
  the parent rock (dark blue shales).

The parent rock, which contains some nodules of iron oxides, is composed of very fine quartz and small amounts of muscovite and chlorite. In the dark grey clay, kaolinite is the dominant clay mineral with very subordinate amounts of micaceous clay. Under the SEM (Fig. 3f) the texture of the kaolinitic clay shows large crystal flakes and expanded rocks of kaolin, with a high porosity.

In another pit (La Croix-des-Landelles, Fig. 2), the kaolinised Ordovician shales are overlain by silicified horizons.

At the bottom of profiles developed on Precambrian shales (Brioverian) kaolinite is associated with chlorite where the clays have retained the texture of the parent rock.

Despite careful study, no aluminium hydroxides (gibbsite or boehmite) could be proved in any of these formations.

Genesis

The weathering profiles have a wide geographical distribution and there is a close association between the kaolinised rocks, the ironstone development and silicificied horizons (Figs 1 & 2). The profiles developed during the Cretaceous and Eocene, under a hot wet climate. The existence of such climates have previously been recognized in various areas of France through the occurrence of Wealden type sediments, bauxite and also through recent studies of the flora in clays of lower Ypresian age (Durand & Ollivier-Pierre 1969). Under such hot wet climates deep lateritisation occurred beneath a peneplain of the Hercynian platform which became emergent at the end of the Carboniferous. These climates continued throughout the Palaeogene, though becoming progressively less humid with longer dry intervals during the Stampian. The pisolitic ironstone crust and the vesicular ironstone are the result of these periodic arid conditions, and some of the silicificied horizons may also be a consequence of such conditions. Indeed, under drier climates with alternate wet and dry seasons, some of the kaolinite profiles have undergone an intensive chemical reworking *in situ*, indicated by the existence of ironstone and by the transformation of the large hexagonal plates of kaolinite into large particles with ragged outlines (Esteoule & Esteoule-Choux 1964).

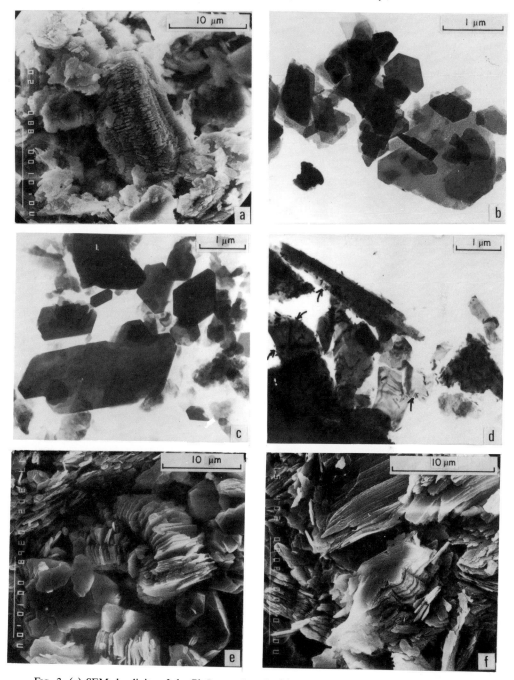

FIG. 3. (a) SEM: kaolinite of the Ploëmeur deposit. (b) TEM: kaolinite of Le Bohu-Robien. (c) TEM: kaolinite from 'Grès Armoricain'. (d) TEM: kaolinite with star-shaped twins of goethite (arrows). (e) SEM: kaolinite from 'Grès Armoricain'. (f) SEM: kaolinite from 'Schistes à Calymènes'.

During the Ypresian and Upper Eocene, part of the cover was stripped away by erosion to be deposited as sedimentary kaolinite in numerous continental basins. It is evident that laterite must have covered Brittany formerly to a much greater extent than at present. The laterite cappings now occur at various altitudes: 160–230 m in Leuhan-Guiscriff-Plouray area, 170 m at Plemet, more than 200 m south of Quessoy, 90 m at St-Hélen and 100 m in the

Châteaubriant area. We must regard these as remnants of ancient peneplanation and later uplift. Tectonic movements at the end of the Tertiary period are well known (Durand & Esteoule-Choux 1977). Uplift led to dissection of the landscape and deposition of important occurrences of loose ironstone gravels, ironstone conglomerates and 'Grès ladères' during the Quaternary and Recent.

Economic importance

At present, the weathering profiles on granitic rocks account for the greater part of the Breton production of kaolin (Table 1), which is obtained from the Ploëmeur, Berrien and Quessoy areas; the kaolin of Plemet is no longer worked. Among the kaolinitic profiles on sedimentary rocks, kaolin is still worked in the Châteaubriant area, at Tertre Rouge, and at La Croix-des-Landelles (Ordovician shales). Ironstone crusts were intensively quarried at the end of the nineteenth century in the Plemet and in the Châteaubriant areas and in many other places. Today, iron ore is still worked near Châteaubriant (Rougé) because of its high assay value (45–50%) and the absence of carbonates.

TABLE 1. *Breton production of kaolin (tonnes)*

Ploëmeur deposit	: in 1980 :	140.000	
Berrien deposit	: in 1980 :	35.000	
Quessoy deposit	: in 1974 :	60.000	(washed kaolin)
		50.000	(rough kaolin)
Tertre Rouge	: in 1980 :	17.000	

Conclusion

In Brittany the weathering mantle is very well developed and is seen overlying sedimentary, granitic rocks and metamorphic rocks. The parent minerals (micas, chlorite, feldspars, quartz) have been replaced by kaolinite, due to lateritic weathering at the end of the Mesozoic and beginning of Tertiary times. The texture under the SEM of all these kaolins is characteristic of a weathering environment. In many places laterite is associated with ironstone and/or silicificied horizons. The abundance of iron and silica is also a consequence of the deep weathering of the parent materials, but the removal of much, if not all of the iron, and of the silica may have occurred during periods of arid climate.

References

BELLION, G. 1979. *Contribution à l'étude du gisement de kaolin de Ploëmeur (Morbihan): caractères, origine*. Thèse, 3ème cycle, Rennes, 189 pp.

CHAROY, B. 1975. Plöemeur kaolin deposit (Brittany): an example of hydrothermal alteration. *Pétrologie*, **1**, 4, 253–66.

DURAND, S. 1960. Le Tertiaire de Bretagne. Étude stratigraphique, sédimentologique et tectonique. *Mém. Soc. géol. Min Bretagne*, **XII**, 389 pp.

DURAND, S. & OLLIVIER-PIERRE, M. F. 1969. Observations nouvelles sur la présence du pollen de palmier Nypa dans l'Eocène de l'Ouest de la France et du Sud de l'Angleterre. *Bull. Soc. géol. Min. Bretagne, C*, **I**, 49–57.

DURAND, S. & ESTEOULE-CHOUX, J. 1977. Sédimentation tertiaire et tectonique dans le Massif Armoricain. *5ème Réunion annuelle des Sc. de la Terre, Rennes, 19–22 Avril 1977*, 212.

ESTEOULE, J. & ESTEOULE-CHOUX, J. 1964. Etude en microscopie électronique de divers types géolo-

giques de kaolins. *C. r. hebd. Séanc. Acad. Sci., Paris*, **259**, 2469–72.

ESTEOULE-CHOUX, J. 1967. *Contribution à l'étude des argiles du Massif Armoricain. Argiles des altérations et argiles des bassins sédimentaires tertiaires*. Thèse Sci., Rennes, 319 pp.

HINCKLEY, D. 1963. Variability in 'cristallinity' values among the kaolin deposits of the coastal plain of Georgia and south Carolina. *Proc. 11th Nat. Conf. clays & clay minerals*, **13**, 229–36.

KELLER, W. D. 1976. Scanning electron micrographs of kaolins collected from diverse environments of origin. I, II. *Clays Clay Miner.* **24**, 107–13, 114–7.

KERFORNE, F. 1918. Sur l'âge des minerais de fer superficiels de la région de Châteaubriant. *Bull. Soc. géol. Fr., 4ème série*, **XVII**, 229–32, année 1917.

NICOLAS, J. 1956. *Contribution à l'étude géologique et minéralogique de quelques gisements de kaolin bretons*. Thèse Sci., Paris, 254 pp.

J. ESTEOULE-CHOUX, Institut de Géologie, Université de Rennes I, Avenue du General Leclerc, 35042 Rennes-Cedex, France.

The origin and occurrence of Devon Ball Clays

A. Vincent

SUMMARY: Valuable deposits of two types of kaolinitic clay (China Clay and Ball Clay), are found in the South-west of England. *China Clay* is comparatively coarse grained, well-ordered kaolinite derived from granite. Kaolinite particles are also derived from weathering of other country rocks, but such particles are fine-grained, disordered kaolinite. The properties of the two types of kaolinite are different. The two sedimentary *Ball Clay* basins of Devonshire occur in association with the Sticklepath Fault system. The properties of the various seams depend on the provenance of the kaolinite particles and the complex sedimentology of the two basins.

Ball Clays have been in general use in pottery compositions in the British Isles since the seventeenth century, and are now extensively exported to Europe and other parts of the World. The continuing popularity of the ball clays in ceramic applications depends upon their unique and consistent combination of physical properties.

However the term 'ball clay' is not derived from any specific property, but from the original method of production. In open pits, the clay was cut into cubes (Scott 1929), the sides of which were approximately 22–25 cm and each cube weighed 14–16 kg. The eventual rounding of the corners with handling before sale gave rise to the term 'ball clay'. Three deposits in South-west England were originally worked in this fashion, and hence are the only true Ball Clays. These deposits are located around Wareham in Dorset, and in the Petrockstowe and Bovey Basins in Devonshire. The positions of these deposits, together with the occurrence of the other kaolinitic deposits, the China Clays, are shown in Fig. 1.

The physical properties of Ball Clay depend mainly upon its mineralogical composition and its particle size distribution, particularly in the lowest size ranges. The controlling factors in these parameters are the provenance and mineralogy of the kaolinite and the environments of deposition of each seam or group of seams. This paper is concerned with these factors in relation to the Petrockstowe and Bovey Basins.

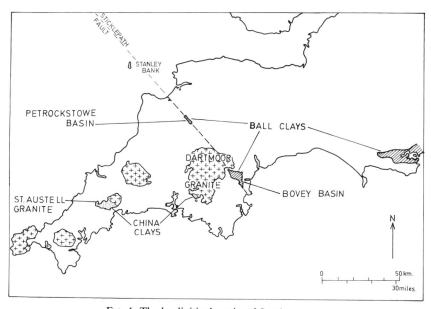

FIG. 1. The kaolinitic deposits of South-west England.

Provenance of the kaolinite and other minerals

The Ball Clays are generally formed from three main components, kaolinite, micaceous minerals and quartz, whilst carbonaceous material is also often present. Marcasite or siderite is also usually found in small quantities, whilst there are occasional minor occurences of tourmaline, felspar, chlorite, montmorillonite, interstratified clay minerals and anatase. In commercially available Ball Clays the ranges of dominant minerals present (Mitchell & Stentiford 1973) are as follows:

kaolinite: 20–95%,
micaceous mineral: 5–45%,
quartz: 1–70%.

The identification of the micaceous mineral is problematic, the mineral being impossible to physically separate from the kaolinite in a pure form. From X-ray diffraction analysis, it would appear that most of the micaceous mineral present is of illitic character, although admixtures with crystalline mica derived from Dartmoor are found in the Bovey Basin (Mitchell & Stentiford 1973). The quartz present in the clays varies from rounded grains in the Petrockstowe Basin, to sub-angular granitic grains at Bovey, and range in size from the clay grade up to 70 μm. Carbon occurs in four forms: (1) as seams of 'lignite', which are generally admixtures of clay and lignitised vegetation; (2) as discrete fragments in the clay (e.g. seeds and twigs); (3) as fine grained carbon (i.e. in the clay grade); and (4) as a colloidal form which exists as a coating on the individual crystallites of the clay mineral.

The proportions and form of the main components of the Ball Clays control the physical properties, and none more so than the kaolinite component. The kaolinite found in the Petrockstowe Basin is of the type often referred to as '*b*-axis disordered kaolinite' (Brindley & Robinson 1946). This type of structural disorder is readily recognized from the appearance of X-ray diffraction patterns. The most widely accepted method for quantifying the 'degree of crystallinity' is that devised by Hinckley (1962). The method involves measurement of peak heights from 20 to 22° & 2θ on the X-ray diffractogram. On the Hinckley scale, a degree of crystallinity of zero represents completely disordered kaolinite, whilst approximately 2 would represent almost perfectly ordered kaolinite. The range in degree of crystallinity of kaolinite is as follows:

china clay: 1.0–1.3,
Bovey Basin: 0.1–0.9,
Petrockstowe Basin– 0.1–0.3.

China Clay kaolinite is regarded as having an ordered structure, the Bovey Basin to contain a mixture of ordered and disordered kaolinite particles, and the Petrockstowe Basin to contain exclusively disordered particles.

The ordered kaolinite of the China Clays was derived *in situ* on the Dartmoor Granite fundamentally, in the author's opinion, by hydrothermal processes. The factors relating to the formation of china clays were reviewed by Bristow in 1977 and work is continuing. The disordered kaolinite of the Petrockstowe Basin was derived from the weathering of Culm Measure Slates (Bristow 1968). The kaolinite of the Bovey Basin was derived from the erosion products of China Clay deposits and the weathering products of the country-rocks during the sub-tropical phases of the Palaeogene (Vincent 1974). The Bovey Basin therefore contains both ordered and disordered varieties of kaolinite usually as admixtures, and hence produces the greatest variety of Ball Clay types.

The differing provenances of the kaolinite in the Petrockstowe and Bovey Basins similarly accounts for the variations in the quartz and micaceous minerals, and also the great difference in the properties of the clays produced from the two basins.

The sedimentary basins

During the early Tertiary, both the ordered kaolinite of the China Clay deposits in the granite and the disordered kaolinite in the sub-tropical weathering mantle must have been widespread in the County of Devon. Fookes *et al.* (1971) and Dearman & Fookes (1972) indicate that considerable weathering took place, but make no direct reference to kaolinite. The formation of suitable kaolinite by weathering processes would probably have been selective and as described by Bristow (1968). The same area, however, has not been a region of extensive deposition since the Permian and kaolinites deposited in a marine environment are generally altered to other clay minerals. Therefore the Ball Clays would be expected to be of fresh water origin and to have been deposited in sedimentary areas within the ancient land mass.

Early writers (Key 1862; Pengelley 1862) considered the Bovey Basin to be a natural lake, basin whilst Jukes-Brown (1909) recognized a probable tectonic origin. It was not, however, until 1957 that the tectonic mechanism for the formation of both basins was recognized. Blyth (1957) produced a paper on 'The Lustleigh Fault in North-east Dartmoor' which he correlated with the Sticklepath Fault of Dearman (1950) and suggested a dextral movement. Blyth recognized the continuation of the Lustleigh Fault to the south-east, passing under the Bovey Basin and continuing to the coast near Torquay. To the north, the fault system cuts off the end of the Crediton Trough, passes under the Petrockstowe Basin and emerges on the coast to the west of Bideford, and can be traced across the Bristol Channel into Pembrokeshire, where at Flimston there are minor clay deposits. Perhaps more significantly, a third Tertiary basin exists on the same fault line under the Bristol Channel, some 4.5 km to the east of Lundy Island at Stanley Bank (Fletcher 1975).

Both Ball Clay areas in Devonshire are known to be deep tectonic basins, but probably the general surface level during sedimentation was not very different from that of today. In both areas, as subsidence occurred along the Sticklepath Fault system, sedimentation relating to the degree of subsidence took place. Settling of the sands, silts and muds found in the sequences is not difficult to understand, but that related to the finer grained clays is not easy to visualise. Many Ball Clays are more than 60% less than 0.5 μm and such particles have been observed to remain in suspension for many years in flooded pits and would not be expected to settle readily (Hazen 1914; Odén 1916). Large thicknesses of fine grained clays, however, show that such clays did accumulate. Under conditions of low pH flocculation occurs, a fact observed in clay technology and described by Dollimore & Horridge (1973). The pH of Ball Clays varies from 3.5 to 7 with the most carbonaceous clays giving the lower values. Olayinska Asseez (1970) describes the effects of organic acids on flocculation, and probably such factors controlled the sedimentation of the Ball Clays. When flocculation is coupled with high concentrations of clay particles, seams of clay of fine-grained character can be laid down (Vincent 1974). Clays laid down in this fashion show no banding or grading and little lateral variation, features on which the quality control processes of commercial extraction are dependent. As the focus and intensity of subsidence varied in the sedimentary areas, then the environment and strata laid down also varied, although with considerable differences between the two basins.

The age of the deposits in the two basins is not precisely known. The only fossil evidence is provided by plant remains in a lignitised condition and the lack of animal remains possibly indicates the difficulties of survival and preservation in the acid and turbid conditions which prevailed. The long-ranging plants do not provide good indications of precise age although the Bovey Basin lignites have attracted considerable attention (Pengelley & Heer 1861, Reid 1910 and Chandler 1957). The general consensus of their opinion based on sections in the 'Old Coalpit' near Bovey Tracey itself points to an Oligocene age for the Bovey lignites. These deposits are higher in the Bovey Formation than the commercial Ball Clay horizons, which could well be of Eocene age. The determination of the precise age of the strata in the two basins will depend upon palynological investigations and some recent work (Freshney et al. 1979) suggests that the Petrockstowe sediments are mainly Eocene with only a little Oligocene strata. There is, however, not complete agreement amongst palynologists and work continues. The author considers that the known strata in the Bovey Basin cover a period in time ranging through the Upper Eocene to the Middle Oligocene, but that the final sediments deposited on the western side of the basin may well be much later in age (Vincent 1974).

The two basins, although related in terms of tectonic origin and age, show few other similarities. The sedimentation patterns and the provenance of the kaolinite are rather different, and each of these two important commercial Ball Clay deposits is, as far as is known, unique.

The Petrockstowe Basin

Sedimentation in the Petrockstowe Basin appears to have been controlled by contemporaneous fault movements, mostly along the northwest to south-east trend line of the Sticklepath Fault system (see Fig. 1). The faulting divides the basin axially into a deep central trough around 670 m deep, with flanking shelf areas to the north-east and south-west (Freshney 1970, see Fig. 2). The earliest deposition, fluviatile in nature, took place in the axial trough areas and consisted mainly of sands and gravels, with silty clays representing occasional overbank deposits. Later, the river that flowed from the south-east, migrated beyond the central trough area so that overbank deposits, particularly in the shelf areas, became more prevalent. Freshney (1970) demonstrated that much of the sedimentation in the trough was cyclical with upward transitional sequences from sandy gravels to smooth clays. Cyclicity is also evident in the shelf deposits, but without the gravels.

A. Vincent

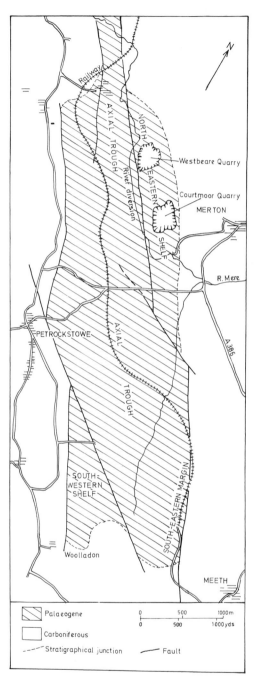

Fig. 2. The Petrockstowe Basin (after Freshney 1970).

Ball Clay is worked by both the major clay companies at three locations, two of which are in shelf areas and the other is an overbank sequence in the trough deposits. E.C.C. Ball Clays Ltd have exploited the South-Western (or Woolladon) Shelf and the trough deposits on the south-eastern margin at Stockleigh Moor and Meeth Claypit. Watts, Blake, Bearne and Co. P.L.C. produces 110,000 tonnes per annum from the North-Eastern (or Merton) Shelf, and it is to this area and the northern half of the Petrockstowe Basin that the author's experience is confined.

The North-Eastern Shelf extends for some 2 km in a north-westerly direction from the Merton to Petrockstowe transverse road (which divides the basin into two halves). The shelf deposits are 4–500 m in width, the easterly boundary being stratigraphic whilst the westerly edge is an internal fault with the dominant north-west to south-east trend. The course of the River Mere has largely been diverted to the south-western side of the internal fault to facilitate the production of Ball Clay.

Changes in the degree and focus of subsidence at the time of deposition were rapid, so that as well as cyclic sedimentation, both minor and larger unconformities are common. Movement in the basement after sedimentation caused folding and faulting (some of which is reversed faulting) and hence a very complicated occurrence of the Ball Clay seams, which makes the deposit most difficult to evaluate.

There are three sequences on the North-Eastern Shelf, each with a different structure and occurrence. The most southerly is that worked in Courtmoor Quarry (GR 5180 1170), and this sequence has a pitching and asymmetric synclinal structure, probably representing the earliest of the clay sediments on the Shelf. To the north, the Courtmoor sequence is overlain transgressively by the plane dipping Westbeare sequence, which in Westbeare Quarry (GR 5130 1215) shows steep dips and monoclinal folding. The Courtmoor sequence is overstepped at the northern end of Westbeare Quarry and is not found further to the north. The area to the north of Westbeare is less well-known, but has some history of working. The relationship to the Westbeare sequence is, as yet, not clear, but the northern area has a broad synclinal structure and is thought to represent the final phases of sedimentation on the shelf, which here merges into the trough deposits on the other side of the internal fault.

The Ball Clays in the various sequences have slightly different properties and have to be exploited accordingly, but basically the strength and plasticity of the disordered kaolinites of the Petrockstowe Basin are of considerable value in ceramic compositions.

The Bovey Basin

The Bovey Basin can be divided into two geographical areas of which the northern and larger area is the most important (see Fig. 3). The main area forms a rough parallelogram stretching in a north-westerly direction from Newton Abbot to beyond Bovey Tracey, a distance of some 11 km with a maximum width of 8 km. The southern area stretches from Decoy to Aller but also contains Aller Gravels, Cretaceous Greensand and a faulted inlier of New Red Sandstone, as well as commercial clays. The Bovey Basin is certainly the deepest of the three Ball Clay deposits in the South-west of England. On the basis of a gravity survey, Fasham (1971) estimated the depth of the Basin to be 1300 m. The lower strata however, are overstepped by the later Bovey deposits and the deepest boreholes only reach 320 m, so the nature and density of the larger part of the Bovey strata is not known. The commercial Ball Clay seams occur in the upper strata.

The sediments of the Bovey Basin consist of clays, sandy clays, sands, siliceous clays, carbonaceous clays, lignitic clays and lignite. The seams are worked selectively for the production of various types of Ball Clay. In contrast to the Petrockstowe Basin, there is considerable variety in the types of kaolinite particle present and in the environments of deposition. In consequence, there is a far greater range of properties in the ball clays produced from the Bovey Basin. Scott (1929) showed a twofold division of the clays of the eastern outcrop which he identified as the 'whiteware' and 'stoneware' clays. The same twofold division was also recognized by Edwards (1976) who distinguished the non-carbonaceous clays of the Abbrook Member (the 'stoneware' clays of Scott) and the clays and lignites of the Southacre Member which coincide with Scott's 'whiteware' clays. Most of the western outcrop of the commercial seams is covered by later, unconformable and sandier Bovey material. There is a modification of the clay lignite sequence of Southacre Quarry (GR 8530 7530) through non-carbonaceous clays at Chudleigh Knighton Quarry (GR 8400 7710) to sand at Little Bradley (GR 8280 7760). In terms of lithology it is perhaps incorrect to continue to call the Southacre Member by the same name at Chudleigh Knighton and Little Bradley, but in terms of clay mineralogy, it is appropriate.

The Abbrook Member consisting of sands, siliceous clays and non-carbonaceous clays show admixtures of kaolinite particles and considerable variety in particle geometry, so that the Ball Clays from the Abbrook Member although not as strong and plastic as the Petrockstowe clays have more valuable firing characteristics. Some seams also have rheological properties which give them great importance in the sanitary ware industry. The clays of the Southacre Member however contain little disordered kaolinite and have probably been derived from the erosion of China Clay deposits along the Lustleigh-Sticklepath Fault System within the Dartmoor Granite. The presence of fine grained and colloidal carbon in the Southacre clays adds strength and plasticity to the ordered kaolinite particles and hence, the clays of the Southacre Member are the only large deposits of white firing Ball Clay so far discovered.

There are also differences in the sedimentology of the two sequences, with that of the Southacre Member being rather more complex than that of the Abrook Member. During Southacre times, most of the material being deposited was entering from the north-west with the development of outwash fans over most of the northern part of the basin, stretching from Bovey Tracey to Chudleigh Knighton Heath. Cessation of deposition on the outwash fans was followed by sedimentation in lakes which formed the clay deposits. The quiet turbid waters of these lakes, with pH reduced by organic acids, gave rise to flocculation. The floccs began to settle and interfere to such an extent that fine particles and even buoyant material, such as twigs, would be entrained in an 'en masse' sedimentation. Carbon itself has numerous seats for the attraction of kaolinite and could often form the nucleus of a flocc. Under such conditions fine grained non-graded and non-banded clays with included plant remains could be formed (Vincent 1974). Meanwhile, floating vegetation collected at the furthest end of the lake, in a back swamp environment, in which formed lignite and lignitic clay. As the focus of subsidence migrated, so also did the environments of deposition, and hence the properties of the Ball Clay vary in a like manner. The three types of environment can be readily recognized in the Southacre sequence, with sand to the north, clay at Chudleigh Knighton, and the first seams of lignites appearing just to the south of the A38 trunk road. The lignites become dominant in the lower part of the sequence through Southacre Quarry to Newton Abbot Clays Quarry (GR

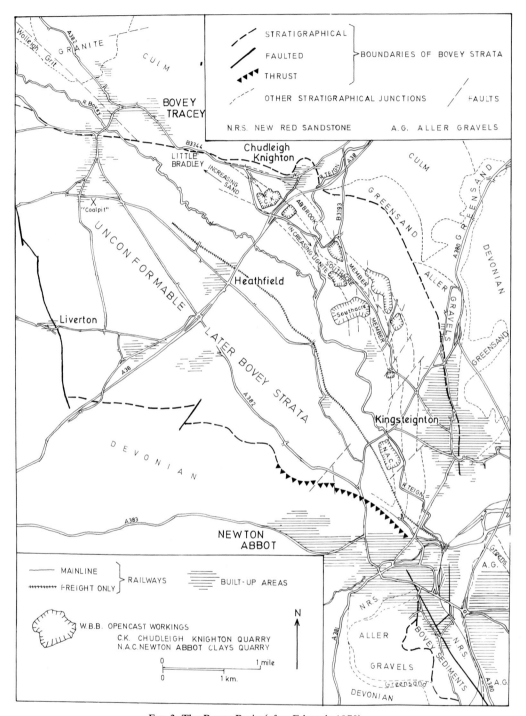

FIG. 3. The Bovey Basin (after Edwards 1970).

8600 7315) in the south of the Basin. The Abbrook Member becomes much sandier towards the north before being overstepped by the Southacre Member, but there is an absence of carbonaceous matter except in the upper part. The lack of lignite could be due to a through circulation existing throughout Abbrook times, but the author speculates that the condition is more likely to be due to climatic conditions with a change occurring towards the end of the period.

From this complex Formation, the two clay companies produce almost half a million tonnes per annum of carefully selected and blended Ball Clays mainly to suit the needs of the ceramic industry.

ACKNOWLEDGMENTS: The author is indebted to the Chairman, Mr C. D. Pike, O.B.E., M.A., LL.B. (Cantab), and to the Board of Directors of Watts, Blake, Bearne & Co. P.L.C., for permission to publish this paper.

References

BLYTH, F. G. H. 1957. The Lustleigh Fault in northeast Dartmoor. *Geol. Mag.* **94**, 291–6.

BRINDLEY, G. W. & ROBINSON, K. 1946. Randomness in the structures of kaolinitic clay minerals. *Trans. Faraday Soc.* **42B**, 198.

BRISTOW, C. M. 1968. The derivation of the Tertiary sediments in the Petrockstowe Basin. *Proc. Ussher Soc.* **2**, 29–35.

—— 1977. A review of the evidence for the origin of the kaolin deposits of S.W. England. *Proc. 8 int. Kaolin Symp. Mtg. Alunite. Madrid-Rome*, pp. 1–19.

CHANDLER, M. E. J. 1957. The Oligocene flora of the Bovey Tracey Lake Basin, Devonshire. *Bull. Br. Mus. nat. Hist.* **3**, 71–123.

DEARMAN, W. R. 1950. *The structure of the Okehampton district*. Unpublished Ph.D. Thesis, University of London.

—— & FOOKES, P. G. 1972. The influence of weathering on the layout of quarries in South-West England. *Proc. Ussher Soc.* **2**, 372–87.

DOLLIMORE, D. & HORRIDGE, T. A. 1973. The dependance of the flocculation behaviour of china clay—polyacrylamide suspension on the suspension pH. *J. Colloid and Interface Sci.* **42**, 581–8.

EDWARDS, R. A. 1970. *The Geology of the Bovey Basin*. Unpublished Ph.D. Thesis, University of Exeter.

—— 1976. Tertiary sediments of the Bovey Basin, South Devon. *Proc. Geol. Ass.* **87**, 1–26.

FASHAM, M. J. R. 1971. A gravity survey of the Bovey Tracey Basin, Devon. *Geol. Mag.* **108**, 119–30.

FLETCHER, B. N. 1975. A new Tertiary basin east of Lundy Island. *J. geol. Soc. London*, **131**, 223–5.

FOOKES, P. G., DEARMAN, W. R. & FRANKLIN, J. A. 1971. Some engineering aspects of rock weathering with field examples from Dartmoor and elsewhere. *Q. J. eng. Geol.* **4**, 139–85.

FRESHNEY, E. C. 1970. Cyclical sedimentation in the Petrockstowe Basin. *Proc. Ussher Soc.* **2**, 179–89.

—— BEER, K. E. & WRIGHT, J. E. 1979. Geology of the country around Chulmleigh. *Mem. geol. Surv. G.B., Sheet 309*.

HAZEN, A. 1914. On sedimentation. *Trans. Am. Soc. civ. Engrs* **53**, 45–88.

HINCKLEY, D. N. 1962. Variability in crystallinity values among the kaolin deposits of the coastal plain of Georgia and South Carolina. *Proc. 11th National Conf. on Clays and Clay Minerals*, p. 299 (W. F. Bradley, ed.).

JUKES-BROWN, A. J. 1909. The depth and succession of the Bovey Deposits. *Jl. Torquay Nat. Hist. Soc.* **1**, 1–23.

KEY, J. H. 1862. On the Bovey Deposit. *Q. Jl geol. Soc. London*, **18**, 9.

MITCHELL, D. & STENTIFORD, M. J. 1973. Die Gewinnung und die Eigenschaften von Plastischem Ton aus dem Devon (The production and properties of Devon ball clays). *Silikat J.* **12**, 185–94.

ODÉN, S. 1915–1916. On the size of particles in deep-sea deposits. *Proc. Roy. Soc. Edinb.* **36**, 219–36.

OLAYINKSA ASSEEZ, L. 1970. Effects of organic acids on the flocculation of suspended sediments in tropical man-made lakes. *J. Hydrology*, **11**, 253–7.

PENGELLEY, W. 1862. *The Lignites and Clays of Bovey Tracey*.

—— & HEER, O. 1861. On the fossil flora of Bovey Tracey. *Phil. Trans*, **112**, 1019–39.

REID, C. & E. M. 1910. The lignite of Bovey Tracey. *Phil. Trans. B*, **201**, 161–178.

SCOTT, A. 1929. Ball clays. *Mem. Econ. geol. Surv. G.B., 31*.

VINCENT, A. 1974. *Sedimentary Environments of the Bovey Basin*. Unpublished M. Phil. Thesis, University of Surrey.

A. VINCENT, Watts, Blake, Bearne & Co. P.L.C., Park House, Courtenay Park, Newton Abbot, Devon TQ12 4PS.

The Ayrshire Bauxitic Clay: an allochthonous deposit?

S. K. Monro, F. C. Loughnan & M. C. Walker

SUMMARY: Since the original study of the Ayrshire Bauxitic Clay by Wilson in 1922, new sections have been examined in both quarries and boreholes. It has also been suggested that the Ayrshire Bauxitic Clay is, in effect, a typical flint clay comparable with those of North America, South Africa, Australia and elsewhere. A re-examination of the outcrop and borehole evidence has been undertaken with a detailed chemical, mineralogical and petrographical study of the core of a bore that penetrated both the Ayrshire Bauxitic Clay and associated altered basalts. The study presents a body of evidence which suggests that, in part at least, the Ayrshire Bauxitic Clay is allochthonous in origin.

The Ayrshire Bauxitic Clay is Carboniferous (Namurian) in age, occurring at the top of the Passage Group in southern Strathclyde (Fig. 1). It was first described in 1895 by John Smith, though it was not until 1922 that a detailed description was published by G. V. Wilson. Wilson (1922) emphasized the massive structure, conchoidal fracture, unusual density and induration of the bauxitic clay and the variety of textures ranging from fine-grained to oolitic, pisolitic and coarsely clastic. He also noted that the rock is 'bauxitic in the chemical sense of containing more Al_2O_3 than can be accommodated with SiO_2 to give kaolinite'. From its association with altered basalts of Namurian age, he concluded that it developed from these lavas, partly as a residual crust and partly as a transported sedimentary deposit derived from erosion of this crust. He believed both types were later subjected to diagenetic changes involving silicification of some of the bauxitic minerals to form a second generation of kaolinite.

Subsequently, Lapparent (1936) identified boehmite and diaspore in selected samples. Although these aluminium oxhydroxide minerals proved subordinate to kaolinite, he agreed with Wilson's use of the term *bauxitic clay*. A number of authors (Bosazza 1946, Keller 1967, Robertson 1971 and Loughnan 1978) have drawn attention to apparent similarities between the bauxitic clay and the flint clays of North America, South Africa, Australia and elsewhere, and have suggested that the bauxitic clay of Ayrshire is in effect a typical flint clay. If this is correct, Wilson's mechanisms for the formation of the bauxitic clay assume greater significance in that the origin of flint clays has long been a contentious issue, with controversy centred primarily on whether they are essentially autochthonous or allochthonous.

The bauxitic clay lithology is peculiar to the top of the Passage Group. Though it may be interbedded with other lithologies, it invariably grades down into fresh basaltic lava. The sequence including the bauxitic clays may be defined as a lithostratigraphic formation and the term Ayrshire Bauxitic Clay Formation is proposed. The Ayrshire Bauxitic Clay Formation is that sequence of strata which immediately overlies fresh basaltic lavas of Passage Group age and is characterized by the presence of bauxitic clay and related rock types. Interbedded coals, seatclays and mudstones are also included within the formation. The formation is followed by cyclical sequences of sediments typical of Coal Measures strata.

The Ayrshire Bauxitic Clay Formation
Geological setting

The Ayrshire Bauxitic Clay Formation crops out along the margin of the Westphalian (Coal Measures) coalfield of Ayrshire (Fig. 1). It is best developed along the northern crop, extending from Saltcoats [NS24.41] on the coast eastwards to Cunninghamhead [NS38.41], with only small isolated outcrops elsewhere, and no development south of the Kerse Loch Fault. Everywhere it occurs above basaltic lavas of Namurian (Passage Group) age.

Tectonically the area is dominated by major north-east to south-west faults which were active during sedimentation (Richey 1930). The variation in sediment pattern and thickness is evident from a comparison of the sections (Fig. 2). At Smithstone, north of the Dusk Water Fault, repeated sequences of bauxitic clay, frequently carbonaceous, and coal are developed up to a thickness of around 20 m, while at Dubbs, south of the Dusk Water Fault, a single bauxitic clay unit occurs around 4 m thick. Lateral changes in the detail of the profile are also evident. The transition from bauxitic

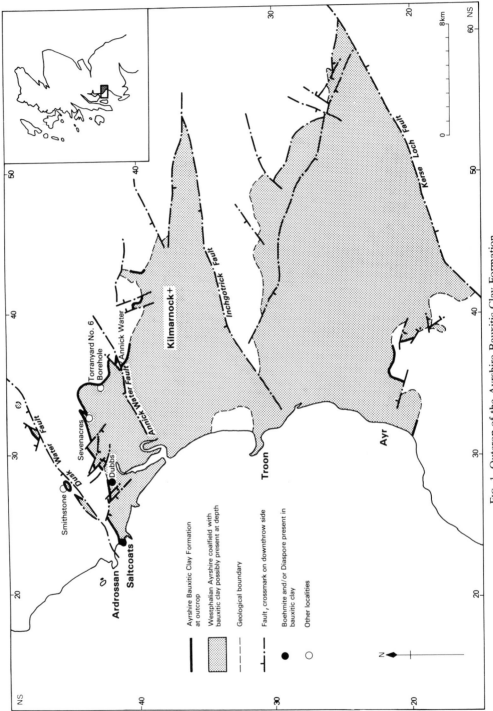

FIG. 1. Outcrop of the Ayrshire Bauxitic Clay Formation.

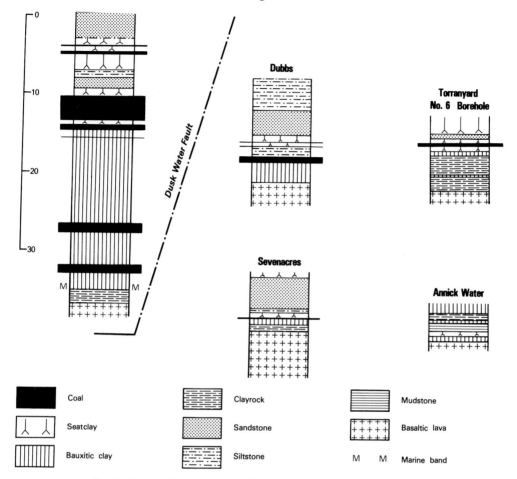

FIG. 2. Comparative sections into the Ayrshire Bauxitic Clay Formation.

clay to basaltic lava is never sharp and is often marked by a transitional clayrock lithology, rich in sphaerosiderite. South of the Dusk Water Fault interbedding with other lithologies occurs locally, as at Annick Water where a black laminated mudstone with plant remains and a seatclay occur within the bauxitic clay.

The bauxitic clay is noted for the occurrence of boehmite and diaspore although, these minerals have not been detected anywhere other than at the original localities examined by Lapparent (1936), at Saltcoats and at Dubbs (Fig. 1).

The Torranyard No. 6 Borehole was located about 12 km east of Dubbs Quarry (Fig. 1). The borehole penetrated basal Coal Measures strata, the Ayrshire Bauxitic Clay Formation, and ended in altered basalts (Fig. 3). The sequence developed here is similar in thickness to

that at Dubbs and typical of most of the northern outcrop of the Ayrshire Bauxitic Clay Formation to the south of the Dusk Water Fault. A chemical and mineralogical study has been undertaken of the core of the borehole to gain a better understanding of the relationship between flint clays and the bauxitic clay and to examine Wilson's proposals concerning the origin of the latter.

Petrography

The Torranyard No. 6 Bore (Fig. 3) was taken to a depth of 14.5 m and terminated in altered basalt in which the feldspar laths are mostly replaced by kaolinite and the groundmass completely altered to green clay material. The green clay varies somewhat in optical properties, being in places very dark coloured

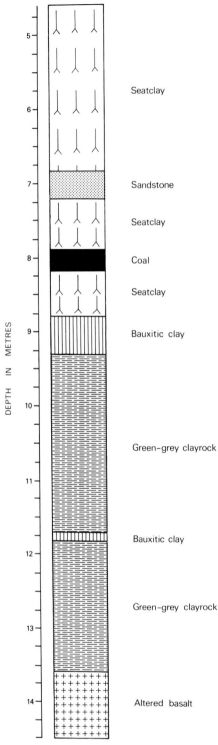

DEPTH IN METRES

Seatclay

Sandstone

Seatclay

Coal

Seatclay

Bauxitic clay

Green–grey clayrock

Bauxitic clay

Green–grey clayrock

Altered basalt

FIG. 3. Detailed section of the Torranyard No. 6 Bore.

and almost isotropic whereas elsewhere it is a much paler shade and the birefringence is considerably higher. Disseminated through it are fine vermicular crystals, which also are characterised by a high birefringence. Upwards the volcanic textures become progressively more obscure and the kaolinite content increases at the expense of feldspar and particularly the green clay.

At a depth of 13.5 m from the surface the altered basalt is overlain by a sequence of green-grey clayrocks which for the most part have clastic textures. Kaolinite is the dominant mineral constituent of these rocks although siderite is generally abundant, whereas the green clay material is either absent or sparse. In several samples from this zone the kaolinite microlites are aligned parallel to the direction of the bedding and this has imparted aggregate birefringence to the rock. The effect may be due to compaction following deep burial but, since the immediately underlying and overlying rocks generally have a massive structure, it would appear more likely that the alignment is a primary feature that has arisen through the settling of kaolinite microlites in a shallow lake or flood-plain environment. Fine vermicular crystals resembling those encountered in the altered basalt but composed of kaolinite, are evident in thin sections of these clayrocks and in one sample obtained from the 13.0 m level, clasts containing residual volcanic textures were observed. Similar clasts have been described from flint clays overlying weathered basalt of Early Permian age in the Sydney Basin, Australia (Loughnan 1973, 1975).

The green-grey clayrocks are overlain at a depth of 11.8 m by a thin bed of dense, massive bauxitic clay comprising abundant relatively clear ooliths and pisoliths, the latter ranging up to 5 mm across, together with angular clasts embedded in a brownish, fine grained kaolinitic matrix. The oolites, pisolites and to some extent the clasts, contain abundant, fine vermicular kaolinite crystals but ferruginous minerals, mostly siderite, are commonly present and tend to depict the concentric structure of the ooliths and pisoliths.

The thin bed of bauxitic clay is succeeded by a second zone of green-grey clayrocks, about 2.4 m thick, that do not differ appreciably in composition and texture from those forming the interval between 13.5 and 11.8 m. Beds composed predominantly of oriented kaolinite microlites interspersed with more massive units are also relatively common in this zone but clasts containing volcanic textures have not been observed.

FIG. 4. Photograph of a thin section of the bauxitic clay at a depth of 8.8 m. Note the presence of coarse clay clasts and small oolites. The section measures 4 cm across.

The upper zone of dense bauxitic clay forming the interval between 9.3 and 8.8 m, has a massive appearance and a grey-brown colour. Toward the base of the zone coarse grains and spherulites of siderite are common and the texture is not dissimilar to that of the underlying green-grey clayrocks. Higher in the sequence, however, ooliths and pisoliths become increasingly more abundant and at the top, tend to predominate. Nevertheless, coarse angular clay clasts, which resemble some of the underlying green-grey clayrocks in that they have aggregate birefringence, are also relatively common (Fig. 4). These clasts are, without doubt, of sedimentary origin and probably are intraformational having been derived from the green-grey clayrocks.

A sequence of typical Coal Measures strata including seatclay, argillaceous sandstone, fireclay and coal, overlies the upper zone of bauxitic clay.

Chemical data

Chemical analyses have been made of representative samples of the core for both major and trace elements (Tables 1 and 2) using a combination of techniques including X-ray fluorescence spectrometry, atomic absorption spectrophotometry and gravimetry, to complement the X-ray diffraction data in evaluating the mineral composition and to ascertain whether trends in depletion and enrichment of the various elements accord with those characterizing weathered profiles generally. Chemical weathering is a process of dissolution and removal of the more mobile elements, e.g. the alkalies and alkaline earths, with the concentration of those less mobile such as aluminium and titanium. Since these processes are most active at the surface it follows that decomposition of a homogeneous parent rock is accompanied by progressive enrichment of the immobile elements as the surface is approached. Complete homogeneity is rarely encountered in nature and consequently, to alleviate difficulties arising in this respect, an immobile element such as aluminium is used as an internal standard.

In constructing of the graphs shown in Figs 5–8, the sample obtained from the base of the altered basalt has been taken as the reference parent material and for each of the elements the concentration ratios, defined as follows, have been plotted.

The Ayrshire Bauxitic Clay

TABLE 1. *Chemical analyses of major elements.*

Depth m	8.5	8.8	9.0	9.2	9.4	9.6	9.8	10.0	10.2	10.4	10.6	10.8	11.0	11.2	11.4
SiO_2	41.5	39.4	41.7	40.2	39.5	38.4	25.3	26.7	32.3	21.8	24.2	35.3	30.7	24.6	29.2
Al_2O_3	36.7	36.5	38.1	35.8	34.3	33.5	22.0	23.4	28.1	19.1	20.0	30.8	27.1	21.4	25.1
Fe_2O_3	2.22	4.44	2.37	4.15	5.22	6.44	27.6	25.6	16.53	32.5	28.9	13.5	20.4	30.0	24.1
TiO_2	4.15	4.38	4.10	5.61	5.32	5.87	3.56	2.84	4.57	3.37	6.22	3.56	3.76	2.46	3.40
MnO_2	0.01	0.02	0.01	0.02	0.03	0.04	0.15	0.16	0.10	0.23	0.15	0.11	0.09	0.11	0.09
CaO	0.46	0.15	0.15	0.18	0.20	0.21	0.36	0.73	0.36	0.65	0.85	0.40	0.91	1.36	0.94
MgO	0.15	0.92	0.21	0.10	0.13	0.17	0.40	0.32	0.22	0.54	0.66	0.66	1.02	1.48	1.40
K_2O	0.28	0.06	0.02	0.05	0.05	0.06	0.03	0.03	0.07	0.02	0.05	0.05	0.05	0.03	0.05
Na_2O	0.57	0.22	0.01	0.02	0.12	0.22	0.03	–	0.11	–	0.08	–	0.10	–	0.03
P_2O_5	0.22	0.08	0.12	0.17	0.11	0.10	0.13	0.43	0.16	0.37	0.14	0.17	0.52	0.67	0.55
SO_3	0.14	0.08	0.14	0.11	0.13	0.10	0.15	0.89	0.14	0.63	0.38	0.10	0.09	0.06	0.09
H_2O	13.03	12.9	13.09	12.79	12.78	12.93	8.01	9.24	10.97	8.18	8.28	11.60	10.33	9.11	10.11
CO_2	1.03	0.44	0.28	0.48	1.52	1.30	13.1	9.24	6.40	11.19	10.19	3.31	5.26	8.31	5.07
Total	100.5	99.6	100.2	100.2	99.4	99.3	100.7	99.6	100.0	99.4	100.1	99.5	100.3	99.4	100.3

Depth m	11.6	11.8	12.0	12.2	12.4	12.6	12.8	13.0	13.2	13.4	13.6	13.8	14.0	14.2	14.4
SiO_2	28.1	41.2	27.0	35.2	33.1	26.7	23.1	24.5	21.8	38.8	34.3	33.8	39.0	40.3	42.3
Al_2O_3	24.1	38.2	24.5	30.0	28.8	22.6	19.6	21.4	17.6	31.7	28.5	25.6	25.2	22.5	19.6
Fe_2O_3	25.3	3.09	26.4	15.0	19.5	27.1	30.9	30.1	31.5	10.8	18.1	19.3	17.1	17.3	17.0
TiO_2	3.37	3.37	3.08	3.50	4.11	2.91	2.57	3.07	2.15	4.20	3.53	3.50	3.40	3.03	2.81
MnO_2	0.12	0.01	0.26	0.11	0.05	0.27	0.35	0.16	0.65	0.01	0.09	0.08	0.05	0.05	0.05
CaO	1.33	0.14	1.11	0.47	0.29	1.24	1.72	1.16	2.08	0.27	0.95	0.97	0.69	1.13	1.67
MgO	1.07	0.41	–	0.84	–	1.89	1.10	1.63	1.09	0.31	1.38	1.82	1.76	2.28	2.68
K_2O	0.04	0.05	0.04	0.06	0.07	0.05	0.03	0.05	0.03	0.13	0.10	0.14	0.89	0.76	1.93
Na_2O	–	0.23	–	0.13	–	0.07	–	0.02	–	0.14	–	0.29	0.71	0.36	1.00
P_2O_5	0.73	0.09	0.46	0.20	0.21	0.14	0.15	0.54	0.13	0.04	0.06	0.09	0.07	0.22	0.57
SO_3	0.11	0.09	0.11	0.69	0.13	0.00	0.05	0.07	0.06	0.05	0.12	0.11	0.07	0.08	0.08
H_2O	9.59	13.46	9.13	11.28	11.33	8.75	7.50	8.86	6.84	10.43	12.51	10.00	10.18	10.62	7.98
CO_2	6.90	0.35	8.06	2.86	1.65	8.69	12.35	8.65	16.15	3.45	0.28	4.47	0.65	1.15	2.00
Total	100.7	99.6	100.2	100.3	99.4	100.4	99.6	100.2	100.1	100.3	99.9	100.1	100.3	99.8	99.7

TABLE 2. *Trace elements expressed as parts per million*

Depth(m)	8.5	8.8	9.0	9.2	9.4	9.6	9.8	10.0	10.2	10.4	10.6	10.8	11.0	11.2	11.4
V	215	400	305	335	325	330	540	490	540	605	565	275	455	345	400
Cr	340	495	405	390	465	525	1025	1430	1140	2130	1170	405	230	275	220
Ba	440	75	40	225	135	120	160	10	290	110	170	280	205	130	340
Nb	110	115	85	90	75	90	55	45	55	30	55	40	40	20	35
Zr	440	690	505	450	355	395	200	175	275	155	225	235	390	135	185
Y	30	20	15	15	15	10	5	5	10	10	10	5	10	5	10
Sr	205	120	180	410	285	125	195	295	185	485	365	315	210	125	225
Rb	20	15	15	20	15	25	45	45	45	20	20	5	10	20	15

Depth(m)	11.6	11.8	12.0	12.2	12.4	12.6	12.8	13.0	13.2	13.4	13.6	13.8	14.0	14.2	14.4
V	390	255	495	355	540	590	600	445	535	360	475	410	340	265	205
Cr	325	395	370	480	805	670	650	235	310	505	565	530	670	1205	510
Ba	230	40	270	520	335	305	155	195	320	580	350	240	870	220	335
Nb	30	80	30	35	40	30	25	30	25	50	40	45	45	35	35
Zr	185	555	195	205	230	165	140	160	95	280	145	215	239	175	165
Y	15	20	10	10	10	5	5	5	15	15	15	55	45	35	40
Sr	460	85	725	660	470	145	115	170	125	85	90	85	130	155	140
Rb	15	5	15	5	5	25	25	20	25	5	10	15	35	10	60

$$\text{Concentration ratio of element} = \frac{\% \text{ element}_{(d)}}{\% \text{ element}_{(ref)}} \times \frac{\% \text{Al}_{(ref)}}{\% \text{Al}_{(d)}}$$

where d = sample at a specific depth and ref = reference parent material.

In these graphs greater enrichment relative to that of aluminium is indicated by values above unity whereas those less than unity show relative depletion.

Of the major elements, sodium and potassium are readily mobile and within the zone of altered basalt there is a marked decline in their concentration relative to that of aluminium with decreasing depth (Fig. 5). Above this zone the values are very low and little variation is evident. Silicon, which is also a potentially mobile constituent, is depleted to some extent in the altered basalt zone, but in the succeeding units it has been stabilized presumably through incorporation in the remarkably resistant kaolinite structure. On the other hand titanium is stable in most weathering environments and as such the plot of the concentration ratios should coincide with that of aluminium. However, the curve shown in Fig. 5 contains noticeable departures from this trend, especially in the upper zone of green-grey clayrocks and also in the two beds of bauxitic clay where, contrary to expectation, the lowest values were recorded. Calcium, magnesium and ferrous iron

resemble the alkalies in being readily mobile but as is evident in Fig. 5, there is no overall depletion of these elements with decreasing depth. Rather, the two curves have much the same erratic trend suggestive of simultaneous precipitation from either surface or ground waters prior to or immediately following shallow burial. Certainly the coarsely crystalline form of the siderite in these rocks gives credence to the concept of authigenic development of the mineral.

Trace elements generally tend to have a more variable distribution in the parent rock than the major elements and consequently their concentration curves rarely appear quite as smooth. This is evident in Figs 6 and 7 where the concentration ratios have been plotted for rubidium, barium and strontium, all of which are readily mobile, yttrium which is mobile under acid conditions and also for zirconium, niobium, chromium and vanadium, which according to Goldschmidt (1954) rank among the most immobile of the elements under reducing conditions.

The curve for yttrium is similar to that recorded for sodium and potassium in that the concentration ratios decrease toward the top of the altered basalt zone and relatively low values persist in the overlying clayrocks. For the remaining mobile elements, however, the concentration curves appear devoid of distinctive trends and, like calcium, magnesium and ferr-

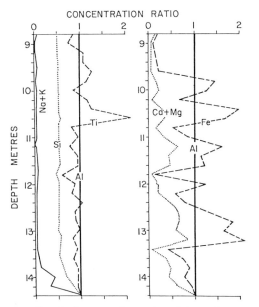

FIG. 5. Concentration ratios of the major elements relative to that of aluminium and plotted against depth.

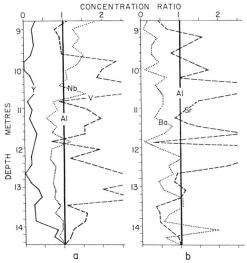

FIG. 6. Concentration ratios for yttrium, niobium, vanadium, barium and strontium relative to that of aluminium and plotted against depth.

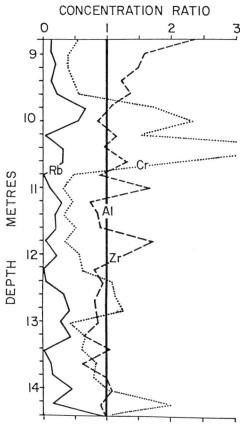

FIG. 7. Concentration ratios for rubidium, chromium and zirconium relative to that of aluminium and plotted against depth.

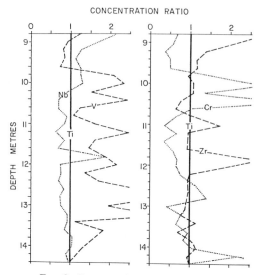

FIG. 8. Concentration ratios for niobium, vanadium, chromium and zirconium relative to that of titanium and plotted against depth.

ous iron, these elements have possibly been introduced by either surface waters at the time of deposition or subsequently by groundwaters. Of the immobile trace elements, zirconium and niobium tend to be concentrated to much the same degree as aluminium although both are somewhat enriched in the bauxitic clay as well as in parts of the upper zone of green-grey clayrocks. On the other hand, the curves for chromium and vanadium appear to reflect random distribution for these elements with the only noticeable trend being relative depletion in the two zones of bauxitic clay. In this respect they resemble the curve recorded for titanium.

In Fig. 8 the concentration ratios for niobium, vanadium, zirconium and chromium have been replotted using titanium as the internal standard. However, the results do not differ appreciably from those obtained with aluminium as the standard.

Mineral composition

Quantitative evaluation of the mineral composition of each of the core samples from the Torranyard bore has been made from the X-ray diffraction and chemical data and the results for the altered basalt, green-grey clayrocks and bauxitic clay are plotted against depth in Fig. 9. For these determinations assumptions with respect to the structural formulae of some of the minerals were necessary and hence, the values shown in Fig. 9 must be regarded as approximations only. Nevertheless, since the objective is the establishment of possible trends, the method is believed satisfactory for the purpose.

Kaolinite is the dominant constituent in all samples of the core from below the sandstone bed (Fig. 3). At the base of the altered basalt it accounts for 40% of the rock but higher in the sequence it becomes progressively more abundant and at the 13.6 m level exceeds 80%. In the overlying strata the kaolinite content varies inversely with that of the ferruginous minerals reaching maxima in the two zones of bauxitic clay where ferruginous minerals are relatively sparse. In the altered basalt and green-grey clayrocks as well as in the coarse birefringent clasts within the bauxitic clay, the kaolinite is fine-grained and of the *b*-axis disordered type similar to that characterizing seatclays and underclays. However, the vermicular crystals comprise a very well-ordered form of the min-

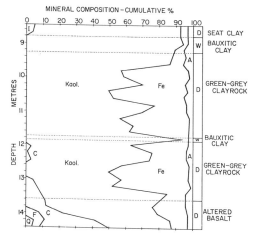

FIG. 9. Mineral composition in relation to depth in the lower part of the section exposed in the Torranyard No. 6 Bore. Q = quartz; F = feldspar; C = chlorite and degraded chlorite; I = illite; Kaol. = kaolinite; Fe = ferruginous minerals, mostly siderite and A = anatase. W = kaolinite is predominantly well-ordered. D = kaolinite is mainly disordered.

eral and consequently, where these are abundant as in the bauxitic clay, the kaolinite yields Hinckley indices in the range of 0.9–1.2 (Fig. 10).

Much of the green clay in the altered basalt and parts of the green-grey clayrocks have the X-ray diffraction pattern of a smectite in that the basal spacing expands readily from about 14–17 Å on glycol saturation and, after heating to 450°C, collapses to approximately 10 Å. Nevertheless, some samples especially from the base of the alterered basalt zone, contain minor amounts of a non-expandable 14 Å mineral, presumably chlorite, and also what appear to be intergrades of chlorite and smectite. Apparently the green clay represents chlorite in varying degrees of breakdown and, since it is either absent or rare in samples from above the altered basalt zone, presumably kaolinite and ferruginous minerals are the end products of this degradation process.

Siderite is by far the most abundant of the ferruginous minerals and in some samples of the green-grey clayrocks it exceeds 40% of the total constituents. Mostly it occurs as coarsely crystalline grains and spherulites, some of which measure more than 3 mm across. Several of the grains and spherulites have a brown rim indicative of incipient oxidation. The siderite is undoubtedly of authigenic origin and apparently formed by slow precipitation in a strongly reducing environment.

Feldspar occurs as remnants in highly kaolinised laths at the base of the altered basalt zone. It has multiple twinning and a mean refractive index at or slightly above that of the associated kaolinite and probably is of labradorite composition.

Quartz is absent from all samples examined from below the sandstone unit (Fig. 3) with the exception of those at the base of the altered basalt zone where a very small amount of the

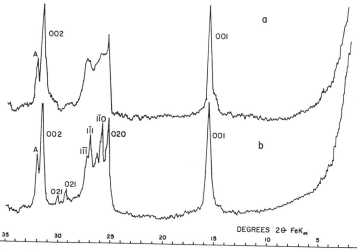

FIG. 10. X-ray diffraction traces. (a) Disordered kaolinite typical of the altered basalt and green-grey clayrocks. (b) Well-ordered kaolinite characteristic of the bauxitic clay.

mineral, present probably in the form of chalcedony, accompanies the feldspar remnants. However, in the sandstone and overlying seatclay it is abundant and not infrequently forms the dominant mineral constituent.

Illite is invariably associated with quartz in the sandstone and overlying strata, and also is present in the seatclays both above and below the thin coal seam where quartz is apparently absent. However, it has not been encountered in the bauxitic clay, green-grey clay rocks and altered basalts.

All X-ray diffraction traces of the altered basalt, green-grey clayrocks and bauxitic clay have a reflection at 3.51 Å corresponding to the 101 spacing of anatase. The intensity of this reflection increases with the TiO_2 content of the sample and, since rutile has not been detected in the strata by either X-ray or petrographic means, presumably most if not all of the titania recorded in the chemical analyses (Table 1) is present as anatase.

Boehmite and diaspore have not been recognized in samples of the core from this bore by either petrographic or X-ray diffraction methods. Nevertheless, chemical analyses of the bauxitic clay at a depth of 11.8 m do indicate a slight excess of alumina over that required to combine with silica to yield kaolinite and possibly one or both of these aluminium oxyhydroxides minerals is present as submicroscopic particles and in amounts below the level of detection by X-ray diffraction.

Discussion

In Table 3 the various features of the bauxitic clay are summarized and compared with those considered by Keller (1967) as characteristic of flint clays. In terms of occurrence, sedimentological structures, mineral composition and textural variation the two are virtually indistinguishable, a fact that supports the proposal that the bauxitic clay is indeed a flint clay.

With respect to the mechanism of formation of the bauxitic clay, the absence of quartz and illite from both the bauxitic clay and green-grey clayrocks leaves little doubt that these units are derived from the basalt. They may, however, have formed in a variety of ways:

(a) they may have formed *in situ* through weathering of the basalt, in which case two superimposed profiles are present and the green-grey clay rocks represent an intermediate stage in the transition from basalt to bauxitic clay;

(b) they may consist of accumulations of detritus eroded from pre-existing soils developed on the basalt, with both the bauxitic clay and the clayrock being of transported origin;

(c) an eroded surface may separate the grey-green clayrocks from the bauxitic clay and only the latter may be of transported origin.

Over the deposit as a whole there is a marked consistency in the bauxitic clay, clayrock, basal-

TABLE 3. *A comparison of various features of the Ayrshire bauxitic clay with those of a typical flint clay*

	Ayrshire bauxitic clay[a]	Flint clay[b]
Occurrence	In non-marine strata associated with coal	In non-marine strata, frequently but not invariably associated with coal
Structures		
Bedding	Coarse and lenticular	Coarse and lenticular
Lamination	Non-laminated, massive	Non-laminated, massive
Induration	Considerable compared with associated strata	Considerable compared with associated strata
Composition		
Dominant	Kaolinite, well-ordered	Kaolinite, generally well-ordered
Accessories	Siderite, anatase (Boehmite and diaspore recorded elsewhere)	diaspore, boehmite, quartz, siderite, anatase, illite
Textures	Cryptocrystalline, oolitic, vermicular, brecciated	Variable. Cryptocrystalline, oolitic, brecciated, vermicular

[a]Exposed in the Torranyard No. 6 Bore.
[b]From Keller (1967).

tic lava profile, a factor which lead Wilson (1922) to conclude that the bauxitic clay developed largely *in situ*. Unlike autochthonous lateritic bauxites of more recent origin, such as those at Weipa (Loughnan & Bayliss 1961) and at Gove (Somm 1975) in northern Australia, there is no clear differentiation of the succession into pallid, mottled and concretionary zones nor is there the marked concentration of ferric oxides and hydroxides along with alumina and other immobile elements in the bauxitic clay. The succession has undoubtedly been subjected to diagenetic alteration involving reduction, mobilization and redistribution of the iron and probably also partial silicification of gibbsite as discussed by Wilson (1922). The latter would seem to be the most likely explanation for the vermicular habit and well-ordered structure of much of the kaolinite in the bauxitic clay as well as the unusual density and induration of the rocks.

The well-ordered structure of the kaolinite of the bauxitic clays is in contrast to the disordered structures of the kaolinite in the clayrocks. Steep synaeresis cracks with films of well-crystallized kaolinite occur at Smithstone and emphasize the role of early diagenesis in the formation of the bauxitic clays.

Wilson (1922) recognized an oolitic or pisolitic variety of bauxitic clay which he considered to be of allochthonous origin being formed on the floor of shallow lagoons. There are, however, a number of other features which support an allochthonous origin for at least some of the bauxitic clays. A relative deficiency of some of the immobile elements, notably titanium and chromium, and the corresponding enrichment of others, e.g. zirconium and niobium, occurs in the bauxitic clay of the borehole section. Cameron (1980) states that TiO_2 levels of around 5.53% may be expected in a residual deposit with an average Al_2O_3 value of 37.31%. Most of the titanium is present within the resistate minerals, rutile, titanomagnetite or illmenite. Cameron suggests that the very few higher TiO_2 values, up to 14.17%, may be due to concentration of these resistate minerals in localized heavy mineral placers. Likewise removal of these minerals may also result in anomalously low TiO_2 values.

The development of bauxitic clay north of the Dusk Water Fault is typified by the succession at Smithstone (Fig. 2). Repeated sequences of coal and bauxitic clay occur with no evidence to suggest that these coals were ever covered by hot lava flows or ash falls. A marine band locally preserved in the Smithstone area at the base of the bauxitic clay confirms that here the bauxitic clay is of allochthonous origin, as part of a cyclical sequence of sediments. At Annick Water (Fig. 2) black mudstones with plant remains occur interbedded with bauxitic clay. If the bauxitic clays at these locations had formed *in situ*, the coal and plant remains would have been destroyed by the intense weathering necessary to convert basalt to kaolinite and bauxitic minerals.

The most convincing argument favouring an allochthonous origin for the section from the Torranyard bore is the occurrence within the bauxitic clay of randomly orientated clasts composed of parallel microlites of disordered kaolinite. These clasts have been derived from a sedimentary source, the most likely being a reworking of the underlying clayrocks, and hence, their presence refutes the concept that the bauxitic clay is here developed *in situ*.

In addition to Wilson's criteria of the presence of ooliths and pisoliths, an allochthonous origin, in part at least, for the bauxitic clay may be postulated when any of the following features are developed;

(a) occurrence of superimposed profiles or other sediments interbedded with bauxitic clay;

(b) presence of randomly orientated clasts of parallel microlites of disordered kaolinite within the bauxitic clay;

(c) anomalously high (or low) values of immobile elements where these result from concentration or depletion of resistate minerals.

On these criteria the bauxitic clay development north of the Dusk Water Fault is regarded as being largely allochthonous in origin with some local autochthonous clayrock grading down into basaltic lava. South of the Dusk Water Fault the evidence would suggest that part of the bauxitic clay is at some localities allochthonous in origin, but elsewhere it may have developed *in situ*.

ACKNOWLEDGMENTS: The work of one of the authors (S. K. Monro) was carried out as part of the programme of the South Lowlands Unit of the Institute of Geological Sciences. He gratefully acknowledges helpful discussions with I. B. Cameron and A. Davies. The paper is published with permission of the Director, Institute of Geological Sciences (NERC).

References

BOSAZZA, V. L. 1946. *The Petrography and Petrology of South African Clays.* Published by author, Johannesburg. 313 pp.

CAMERON, I. B. 1980. Titanium dioxide in the Ayrshire Bauxitic Clay. *Miner. Reconnaissance Prog. Rep. Inst. Geol. Sci.*

GOLDSCHMIDT, V. M. 1954. *Geochemistry.* MUIR, A. (ed.) Clarendon Press, Oxford. 730 pp.

KELLER, W. D. 1967. Flint clay and flint-clay facies. *Clays Clay Miner.* **16**, 113–28.

LAPPARENT, J. de 1936. Boehmite and diaspore in the bauxitic clays of Ayrshire. *Summ. Prog. geol. Surv. Gt. Br.* for 1934, pt 2, 1–7.

LOUGHNAN, F. C. 1973. Kaolinite clayrocks of the Koogah Formation, New South Wales. *J. geol. Soc. Aust.* **20**, 329–41.

LOUGHNAN, F. C. 1975. Laterites and flint clays in the early Permian of Sydney Basin, and their palaeoclimatic implications. *J. sedim. Petrol.* **45**, 591–8.

LOUGHNAN, F. C. 1978. Flint clays, tonsteins and kaolinite clayrock facies. *Clay Miner.* **13**, 387–400.

LOUGHNAN, F. C. & BAYLISS, P. 1961. The mineralogy of the bauxite deposits near Weipa, Queensland. *Am. Mineral.* **46**, 209–17.

RICHEY, J. E. 1930. The geology of North Ayrshire *Mem. geol. Surv. U.K.* 197–220.

ROBERTSON, R. H. S. 1971. Ayrshire bauxitic clay, a flint clay. *Clays Clay Miner.* **19**, 341.

SOMM, A. F. 1975. Gove bauxite deposits, N. T. *Aust. Inst. Min. Metall., Mono 5,* **1**, 964–8.

SMITH, J. 1895. On a bed of ironstone occurring on trap tuff in the parishes of Stevenston, Dalry and Kilwinning. *Trans. geol. Soc. Glasg.* **10**, 133–6.

WILSON, G. V. 1922. The Ayrshire bauxitic clay. *Mem. geol. Surv. U.K.* 28 pp.

S. K. MONRO, Institute of Geological Sciences, Murchison House, West Mains Road, Edinburgh EH9 3LA.

F. C. LOUGHNAN & M. C. WALKER, School of Applied Geology, University of New South Wales, P.O. Box 1, Kensington, New South Wales 2033, Australia.

Base metal concentrations in kaolinised and silicified lavas of the Central Burma volcanics

T. R. Marshall, B. J. Amos & D. Stephenson

SUMMARY: Detailed mapping and limited geochemical sampling of a Lower Pliocene trachyandesite-sediment-tuff sequence reveals two juxtaposed but distinct kaolinised sequences caused by tropical weathering and hydrothermal activity. The stratigraphy and sequence of alteration in the lower Taungni section is difficult to establish. The kaolinised rocks are characterized by high lead and arsenic content and contain large irregular masses of silica. Kaolinisation resulted either directly from metalliferous hydrothermal activity or from acidic fluids released by oxidation of sulphides. The upper, well-bedded Kyauktaga section shows regular redistribution of silica resembling that of the silcrete profile and vertical base metal zoning attributable to groundwater movement.

Tertiary sediments of the Central Lowlands of Burma are bounded on the west by the folded late Cretaceous to Eocene flysch succession of the Indoburman Ranges (Brunnschweiler 1966), and to the east by the Shan Plateau, part of a cratonic unit extending into Thailand (Baum *et al*. 1970). The volcanic line of Central Burma (Chhibber 1934) forms the axis of the Central Lowlands extending from Mt Popa volcano northwards to Taungthonlon volcano. South of Mt Popa its probable extension is indicated by doleritic sills in the Pegu Yoma and volcanics on Narcondam and Barren Islands. Along this arc, significant copper deposits, possibly of porphyry type, are known in the Banmauk area and around Monywa (Goossens 1978).

In this paper we report on an occurrence in the Mt Popa area of kaolinised and silicified volcanics characterised by high base metal content and discuss the role of surface and hydrothermal processes in their formation. Details of a complete economic investigation in this area are contained in an unpublished IGS report (Amos *et al*. 1981).

Stratigraphy and structure

The calc-alkaline volcanics of the Mt Popa area are associated with two major sedimentary units (Amos *et al*. 1983). The older unit is the Miocene Myinde group of interbedded shales and fine grained, often calcareous sandstones, overlain in some areas by the transitional Kadetkon formation. Together these form the upper part of the Peguan supergroup of Central Burma. The younger unit is the Pliocene Irrawaddy formation, a shallow water, fluvial, clastic sequence ranging from conglomerates to siltstones, but predominantly poorly lithified,

massive, cross-bedded sandstone. Chhibber (1934) recognized a major division between an 'Older Volcanic' group interbedded and folded with the upper Tertiary sediments and a 'Younger Volcanic' group comprising the 1518 m composite volcano of Mt Popa and associated deposits which rest unconformably on top of the sediments and 'Older Volcanics'. The term 'Older Volcanics' describes a number of isolated outcrops of trachyandesite, rhyodacite and tuff which cannot be correlated as a single unit (Stephenson, Marshall & Amos 1983). However much of the 'older' volcanic activity took place within a narrow stratigraphic range and may be considered almost contemporaneous with an early Pliocene age.

The rocks of Taungni Range (Fig. 1) are composed of a series of 'older volcanics' intercalated with sandstones and minor shales of the Irrawaddy formation. The volcanics are predominantly trachyandesite flows with rubbly tops but agglomerates and minor intercalations of pyroclastics have also been observed. Where fresh, the trachyandesites are invariably fine grained lavas, containing plagioclase, hypersthene, augite and biotite microphenocrysts. In the lower part of the succession, the volcanics form relatively thin bands and lenses intercalated with a high proportion of Irrawaddy sandstones. East of the summit of Taungni Hill these rocks are overlain by a predominantly volcanic unit, the Myage volcanic member, containing only minor intercalations of sediment. This in turn is unconformably overlain by a distinctive series of heavily altered lavas, the Kyauktaga volcanic member (Fig. 2).

Exposures on Taungni Hill are very poor, much of the surface between crags of silica rock being covered by debris of resistant siliceous material which totally masks the underlying bedrock. Although this precludes a detailed

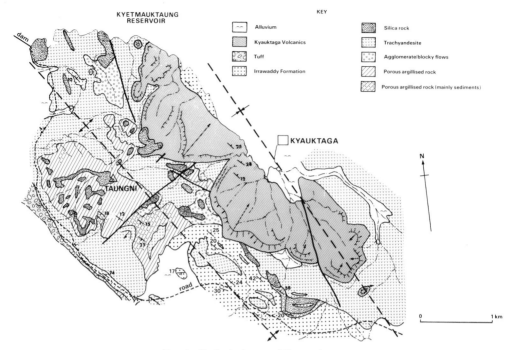

FIG. 1. Geological map of Taungni range.

knowledge of the stratigraphy or structure, it can be determined that the rocks are disposed in a broad, faulted anticline trending southeast through the summit of the hill. The fold continues towards the south-east in a series of offset, faulted segments; volcanics thin out in this direction, and although areas of alteration occur there is no evidence of mineralization. North-west from Taungni Hill the fold reappears, slightly offset, affecting tuffs of the 'older volcanics'. North-east of the Taungni anticline the Kyauktaga syncline is well-defined by the Kyauktaga volcanic member, the dip surface of which occupies the core of the syncline, overlain in places by small, remnant patches of pebbly Irrawaddy sandstone.

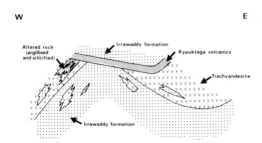

FIG. 2. Diagramatic stratigraphy of Taungni Hill indicating areas of alteration.

Alteration products

The main end products of alteration in the volcanics and associated sediments are a hard, flinty silica rock and a soft, porous, chalky, low density argillised rock containing a high proportion of clay minerals.

The flinty rock, composed of almost pure silica, occurs in a variety of colours and displays a wide range of textures. It ranges from white through pale cream, grey, pink and brown to, more rarely, dark blue or black. In some areas the pale coloured varieties have developed concentric colour banding. Most frequently it is completely amorphous or shows a 'flocculated' texture, but many silicified trachyandesites retain relict feldspars (up to 2–3 mm) and in some places these form a crowded mass of euhedral crystal outlines. The blue varieties of silica rock almost invariably preserve these relict textures, while the majority of paler varieties do not. In thin section, plagioclase laths and relics of rectangular orthopyroxene and irregular clinopyroxene are often still visible and in some samples unaltered cores with silicified rims are present. Cryptocrystalline cherty silica and brown amorphous opaline silica occur in varying proportions, but crystalline quartz is never present. With increased silicification, silica begins to segregate, forming cuspate-bordered areas of banded opaline silica within a cherty matrix. Disseminated haematite within the silica gives many of the hand specimens a pink or red colouration. Occasionally the silicified rock can be seen to be full of lithic fragments, equally silicified, but varying in colour and texture from the matrix and from each other. Voids in the silica rock are extremely rare; where they occur they are lined with small crystals of quartz, or more rarely by botryoidal haematite.

Evidence of the continuity of the silicification process is provided by intraformational breccias containing clasts of andesite which were clearly silicified before incorporation into the breccia. These clasts are set in a matrix of well-sorted opaline silica grains surrounded by cryptocrystalline cherty silica, or sometimes cherty grains surrounded by opaline silica. Some clasts are rimmed by fibrous chalcedony suggesting that they were originally opaline silica and were subsequently converted to cherty silica with consequent dehydration and shrinkage. Certain white sandstones consist of well-sorted, rounded clasts of opaline silica in an unsilicified fine grained matrix.

Finely-disseminated pyrite, rarely accompanied by galena and chalcopyrite, is usually visible in diffuse patches in the dark blue variety of silica rock, and occasionally forms discrete bands or veinlets. Pyrite is very rare in any of the other varieties of silica rock, though small spherulites of pyrite have been observed in a few samples of white silica rock. Granular aggregates of jarosite and alunite, characteristic of the argillic alteration on Taungni Hill, have been found in some thin sections.

The other main type of alteration product is a porous argillised rock, typically brilliant white, soft and friable, with a very low bulk density. It has a chalky appearance, soils the fingers and sticks to the tongue. When completely white it is amorphous, but violet or purple staining sometimes makes visible a relict andesitic texture marked by unstained crystal relics. Colour staining usually occurs in diffuse patches up to a few metres across, but can be localized along joints or in bands; it may be violet, purple (both caused by haematite), yellow or brown, but the colouration is never intense. At some localities the rock is banded by layers of different colour or texture, or both, but commonly the rock is white, amorphous and structureless.

Analyses indicate that this rock contains considerable quantities of kaolinite, confirmed by strong but diffuse kaolinite peaks on DTA charts. Pale pink crystals of alunite are not uncommon within the white matrix. Sediments

may also be affected by this type of alteration and may be identified by relict bedding planes which show up well on weathered surfaces. Original shale and sandstone can be distinguished by their contrasting fine and coarse texture.

Alteration patterns

Two distinct areas of argillisation and silicification may be recognized on Taungni Hill: the Kyauktaga volcanic member and the main bulk of Taungni Hill proper together with ridge extensions to the north. Although both show the same end products, differences in pattern and relationships probably reflect differing processes of alteration.

Kyauktage volcanic member

The alteration that has affected the Kyauktaga volcanics differs fundamentally in at least two aspects from that observed elsewhere on Taungni Hill: (i) The Kyauktaga volcanics show a uniform, near-complete alteration throughout the whole of their outcrop; and (ii) Within this alteration silicification is less intense than elsewhere and affects a relatively small proportion of the unit.

Where the unit unconformably overlies outcrops of massive silicified rock contained within the Myage volcanic member, these terminate at its base suggesting that the period of massive silicification predated the deposition of the Kyauktaga volcanics. The outcrop of the Kyauktaga volcanic member is bounded on all sides by steep scarps and has a maximum thickness of 27 m. At a number of localities fresh trachyandesite at the base of the unit is succeeded upwards by rotted, grey andesitic rocks, with white feldspar and black biotite still visible in a soft, crumbly matrix. Commonly within six metres of the scarp base the rocks have given way to a white, porous, low-density rock, usually amorphous, or with relict andesitic texture. In some localities there are textural variations; the rock may contain bands of differing grain size or colouration, or may contain clasts of equally altered rock, distinguishable by differences in texture or colour. These textural differences show best on weathered surfaces, where they closely resemble bands of tuff but are more likely to represent flow tops.

Towards the top of the unit, the porous rock begins to contain small blebs and stringers of white amorphous silica, and these become more numerous upwards. In some places the porous rock takes on a porcellanous texture, very fine grained, totally amorphous, less easily scratched than the normal porous rock but with little or no apparent loss in porosity, or increase in density. The porcellanous variety is commonly characterized by closely spaced sub-parallel joints, individually impersistent but collectively pervasive.

The topographic upper surface of the unit appears to correspond closely to its stratigraphic upper boundary, with a few isolated patches of the overlying Irrawaddy sandstones preserved perched upon it. On this upper surface the rock is predominantly white soft porous rock, but there are a number of scattered outcrops of white or cream silica rock ranging from 1.5 to 9 m across, which may have resulted from local coalescence of the increasingly numerous silica segregations already described. There is no evidence of silicification in the overlying sandstones.

Pits dug in the Kyauktaga volcanics reveal much the same picture as that derived from the scarp face, with soft porous rock at the surface passing down into grey or pale blue, glutious, clayey rock with a clear andesitic texture at a depth of about 8 or 9 m. In more deeply-weathered sections further east on the same scarp, the alteration extends down into sediments beneath the lavas. Bleached sandstones pass upwards into silicified lavas via a zone of strongly leached, rather rubbly rock, including sediments and volcanics, which shows a marked resemblance to the typical silcrete profile, with a rubble layer separating the silcrete from an underlying pallid zone (Mabbut 1967; Langford-Smith & Drury 1965).

Analyses of major elements in samples taken from the scarp face show a fairly complex redistribution of elements. Five samples were analysed, taken from the upper half of the scarp face; only the one nearest the top of the sheet is visibly siliceous. All the iron is in the ferric state, not exceeding 3%, and normally around 0.5%. Calcium, magnesium and manganese are only present at ppm levels, as is sodium in some samples (Table 1). Table 1 also includes semi-quantitative, partial analyses of similar lithologies from the Taungni Hill area.

DTA analyses of four samples indicate kaolinite and altered feldpathic rock with disordered kaolin. Other aluminous phases were not identified, although a veinlet of gibbsite was identified optically in one specimen. It is assumed that the reaction stopped short at the production of kaolinite and the release of silica from K-feldspar, which is inferred to be present in considerable proportions in the groundmass of the trachyandesites (normative orthoclase ranges from 14 to 16% in fresh trachyandesites).

The existence of kaolinite in preference to gibbsite has been discussed by Curtis & Spears (1971) who concluded that gibbsite is stable relative to kaolinite only at very low dissolved silica concentrations. It is possible that tropical weathering of the Kyauktaga volcanics produced gibbsite bearing profiles such as are found in Hawaii at present (Bates 1962). However subsequent deposition of the Irrawaddy formation produced a cover which may have reduced soil pore-water circulation and allowed silica to dissolve to a level where gibbsite became unstable.

TABLE 1. *Analyses of kaolinised/silicified trachyandesites from Kyauktaga and Taungni Hill*

	1	2	3	4	5	6	7
SiO_2	58.58	49.08	77.34	65.36	52.78	64.4	84.5
TiO_2	0.10	0.10	0.10	0.10	0.01	0.76	2.05
Al_2O_3	16.82	29.98	6.73	19.18	30.70	2.0	0.65
Fe_2O_3	3.09	0.56	2.06	0.58	0.04	1.85[a]	5.78[a]
Na_2O	0.25	0.11	0.16	0.05	0.05	nd	nd
K_2O	2.08	0.36	1.17	0.45	0.63	0.14	0.02
P_2O_5	0.60	0.15	0.20	0.36	0.26	nd	nd
Loss on ign.	17.32	18.84	11.48	13.28	15.56	nd	nd
Total	98.91	98.24	99.28	99.40	100.05	89.15	93.00

[a]Total iron.
1–5 Analyses by classical methods, DGSME, Rangoon. FeO, MnO, MgO, CaO present in trace amounts.
6 & 7 Analyses by direct reading emission spectrometry, IGS, London. nd = not determined.
1, 2 & 3 Samples taken at increasing heights up the scarp face of the Kyauktaga volcanics.
4 & 5 Samples from the dip slope surface Kyauktaga volcanics.
6 Kaolinised rock, Taungni Hill.
7 Silicified rock, Taungni Hill.

The analyses confirm the visual impression that silica has segregated and moved generally upwards. Other elements have been efficiently leached from the rock, presumably migrating downwards (see next section). The combination of this pattern of upward migration of silica, together with the widespread nature of the alteration throughout the unit, suggests a process of weathering acting upon a subaerial surface to produce an incomplete development of the silcrete profile.

Taungni Hill

Fresh trachyandesites interbedded with unaltered sandstones and shales are exposed at the level of the main road around the western and southern margins of the hill, while below the road there are scattered exposures of tuff. Above the road silica rock is exposed at numerous localities around flanks, on some of the ridge crests and on the summit itself. This very resistant rock forms several large outstanding crags up to 20 m high and as much as 100 m long. It also forms extensive outcrops at ground surface, small conical hillocks and bouldery linear features. Its relationship to the other rock types is not easily determinable at the surface as the float of siliceous material masks the less resistant rock types.

Trenching has revealed that much of the rock between the outcrops of silica rock is composed of varieties of porous rock, predominantly white but colour-stained in diffuse patches in many areas. Contacts between porous rock and silica rock are generally sharp but irregular. Isolated cores of silica rock are often found enclosed within the porous rock, and these cores range from sub-spherical or lenticular pods of about 10 cm diameter to irregular lenses and anastomising masses a few metres across. Where the general stratification can be determined, the boundaries of the siliceous masses seem to cut across, suggesting that on Taungni Hill the silicification is not stratiform. Most silicified areas are irregular in outline, but some are distinctly linear, suggesting a possible fracture control (Fig. 1).

Exposures of porous rock can be found on Taungni Hill, though they do not form distinct topographical features and comprise much less of the total outcrop.

Distribution of base metals in soils

Orientation studies revealed the ground surface to be underlain directly by weathered bedrock with almost no development of A and B soil horizons. Depth to bedrock ranged from nil to six metres with an average between 50 cm and 1 m. Since sampling at various depths did not reveal any variation in trace element values an arbitrary depth of 25 cm was chosen for the survey. Samples were taken at 50 m intervals along lines cut at 200 m intervals perpendicular to the trend of the main area of alteration. Chemical determinations were made by atomic absorption spectrometry and by colorimetry and the results contoured at intervals suggested by frequency histograms.

The lead distribution pattern (Fig. 3) shows a zone of values in excess of 660 ppm covering most of Taungni Hill proper and extending a NNE direction to the Kyauktaga volcanics. Within this zone, areas with lead in excess of 0.2% occur on Taungni Hill and on the dip slope of the Kyauktaga volcanics. An outer zone with lead in excess of 37 ppm has the same general trend. These values (>37 ppm) also extend south-eastward along the outcrop of the Kyauktaga volcanics, in which they are widespread on the upper dip slope but absent from the base of the unit. Outside the areas of alteration on Taungni Hill and the Kyauktaga volcanics, lead values are uniformly low (<37 ppm).

Most of the area shows zinc values in the range 30–117 ppm (Fig. 3). Values greater than 118 ppm are found in many localized areas, most of which fall within the zones of high lead. The most noticable feature of the zinc distribution is the strong correlation between a well-defined area of low zinc values (below 30 ppm) and the upper surface of the Kyauktaga volcanics. Scattered high zinc values are associated with the base of this unit.

Copper values are low (<85 ppm) and show no regular distribution, although occasional, slightly higher values (85–117 ppm) occur within the zone of high lead values.

Taking all three metals (Pb, Zn, Cu) together, coincident maxima area found in a belt running NNE from Taungni Hill and into the Kyauktaga volcanics. Outside of this belt the upper surface of the Kyauktaga volcanics contains moderately high lead and low zinc values whilst a narrow band at the base of the unit shows higher zinc and lower lead values possibly due to redeposition of material preferentially leached from above.

More detailed sampling was conducted along lines spaced at 50 m intervals within the area of high metal values outlined above. Results are shown in Fig. 4. High lead values (up to 0.9%) and arsenic values (up to 1300 ppm) define an

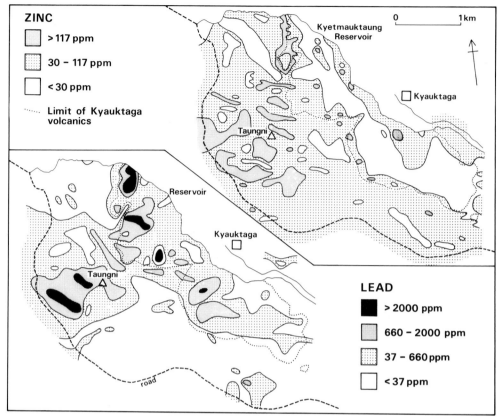

FIG. 3. Distribution of lead and zinc in soil over Taungni Range.

irregular-shaped area which does not show any obvious relationship to the lithology or structure of the rocks. This area also contains slightly higher copper (>58 ppm) than the surrounding areas.

Although zinc seems to correspond with the area of lead concentration in the broad survey of Fig. 3, the more detailed survey (Fig. 4) reveals that, within this area, high lead and high zinc are almost mutually exclusive. Areas of

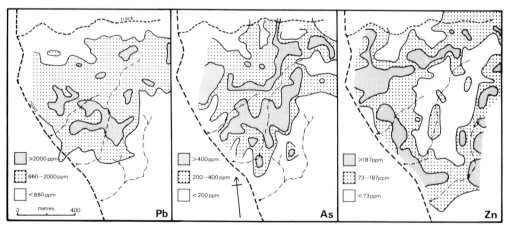

FIG. 4. Distribution of lead, arsenic and zinc in soil within the area of altered rocks, Taungni Hill proper.

relatively high zinc (>187 ppm) enclose an area of lower zinc values which is coincident with the zone of high lead values. High zinc values may in some instances be correlated with geological features such as the main WSW ridge of Taungni Hill, composed entirely of silicified rock, and the lower south-western slopes, underlain by unaltered trachyandesites. Both areas are unaffected by intense, pervasive argillisation.

Bedrock base metal variation

Two trenches were excavated in argillised and silicified rock underlying the zones of highest base metal values on Taungni Hill. Chip samples were taken at 1 m intervals and analysed by atomic absorption spectrometry and semiquantitative optical emission spectrography. Samples of silica rock usually contain the lowest values of most elements. Samples of argillised rock range in lead content from 200 ppm to values in excess of 0.4% but distribution is extremely erratic even within zones of overall high values (Fig. 5). High lead

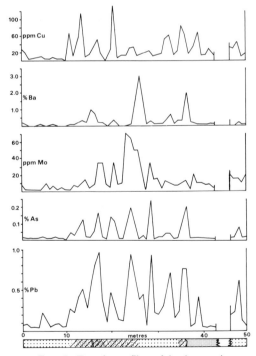

FIG. 5. Trench profiles of lead, arsenic, molybdenum, barium and copper in altered rocks, Taungni Hill proper. Open stipple, argillised rock; shading, stained argillised rock; close stipple, silica rock.

values are associated with high levels of arsenic, barium and molybdenum with generally low copper values and with uniformly low values of zinc, manganese, cobalt and nickel.

Five pits, dug to depths of 7 m in highly altered rocks, were sited on the highest surface metal anomalies of Taungni Hill and the Kauktaga volcanics. Lead, zinc and copper analyses of samples taken at 1 m intervals from each pit face allow vertical profiles of base metal distribution to be drawn (Fig. 6). All three pits on Taungni Hill show high lead values but the distribution varies from strongly banded (Fig. 6), through an overall decrease with depth, to constant levels showing little evidence of any pattern. Copper and zinc show irregular distribution, usually with coincident maxima. The two pits in the Kyauktaga volcanics contain appreciably different levels of lead and zinc although copper values are similar. Stratification of all three metals is pronounced in one pit (Fig. 6) but is shown by lead alone in the other.

Despite the very high values of lead accompanied by other base metals in both soils and bedrock, no lead minerals have been identified during exhaustive field surveys in the area. Blue varieties of silica rock commonly contain a little disseminated pyrite, but preliminary laboratory studies have identified only traces of galena and chalcopyrite in such rocks. Highest lead values occur in highly argillised rocks and investigations by XRD and electron microprobe of earthy, ochrous material containing up to 15% Pb, reveal the presence of complex Pb sulphates and chlorides, especially plumbojarosite Pb $Fe_6(SO_4)_4(OH)_2$ and jagoite $Pb_6Fe_2(Si_2O_3)_3Cl_2$ (J. F. W. Bowles, pers. comm.). However, other more typical samples of argillised material, with 600–1000 ppm Pb, failed to reveal any lead phases, suggesting that much of the lead is evenly distributed throughout the vermiform kaolinitic matrix of the rocks.

Nature of the alteration

The alteration observed on Taungni Hill is likely to have been produced either by advanced hydrolytic leaching, preceding and accompanying hypogene mineralizing fluids or by intense acid leaching resulting from the oxidation of an existing sulphide ore body. In either case field relationships indicate that most of the alteration predated the deposition of the Kyauktaga volcanics where the alteration is of a different type, being akin to supergene duricrust development. The argillic/silicic alteration products associated with these processes are

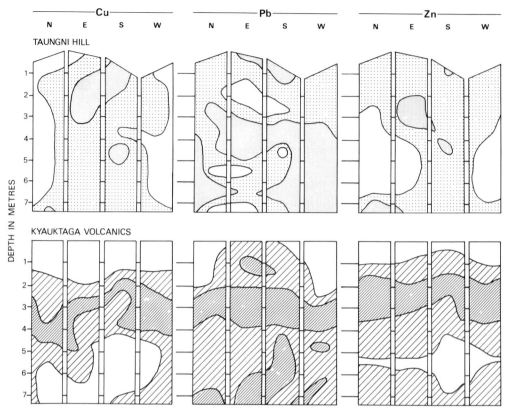

FIG. 6. Comparative pit profiles, Taungni Hill (above) and Kyaukataga volcanics (below). Taungni Hill: heavy stipple, Cu >20 ppm, Pb >2000 ppm, Zn >15 ppm; light stipple, Cu 5–20 ppm, Pb 1000–2000 ppm. Zn 5–15 ppm. Kyauktaga volcanics: heavy shading, Cu >30 ppm, Pb >5600 ppm, Zn >50 ppm: light shading, Cu 10–30 ppm, Pb 1800–5600 ppm, Zn 10–50 ppm.

often very similar and difficult to differentiate (e.g. Burbank, Luedke & Ward 1972).

The alteration on Taungni Hill has many similarities to the advanced hydrolytic type discussed by Hemley & Jones (1964) and described the world over from such Tertiary epithermal districts as the San Juan Mountains (Steven & Ratté 1960), Cochiti (Bundy 1958) and El Queva (Sillitoe 1975). In most of these occurrences an innermost zone of intensely altered, highly siliceous rock, (mostly quartz with some alunite with or without dickite, kaolinite, jarosite, pyrite, haematite and leucoxene), passes outwards into an intermediate zone of advanced argillic alteration (characterized by kaolinite/illite with alunite, sericite, quartz, montmorillonite, rutile and leucoxene). This is usually separated from unaltered rock by widespread propylitic alteration. Much of the inner zone of quartz/alunite rock is gradational into quartz rock having a very porous vuggy texture. Cavities are produced by leaching of feldspar phenocrysts or alunite pseudomorphs after feldspar.

The rocks cropping out on Taungni Hill are similar in physical properies and mineralogical composition to these advanced and intermediate zones of alteration. A major difference is the absence of exposed vuggy quartz rock and the surrounding quartz rock with alunite pseudomorphs. Propylitic alteration of the volcanics is not widespread and a great deal of fresh unaltered andesite crops out between areas of alteration (Fig. 1). It may be that the failure of the quartz/alunite rock to develop was due to an influx of silica at an early stage which effectively precluded the replacement of the majority of feldspars by alunite. Such features have been reported from Summitville (Steven & Ratté 1960). However, the restricted area of propylitic alteration on Taungni Hill suggests the rocks were not saturated with fluids to the extent that they were at Summitville.

The fact that alteration is centred on Taungni

Hill suggests that this was an active centre of solfataric activity. A continuing history of such activity, in a subaqueous environment interspersed with periods of emergence, is deduced from numerous samples of sedimentary rock which exhibit evidence of several cycles of silicification and redeposition. Oxidation of sulphide-rich rock results in highly acid conditions causing intense base leaching (Meyer & Hemley 1967; Hemley *et al.* 1980). On Taungni Hill sulphides are only observed in blue silica rock, the setting of which suggests that the silica rock has been argillised, the original texture lost, the silica remobilised and possibly redeposited as the amorphous silica crags, and all the sulphides oxidized to sulphates.

The generally high base metal values; the close association of lead and arsenic in basic sulphates such as beudantite $PbFe_3(AsO_4)(SO_4)(OH)_6$ (P. R. Simpson, pers. comm.), and occasional high values of other metals such as barium and molybdenum, suggest that the original mineralization took the form of hydrothermal sulphides, later oxidized to sulphates. Similar associations are noted in the San Juan Mountains, Colorado by Burbank *et al.* (1972). If this model is assumed, the almost total lack of relict sulphides in the Taungni area is puzzling. It is necessary to assume that either the acid leaching process has been unusually efficient in removing almost all traces of primary sulphides, or that physicochemical conditions were such that the base metals were precipitated as primary sulphates during the hydrolytic activity. Further clarification of these possibilities by optical or X-ray methods is difficult on account of the finely disseminated nature of the products of mineralisation.

The pattern of base metal distribution supports the different origins of alteration on Taungni Hill and the Kyauktaga volcanics but is unable to provide criteria to discriminate between the models of supergene and hypogene argillic/silicic alteration outlined. The upper surface of the Kyauktaga volcanics, depleted in zinc, indicates supergene leaching and the thin zone of relative concentration at the base of the unit could indicate a limit of leaching or redeposition of surface-derived material. Concentration of lead on the upper surface of this unit,

particularly where it overlies the alteration zone of Taungni Hill, is problematic considering the relative immobility of lead under supergene conditions. It could be caused by redeposition from solution of lead derived from Taungni Hill or by hypogene hydrolytic activity continuing after the deposition of the Kyauktage member.

The nature of the argillic/silicic alteration on Taungni Hill is less obvious than is the essentially supergene alteration of the Kyauktaga volcanics. Much of the alteration must be associated with the hypogene hydrolytic activity which was responsible for the introduction of base metals as discussed above. However, the low values of copper and zinc may be attributed to leaching by acid surface waters which must be assumed to have affected the whole area. Higher zinc values are restricted to unaltered rocks and to areas where metal values may have been preserved by early incorporation into massive silica rock now exposed on ridge crests. In such shallow environments the admixture of supergene and hypogene processes is inevitable and the circulation of downward-percolating meteoric waters through regions affected by the hydrolytic action of rising hydrothermal fluids must produce a variety of effects. It seems probable that the distribution of base metal concentrations in the Kyauktage volcanics are the effect of such an interaction and that the complex argillic/silicic products on Taungni Hill are a result of initial hypogene hydrolytic activity modified by later supergene acid leaching.

ACKNOWLEDGMENTS: The field observations described were made during a Technical Cooperation Programme carried out by the Department of Geological Survey and Mineral Exploration (Rangoon) and the Institute of Geological Sciences (London) and funded by the Ministry of Overseas Development (London) under the auspices of the Colombo Plan. The authors thank their many field counterparts for assistance with geochemical sampling and colleagues in the DGSME laboratories for most of the chemical determinations. Additional identifications and determinations were made by Miss H. Auld, D. Peachey and N. Breward (all IGS).

This paper is published with the permission of the Director of the Institute of Geological Sciences (a constituent body of the Natural Environment Research Council).

References

AMOS, B. J., MARSHALL, T. R. & STEPHENSON, D. 1983. Geology of the country between Mt. Popa and Taungdwingyi, Northern Pegu Yoma, Burma. *Inst. geol. Sci. London, Overseas Div. Rep. No. 40 (in preparation)*.

——, ——, ——, U TUN AUNG KYI, U BA THAW & U

NYUNT HAN 1981. Economic geology and geochemistry of Taungni Hill and of the area between Mt. Popa and Taungdwingyi, Northern Pegu Yoma, Burma. *Inst. geol. Sci. London, Overseas Div. Rep. No. 38 (unpubl.).*

BATES, T. F. 1962. Halloysite and gibbsite formation in Hawaii. *Clays Clay Miner.* **9,** 312–28.

BAUM, F., V. BRAUN, E., HAHN, L., HESS, A., KOCH, K-E., KRUSE, G., QUARCH, H. & SIEBENHUNER, M. 1970. On the geology of northern Thailand. *Beih. geol. Jb.* **102,** 23 pp.

BRUNNSCHWEILER, R. O. 1966. On the geology of the Indoburman Ranges (Arakan Coast and Yoma, Chin Hills, Naga Hills). *J. geol. Soc. Aust.* **13,** 137–94.

BUNDY, W. M. 1958. Wall rock alteration in the Cochiti Mining District, New Mexico. *Bull. New Mexico Bur. Mines Min. Res.* **59,** 1–71.

BURBANK, W. S., LUEDKE, R. G. & WARD, F. N. 1972. Arsenic as an indicator element for mineralised volcanic pipes in the Red Mountains area, western San Juan Mountains, Colorado. *Bull. U. S. geol. Surv.* **1364,** 31 pp.

CHHIBBER, H. L. 1934. *The Geology of Burma.* Macmillan, London. 538 pp.

CURTIS, C. D. & SPEARS, D. A. 1971. Diagenetic development of kaolinite. *Clays Clay Miner.* **19,** 219–27.

GOOSENS, P. J. 1978. The metallogenic provinces of Burma: their definitions, geologic relationships and extension into China, India and Thailand. *In:* NUTALAYA, P. (ed.) *Regional Conference on Geology and Mineral Resources of southeast Asia.* Asian Institute of Technology, Bangkok.

HEMLEY, J. J. & JONES, W. R. 1964. Chemical aspects of hydrothermal alteration with emphasis on hydrogen metasomatism. *Econ. Geol.* **59,** 538–69.

——, MONTOYA, J. W., MARINENKO, J. W. & LUCE, R. W. 1980. General equilibria in the system Al_2O_3–SiO_2–H_2O and some implications for alteration/mineralisation processes. *Econ. Geol.* **75,** 210–28.

LANGFORD-SMITH, T. & DRURY, G. H. 1965. Distribution, character and attitude of the duricrust in the northwest of New South Wales and the adjacent areas of Queensland. *Am. J. Sci.* **263,** 170–90.

MABBUT, J. A. 1967. Denudation geochronology in Central Australia: structure, climate and landform inheritance in the Alice Springs area. *In:* JENNINGS, J. N. & MABBUT, J. A. (eds) *Landform Studies from Australia and New Guinea.* Cambridge University Press. 434 pp.

MEYER, C. & HEMLEY, J. J. 1967. Wall rock alteration. *In:* BARNES, H. L. (ed.) *Geochemistry of Hydrothermal Ore Deposits.* Holt, Rinehart & Winston, New York. 670 pp.

SILLITOE, R. H. 1975. Lead-silver, manganese and native sulphur mineralisation within a stratovolcano, El Queva, northwest Argentina. *Econ. Geol.* **70,** 1190–201.

STEPHENSON, D., MARSHALL, T. R. & AMOS, B. J. 1983. Geology of the Mt. Popa volcano and associated post-Palaeogene volcanics, central Burma. *Inst. geol. Sci. London, Overseas Div. Rep. No. 39 (unpubl.).*

STEVEN, T. A. & RATTÉ, J. C. 1960. Geology and ore deposits of the Summitville District, San Juan Mountains, Colorado. *Prof. Pap. U.S. geol. Surv.* **343,** 70 pp.

T. R. MARSHALL, CRA Exploration Pty Ltd, P.O. Box 175, Belmont, W. Australia 6104.

B. J. AMOS, Institute of Geological Sciences, Keyworth, Nottingham NG12 5GG.

D. STEPHENSON, Institute of Geological Sciences, Murchison House, West Mains Road, Edinburgh EH9 3LA.

A low level laterite profile from Uganda and its relevance to the question of parent material influence on the chemical composition of laterites

M. J. McFarlane

SUMMARY: A 78 foot (24 m) 'low level' laterite profile from Tira, in Busia District, Uganda, is described. Mineral and chemical progressions, from the parent rock through saprolite and into the laterite, indicate its residual development. The chemical compositions of the residuum and the parent material are compared with further data from Uganda and other published results in order to assess the importance of the Fe, Al and quartz content of parent rocks in determining the chemical composition of a laterite. Quartz-richness of parent material does not appear to be disadvantageous for bauxite formation. Neither is Al-richness advantageous. This leads to the conclusion that the absence of bauxite in Uganda may be attributed to unfavourable leaching conditions rather than, as formerly believed, to the absence of suitable parent rocks.

Because laterite-capped planation surfaces are very well developed in Uganda, it was early anticipated (Fox 1932) that bauxite would be found there. In 1970, however, a review based on extensive mapping by the Geological Survey of Uganda (Kafol 1970) indicated that the high level laterites are unbauxitised. They are generally kaolinitic, Fe-rich and with a variable, low content of primary quartz. The quartz-richness of the metasedimentary rocks, on which the majority of the high level laterites survive, was generally believed to be responsible for their siliceous nature. This paper presents some data on a low level laterite profile developed from parent rocks ostensibly more favourable for bauxite formation (i.e. rocks relatively poor in quartz and richer in aluminous reactive minerals).

The Tira profile

Location

The profile (Core No. 604, Geological Survey of Uganda drilling records) is located in Tira, Busia District, near the Victoria/Kyoga watershed, close to the Kenyan border (Fig. 1a).

Geology

The quartz-rich Buganda Series, topped by the high level laterites, form uplands some 40 miles (64 km) to the west (Fig. 1b). They are bounded to the east by lowlands underlain by granite, and further east the Nyanzian-Kavirondian Series of metasediments also forms lowlands above which rise narrow quartzite ridges. To the east of Tira rise the prominent carbonatite plugs of Sekulu, Tororo and Akure.

Geomorphology

In the Kampala area (Fig. 1a) Bishop (1966) recognized two low level landsurfaces, the Kasubi Surface at about 4000 feet (1219 m), in addition to the already recognized Tanganyika Surface at about 3850 feet (1173 m). These surfaces are traceable undisturbed eastwards along the watershed (McFarlane & Brock 1983). They are barely recognizable by summit frequency, using the highest closed contours (Fig. 2a) but emerge more clearly if the area within the highest closed contours is plotted against altitude (Fig. 2b), a technique which separates the numerous small quartzite and carbonatite summits from the much more extensive laterite covered interfluves. The Kasubi and Tanganyika Surfaces comprise the oldest and highest members of a flight of low level erosion surfaces called PIII by Wayland (1931, 1933, 1934), which are extensively developed northwards into the Kyoga area. Their formation may be related to the changes in the local base level of erosion provided by the evolution of the Western Rift Valley (Bishop & Trendall 1967). The site of the core, shown in Fig. 2(c), is just below the change of slope which delimits the Tanganyika Surface (Fig. 2d). In the vicinity of the site the laterite, a pedogenetic laterite is continuous from over 4050 feet (1234 m) to below 3700 feet (1128 m). Although there is some small-scale incision along the streams, the blanket is essen-

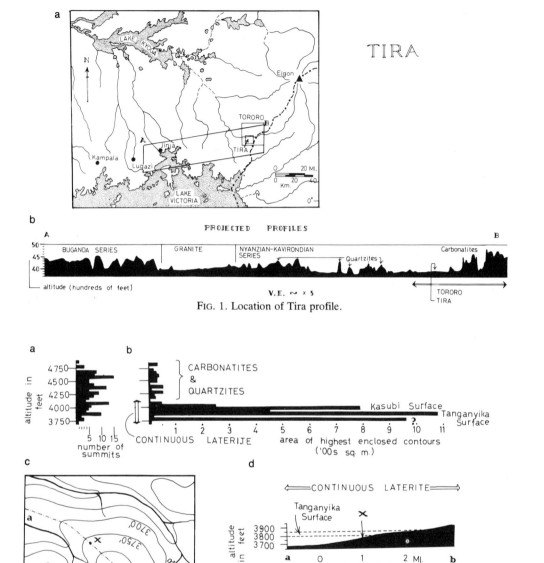

FIG. 1. Location of Tira profile.

FIG. 2. The Tira site and its geomorphological context. (a) Summit frequencies on the 1:50,000 topographic map of Tororo (location in Fig. 1). (b) Summit areas on the same map. This technique separates the numerous small summits attributed to carbonatite and quartzite outcrops from the laterite-blanketed Kasubi and Tanganyika Surfaces, the two oldest and highest components of the polycyclic 'PIII'. (c) The site of the core in Tira (the location of this area is shown in Fig. 1). (d) Profile (location shown in c) showing the site of the core, below the change of slope delimiting the Tanganyika Surface. (e) The core location in relation to the Tanganyika Surface is more clearly shown by the technique of plotting inter-contour distances along the same profile.

tially unincised and there is no reason to believe that it is not still forming. Since the development of pedogenetic laterites in vadose profiles is not terminated by landsurface incision (McFarlane 1983), continued formation has provided a continuous cover across, in this case, three low relief land surface elements, making their differentiation difficult. Both the polycyclic nature of the continuous laterite sheet and the geomorphological context of the core site can be more clearly demonstrated by the technique of plotting inter-contour distances along crest or spur profiles against altitude (Fig. 2e).

The laterite

Parent material

The core penetrated about 100 feet (30 m) of fresh rock below the laterite profile, showing that the immediate parent material is dominantly amphibolite (a metamorphosed calcareous sediment or tuff). X-ray diffraction (Table 1) of a typical sample showed hornblende dominant (*ca.* 90%) with traces of calcite, mica and quartz. Three typical samples average 21.34% Al_2O_3, 7.05% Fe_2O_3 and 38.4% SiO_2. X-ray diffraction of three saprolite samples, however, also identified quartz, feldspar and mica in varying proportions (Table 1).

The profile

The weathering profile consisted of 14.7 feet (4.4 m) of laterite with typical pedogenetic structures (McFarlane 1976). The irregular pisoliths were commonly coated with mangan-

ese, especially towards the base of the profile, where pisolith cementation results in rubbly or lumpy laterite. Overall colouring was yellowish-red (5YR 4/6-8) or red (2.5YR 4/6-8), imparted essentially by the soil-like material between the pisoliths. Some 64 feet (19.2 m) of saprolite, generally pale coloured, cream to grey, underlay the laterite with core stones encountered at 46.8 and 58.8 feet (14.2 and 17.9 m). The water table occurred at about 30 feet (9.1 m) below the surface. Fresh rock was reached at 78.8 feet (24 m). Drilling reached nearly 174 feet (53 m), weathered partings being encountered at 83.3, 95.7 and 145.9 feet (25.3, 29.1, 44.0 m). (The core deviated from the vertical, and so these depths are corrected to vertical depths.)

Mineralogy

Differential thermal analyses (Fig. 3) indicate the progressive mineral transformations. Up the profile, early weathering products (2:1 type minerals) are replaced progressively by the late weathering products kaolinite and goethite. X-ray diffraction (Table 1) confirmed this result.

Chemistry

The distribution of Fe_2O_3, Al_2O_3, SiO_2, Ca, MgO and K_2O in the profile is shown in Fig. 4. In comparison with the fresh rock, the laterite shows considerable increase in Fe_2O_3 and some loss of Al_2O_3. SiO_2 is extensively lost from the laterite but with some slight increase near the surface. MnO accumulates in the lower part of the laterite. Ca and MgO are progressively lost and above the water table K_2O loss is marked.

TABLE 1. *X-ray diffraction of laterite, saprolite and parent material in the Tira profile*

	Depth	Quartz	Mica	Calcite	Feldspar	Hornblende	Smectite	Kaolins	Goethite
Laterite	8 ft (2.44 m)	**						**	**
Saprolite	20 ft (6.1 m)	**					**		
	39 ft (11.89 m)	**	*		*	**	***		
	70 ft (21.34 m)	**	*			***	*		
Parent material	120 ft (26.58 m)	*	*	*		***			

Dominant ***. Present **. Trace *. Analyst G. Stewart.

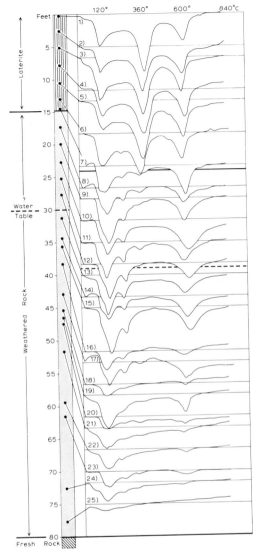

FIG. 3. Mineral progressions in the weathering profile (differential thermal analyses). Low temperature endotherms attributable to smectites dominate the lower part of the profile, kaolinite endotherms becoming more pronounced up-profile. The laterite is dominated by goethite and kaolinite.

Formation

There is no pallid zone in the sense of an Fe-depleted zone, and no indication that enrichment is the result of up-profile movement of the concentrated materials. Rather, progressive loss of the more mobile components allows the relatively less mobile materials to accumu-

late, with no doubt some absolute addition of residua, mechanically and in solution, from upslope locations. Smectites dominate the lower part of the saprolite where Ca, MgO and K_2O survive. Leaching of K_2O corresponds with kaolinite formation and goethite formation follows. Goethite, being stable in the oxidising conditions of the vadose profile, accumulates progressively as leaching continues. The water table plays no special part in the laterite formation, unlike in groundwater laterite where it has significant initial role in the segregation of Fe to form pisoliths (McFarlane 1976).

Discussion

Schellmann (1974, 1977) has provided data to support the view that laterite weathering of quartz-rich rocks differs from that of quartz-free rocks. In the case of the former, weathering trends are away from the SiO_2 side of the ternary diagram, which represents the varying proportions of Al_2O_3. Fe_2O_3 and SiO_2, as shown in Fig. 5(a). In the case of quartz-free rocks the trend is away from the SiO_2 apex. The trend for quartz-rich rocks implies their unsuitability for bauxite formation, gain of Fe_2O_3 exceeding that of Al_2O_3.

Laterite weathering trends in Uganda do not appear to conform with these general trends. At Tira, where the parent rocks are quartz-poor, the trend from parent rock to saprolite and laterite is sharply away from the SiO_2 side (Fig. 5d) and in Buganda, where parent materials are much more quartz-rich, the weathering trend from fresh rock to massive vermiform groundwater laterite is more from the apex (Fig. 5e, Table 2a). Furthermore, this massive vermiform groundwater laterite, almost or entirely free from quartz, is parent material for the formation of a much younger pedogenetic laterite (Table 2b) and the trend from groundwater to pedogenetic laterite, shown in Fig. 5(f), is almost perpendicular to the SiO_2-Al_2O_3 side of the triangle. If these trends are to be related to original quartz content, then the lower the content the smaller the gain in Al_2O_3 in comparison with Fe_2O_3.

The possibility that the Ugandan weathering trends are atypical was rejected in the light of comparisons of other published data with Schellmann's trends. The following conclusions emerged:

(1) Sandstone weathering trends, provided by Lelong (1978) and others lead generally from the SiO_2 apex (Fig. 5c).
(2) Superimposition of these sandstone weathering trends and those of Schell-

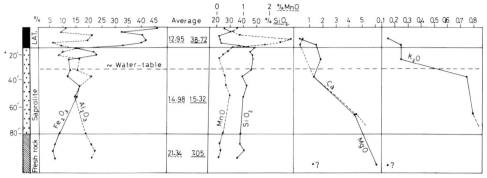

• depths corrected to vertical

FIG. 4. Chemical progressions in the Tira weathering profile.

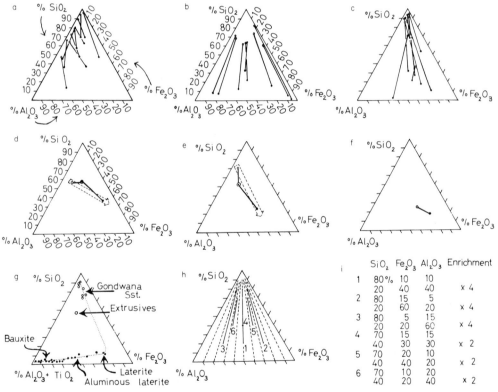

FIG. 5. Ternary diagrams of weathering trends. (a) Weathering trends (parent rock to laterite) for quartz-rich rocks (after Schellmann 1974, 1977). (b) Weathering trends (parent rock to laterite) for quartz-free rocks (after Schellmann 1974, 1977). (c) Weathering trends (parent rock to laterite) for sandstones (after Lelong 1976; Ghosh & Dutta 1978). (d) Weathering trend for amphibolites at Tira. The pecked arrow shows the general trend from parent rock to laterite. Solid line shows the trend from parent rock to saprolite and to laterite. (e) Weathering trend for the quartz-rich Buganda Series. The pecked arrow shows the general trend from parent rock to laterite. Solid line shows the trend from parent rock to saprolite and to laterite (data in Table 2a). (f) Weathering trend for the formation of pedogenetic laterite from a parent material of vermiform laterite in Buganda (data in Table 2b). (g) Weathering trend (parent rock to laterite, aluminous laterite and bauxite) from Phutkapahar, Central India (after Ghosh & Dutta 1978). The extrusives, earlier suggested as parent material for this bauxite, are also shown (data in Table 3c). (h) Trends for equal gain of Al_2O_3 and Fe_2O_3, radiating from the SiO_2 apex of the triangle. (i) Hypothetical construction points for h.

TABLE 2. *Parent materials and laterites*

	Samples	Average weight (%)		
		Al_2O_3	Fe_2O_3	SiO_2
(a) (see Fig. 5e)				
Massive vermiform laterite	5	25.5	37.3	22.5
Saprolite	11	24.81	11.38	44(est.)
Buganda Series				
(parent material)	14	18.4	5.09	60(est.)
(b) (see Fig. 5f)				
Pedogenetic laterite	3	18.6	51.1	15.9
Massive vermiform laterite				
(parent material)	5	25.5	37.3	22.5
(c) (see Fig. 5g)				
Decann extrusives[a]	2	14.21	12.81	41.31
		$(+2.25TiO_2)$		

[a] G. K. Sastri (pers. comm.)

mann shows that chemically dissimilar parent rocks may yield similar laterites and chemically similar rocks may yield very different laterites.

(3) Weathering trends from the Gondwana Sandstone, through laterite, aluminous laterite to bauxite, reported from India by Ghosh & Dutta (1978), are shown in Fig. 5(g). Although the laterite, aluminous laterite and bauxite are chemically very different, the trends from fresh rock to each of these weathering products is generally the same (i.e. from the SiO_2) apex.

Moreover, if the parent rock is the Deccan extrusives (Table 2c) as was earlier believed (Roychowdhury 1956), the trends are essentially similar but from a starting point below the SiO_2 apex (Fig. 5g).

It seems therefore that the significance of the quartz-rich and quartz-free weathering trends is open to question. In part these general trends are artefacts of the cartographic technique. A high proportion of SiO_2 in the rocks inevitably produces a weathering trend from the apex. Rocks with Al_2O_3 in excess of Fe_2O_3, that is those placed along the SiO_2 side of the triangle, are to some extent predisposed to weathering trends away from this side even when, for example, Fe_2O_3 and Al_2O_3 gains are equal (Fig. 5h).

Significant weathering trends, as far as iron and alumina richness of the product are concerned, are those which swing to the left or right of the equal gain trends, that is in the direction of preferential Al_2O_3 or Fe_2O_3 gain. Equal gain trends, radiating from the apex, are shown in

Fig. 5(h) and values for the construction are shown in Fig. 5(i). If the data from Schellmann, Lelong, Ghosh and Dutta are re-examined in terms of divergence from these equal gain trends, the results are as shown in Table 3. From this limited data (27 values), it appears that quartz-richness or poverty cannot be related to relative removal or preferential accumulation of Al_2O_3; both rock groups yield a similar proportion of Al_2O_2 and Fe_2O_3 enriched products, with a preference in both cases for Fe_2O_3 enrichment. In some agreement with the Ugandan data, it seems that rocks with an initially greater Al_2O_3 than Fe_2O_3 content favour relatively greater gain of Fe_2O_3 than Al_2O_3. Since quartz-rich rocks usually contain Al_2O_3 in excess of Fe_2O_3, Schellmann's 'quartz-rich weathering trend' (Fig. 5a) appears to show this.

Conclusions

The apparent lack of importance of quartz content in influencing bauxitisation and the absence of any indication that an initially high Al_2O_3 content is advantageous indicate that factors other than parent rock chemistry are most important in determining the nature of the product of lateritisation. These chemical trends appear to relate to variations in saprolite permeability which offer the more direct control on bauxitisation, determining whether kaolinite dissolution is congruent or incongruent (Garrels & Christ 1965). Thus a quartz content sufficient to provide a supportive skeleton for the saprolite may be an asset, prevents collapse and

TABLE 3. *Preferential enrichment of Al$_2$O$_3$ or Fe$_2$O$_3$ in laterites developed from Al-rich, Fe-rich, quartz-rich and quartz-free parent rocks*

Parent rock	Al$_2$O$_3$ > Fe$_2$O$_3$		Fe$_2$O$_3$ > Al$_2$O$_3$		Fe$_2$O$_3$ = .Al$_2$O$_3$	
	No. of samples	% of total number	No. of samples	% of total number	No. of samples	% of total number
Al$_2$O$_3$ < Fe$_2$O$_3$ (18 samples)	1	5.6	14	77.8	3	16.7
Fe$_2$O$_3$ > Al$_2$O$_3$ (eight samples)	3	37.5	2	25.0	3	37.5
Quartz-rich (19 samples)	3	15.8	12	63.0	4	21.0
Quartz-free (eight samples)	1	12.5	5	62.0	2	25.0

maintains permeability. Conversely, as Gaskin (1975) found, a high proportion of aluminous reactive minerals may yield a comparatively poorly permeable weathering product characterized by kaolinite rather than gibbsite. It follows that even more important than saprolite permeability is the permeability of the more immediate parent material of a bauxite, that of the laterite protore. The relatively poor permeability of the mature massive vermiform groundwater laterite in Uganda appears to be an important factor in the absence of bauxitisation there. By contrast, the more permeable, immature, pisolithic groundwater laterites which overlie essentially similar parent rocks in the Darling Ranges (McFarlane 1981) are well bauxitised. The high level packed pisolithic groundwater laterites in Uganda are usualy well cemented and even less permeable than the massive vermiform laterite, and the spaced pisolithic laterite is sufficiently incohesive for erosion to outpace bauxitisation. Prospects for the discovery of bauxite in Uganda do not appear to be good. It is highly doubtful, however, if the reason for this lies, as was formerly thought, in the nature of the parent rocks.

ACKNOWLEDGMENTS: I am greatly indebted to the Department of Geology and Mines of Uganda, and in particular to Drs J. Almond and P. Nixon for providing me with the Tira samples and for various assistance with them. Partial analyses (wet silicate) were also carried out by the Department. XRD analyses were by Dr A. Parker and Miss G. Stewart, Sedimentology Laboratory, Reading University. The National Agricultural Laboratories, Nairobi, permitted my use of their DTA equipment. The projected profiles (Fig. 1b) were by Dr P. Brock. Financial assistance came from the Science Research Council (NATO), the Goldsmith Company and the British Council.

References

BISHOP, W. W. 1966. Stratigraphical geomorphology. *In:* DURY, G. H. (ed.) *Essays in Geomorphology,* pp. 139–76. Heinemann, London.

BISHOP, W. W. & TRENDALL, A. F. 1967. Erosion-surfaces, tectonics and volcanic activity in Uganda. *Q. Jl geol. Soc. Lond.* **122**, 385–420.

FOX, C. S. 1932. *Bauxites.* Crosby, Lockwood & Son, London.

GARRELS, R. M. & CHRIST, C. L. 1956. Solutions, minerals and equilibria. *Am. Miner.* **55**, 1380–9, Washington.

GASKIN, A. R. J. 1975. Investigation of the residual iron ores of the Tonkolili District, Sierra Leone. *Trans. Inst. Min. Met. Sec. B, Applied Earth Sciences,* B98–B119.

GHOSH, K. P. & DUTTA, B. C. 1978. Mineralogy and genesis of Phutkapahar bauxite deposits of Eastern Madha Pradesh, India. *4th int. Congr. for the*

study of bauxites, alumina and aluminum. **1**, Bauxites, 204–55. Athens.

KAFOL, N. 1970. An investigation into the possible sources of raw material for an aluminium industry in Uganda. *Rep. geol. Dep. Uganda,* NK/1, 31 pp.

LELONG, F. 1976. See Ghosh & Dutta (1978).

McFARLANE, M. J. 1976. *Laterite and Landscape.* Academic Press, London. 151 pp.

—— 1981 Morphological mapping in laterite areas and its relevance to the location of economic minerals in laterite. *In: Lateritisation Processes,* pp. 308–17. Oxford and IBH Publishing Co.

—— 1983. Laterites. *In:* GOUDIE, A. & PYE, K. (eds) *Chemical Sedimentation and Geomorphology.* Academic Press, London.

—— & Brock, P. W. G. 1983. Cartographic analyses of 'high level' laterites—an example from Uganda—and the relevance of such techniques to lateritic mineral prospecting and to geochemical prospecting through laterites (in preparation).

ROYCHOWDHURY, M. K. 1956. Bauxite in Bihar, Madhya Pradesh, Vindhya Pradesh, Madhya Bharat and Bhopal. *Mem. geol. Surv. India,* **85**, 1–271.

SCHELLMANN, W. 1974. Kriterien für die bildung, prospektion und bewertung lateritischer silikat-bauxite. *Geol. Jb.* **D7**, 3–17.

—— 1977. The formation of lateritic silicate bauxites and criteria for their exploration and assessment of deposits. *Natural Resources & Development,* **5**, 119–34.

WAYLAND, E. J. 1931. Summary of progress of the Geological Survey of Uganda. *Summ. Prog. geol. Surv. Uganda*, pp. 1919–29.

—— 1933. The peneplains of East Africa. *Geogrl J.* **82**, 95.

—— 1934. The peneplains of East Africa. *Geogrl J.* **83**, 79.

M. J. McFARLANE, 32 Northcourt Avenue, Reading RG2 7HD.

Palaeoenvironment of lateritic bauxites with vertical and lateral differentiation

Ida Valeton

SUMMARY: Formation of lateritic bauxites of the type described in this paper occurs world-wide in Cretaceous and Tertiary coastal plains. The bauxites form elongate belts, sometimes hundreds of kilometres long, parallel to Lower Tertiary shorelines in India and South America and their distribution is not related to a particular mineralogical composition of the parent rock. The lateral movement of the major elements Al, Si, Fe, Ti is dependent on a high level and flow of groundwater. Varying efficiency of subsurface drainage produces lateral facies variations. Interfingering of marine and continental facies indicate a sea–land transition zone where the type of sediments also varies with minor tectonic movements or sea-level changes. A typical sediment association is found in India, Africa, South and North America. It consists of (i) red beds rich in detrital and dissolved material of reworked laterites, (ii) lacustrine sediments and hypersaline precipitates, (iii) lignites intercalated with marine clays, layers of siderite, pyrite, marcasite and jarosite, and (iv) marine chemical sediments rich in oolitic iron ores or glauconite. A model is developed to account for element distributions in lateritic bauxites in terms of groundwater levels and flow. Finally it is shown that many high-level bauxites are formed in coastal plains and that they are subsequently uplifted to their present altitude.

Bauxite deposits can be subdivided into lateritic bauxites and karst bauxites. Bardossy (1982) estimated that 85% of the world's bauxite reserves belong to the lateritic type. They occur today at altitudes from sea level to about 2000 m above sea-level. Following Fox (1923), it was believed that bauxite-bearing laterites are formed on high plateaus on top of specific rock types under a tropical climate. However, extensive alteration blankets showing a marked lateral and vertical zonation with separation of iron and aluminium have not been found to develop in the interior of continents. This paper demonstrates that there are distinct patterns in the geographic distribution, mineralogical and geochemical composition and sediment association of such lateritic bauxite blankets. The paper discusses the following four topics with specific reference to Lower Tertiary lateritic bauxite deposits of India and South America, and by comparison with other localities of the world:

— the geographic relationship between *in situ* lateritic bauxites and the shoreline at the time of bauxite formation;
— the vertical and lateral geochemical and petrographical variation within the alteration blanket;
— the nature of sediments associated with *in situ* lateritic bauxites;
— a model for the development of lateritic bauxite belts in space and time.

In this paper, lateritic bauxites are considered to be part of an alteration blanket, which is formed by *in situ* pedogenic processes leading to extremely intensive geochemical separation of Si, Al, Ti and Fe. This process always leads to a vertical division into three major soil horizons:

horizon rich in oxides	$B_{ox,fe,al}$
horizon rich in silicates (saprolite)	B_v
horizon of fresh parent rock	C

Normally the soil sections are truncated. The A-horizon is always eroded. After the Soil Taxonomy of the US Department of Agriculture (1965), this type of lateritic bauxites belong to the subgroup of aquox in the group of oxisols which are soft during time of formation. During uplift above the groundwater level, the Fe-rich parts form hard ferricretes, whereas the Al-rich parts become hard alucretes (Goudie 1973).

Geographic relationship between *in situ* lateritic bauxite and shoreline at the time of bauxite formation

In India and around the Guyana Shield of South America, the distribution of alteration blankets of lateritic bauxite is restricted to elongated belts, which are sometimes hundreds of kilometres long. Their distribution is not related to a particular mineralogical or chemical composition of the parent rock. Root systems penetrating the lateritic bauxites in India and Surinam demonstrate the pedogenic character of the material.

77

India

Lateritic bauxites are widespread in India (Fig. 1). They occur from Kashmir in the north, to Kerala in the south, and from Andhra Pradesh in the east to Kutch in the west. A reconstruction of Lower Tertiary shorelines show that most Indian occurrences are situated on Palaeocene or Eocene coastal plains.

In Kutch, the lateritic bauxites are developed in two distinct stratigraphic horizons, the older one belonging to the Palaeocene, the younger one to the Eocene (Fig. 2). The older and major laterites are formed on the peneplained clastic sediments of the truncated Upper Bhuj Formation of Lower Cretaceous age (Biswas 1977) or on Trap Basalts (Upper Cretaceous to Palaeocene). The peneplained surface of the Bhuj Formation and the flow surface of the basalt have a weak relief of a few metres (Figs 4 & 7).

The preserved bauxite belts are a few hundred metres to several kilometres wide, and the alteration blanket is generally 10–20 m thick. The belts extend parallel to the Lower Tertiary shore line (Fig. 2) in Kutch and the belt continues south-eastwards around the Kathiawar peninsula following the northward indentation of the Gulf of Combay; north of Ahmedabad it turns southward following roughly the outline of the west coast of India (Fig. 1). The distribution pattern of lateritic bauxites resembles a string of pearls around the Deccan peninsula.

The morphology of the alteration blanket reflects the original surface of the horizontal basalt flows or the peneplained sediments of the Upper Bhuj Formation. The absence of deeply incised valleys and the lateral facies variation (see below) indicate a dominantly subsurface seaward directed drainage. The 10–15 m deep present-day valleys were incised during later erosional cycles. The uppermost ferricretes or alucretes of the main alteration blanket are partly removed by erosion before the transgression of the Lower Eocene Laki Formation. A second and younger lateritic bauxite horizon is formed on the top of the Laki Formation (Figs 4 & 7).

The end of the lateritisation process is dated by the marine transgressions of either the Eocene Nummulitic Limestone or the Miocene Gaj. Limestone (Rao, pers. comm.), which cover large parts of the alteration blanket.

Other localities

Another high-quality bauxite belt extends around the eastern part of the Guyana Shield (Figs 3 & 5). Starting in Maestrichtian times and continuing through Lower and Upper Tertiary times, the deeply weathered rocks of the Precambrian hinterland are eroded and redeposited in the eastern Guyana foreland forming an association of sands, sandstones, clays and claystones. During an Eocene break of sedimentation (see pollen dating by Van der Hammen & Wymstra 1964) bauxitisation occurred in the coastal plain. The geomorphology of the Eocene bauxitisation surface consisted of very shallow drainage channels (Aleva 1965, 1979).

The sedimentary parent rocks from which the bauxites in Surinam and Guiana were formed are tidal flat and tidal channel sediments with bioturbation and root horizons characterizing an intertidal environment associated with mangrove vegetation (Valeton 1971, 1973a). The much younger Plio-Pleistocene bauxite deposits around an embayment of the Amazonas River are formed in a similar environment (Klammer 1976).

In North America and in Africa bauxite formation in similar palaeoenvironments are known. Gordon, Tracy & Ellis (1958) described the marine-terrestrial interaction in the nepheline-syenite area of Arkansas during the Eocene of bauxitisation. It is characterized by bauxite formation over a flat topography, a contemporaneous reworking and redeposition of detrital lateritic and bauxitic material as colluvial deposits, fanglomerates or river sediments, associated with coaly beds, lignitic or bituminous shales and marine clayey or sandy sediments.

By reconstructing the coastline of the Eocene Mississippi embayment in the gulf coastal plain of the southern United States, a clear relationship between bauxite formation on different kind of clastic rocks and a coastal plain environment was revealed by Overstreet (1964), Zapp (1965) and others (modified in Fig. 3). A similar relationship for a Lower Tertiary bauxite belt in Equatorial Africa was described by Zanone (1971).

Bauxites on karst

The fact that bauxites are formed in elongated belts parallel to ancient coastlines is well documented for the karst bauxites of the Mediterranean. These formed as terrestrial intercalations in marine Upper Jurassic to Lower Tertiary sediments. Lenses or banks of limestones with marine fossils within the bauxite horizons and associated lignites demonstrating their near shore position were described by

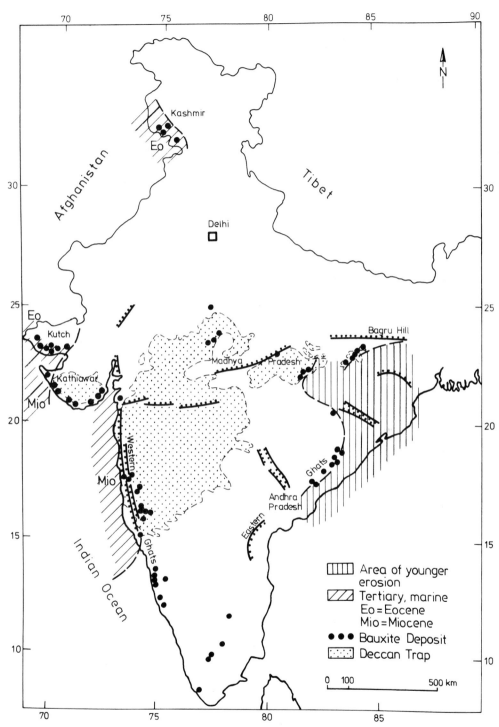

FIG. 1. Distribution of bauxite deposits in India.

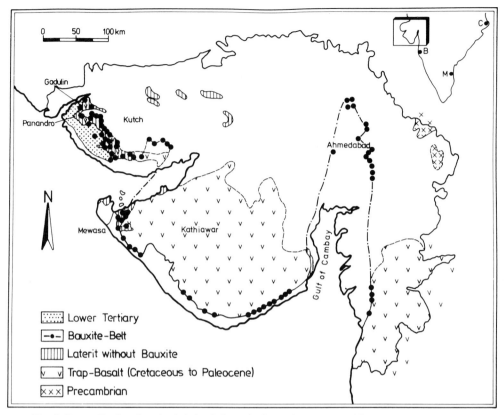

FIG. 2. Lower Tertiary bauxite belt in Gujerat (redrawn from an unpublished map of the Department of Geology and Mining, Gujerat, 1978).

Roch & Deicha (1966), Bonte (1965), Nicolas & Esterle (1965), Nicolas (1968) , Nia (1968) and Combes (1969, 1979). The location of elongated bauxite belts parallel to an ancient coastal line and their association with underlying and overlying shallow-water carbonate facies was described by Combes (1969, 1978, 1979) for southern France and Central Greece. Grubic (1972) describes the same palaeogeographic interpretation for the karst bauxites of the Yugoslavian Dinarids. In America, Keller, Westcott & Bladsoe (1954) showed a lateral transition, from terrestrial diaspore rocks via flint clays and semi-plastic clays, to marine shales in the Pennsylvanian Sheltonham Formation of Missouri.

Vertical and lateral petrographic and geochemical variation within the alteration blanket

The vertical variation in sections of *in situ* lateritic bauxites is well known since its first description by Walther (1915) and later by Millot (1964).

Some typical lateral variations are described below.

India

At Mewasa/Kathiawar (Figs 2 & 4,2) a definite pattern of lateral petrographic and geochemical differentiation is developed. The lateral sequence shows enrichment of Fe to the northeast and enrichment of Al in the opposite direction (Valeton 1966). Since the well-preserved relic textures indicate that the original volume of the parent rock is also preserved, the isovolumetric method of Millot & Bonifas (1955) may be used for calculating the geochemical balance. Thus it is possible to prove that besides the relative enrichment, there is absolute enrichment of Fe in the north-eastern part, whereas Al and Ti are enriched absolutely in the south-western area. The only explanation for this phenomenon is the lateral movement of all major elements (Al, Si, Fe, Ti) leading to a

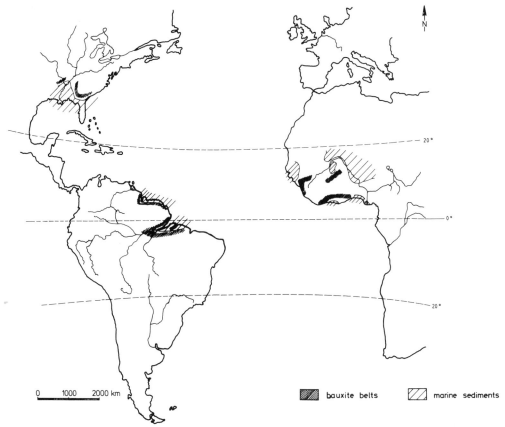

FIG. 3. Bauxite belts: Lower Tertiary around the Mississippi embayment, east of the Guyana shield, and Equatorial Africa; Pleistocene in the Amazonas embayment (after Valeton 1980).

lateral facies differentiation from a kaolinitic laterite to an Al-rich bauxite.

An equivalent Palaeocene alteration blanket in Kutch has well-preserved relic textures, and shows imposing facies variations along both the strike and dip directions of the belt with swells and basins caused by synpedogenetic movements. The differentiation along the strike of the belt (Fig. 4,3, 4,4) is marked by two facies types:

I low-silica type with kaolinitic saprolite (LST on Fig. 4).

II high-silica type with bentonitic and kaolinitic saprolite (HST on Fig. 4).

From the high-silica type, Saharsrabudue (1961) and Talati (1968, 1970) described the bentonitic variation of the saprolite. In addition to this very pronounced vertical geochemical differentiation, there is a lateral transition from nodular to spongy textured gibbsite bauxite into a pisolitic bauxite rich in boehmite and diaspore (Fig. 7). The petrographic and geochemical aspects of the Kutch bauxites are being investigated at present by a German-Indian research team.

On the Udagiri Plateau in the Western Ghats, lateral facies variations of the equivalent horizon to that at Mewasar/Kathiawar are found today in a high level position (Fig. 4,1) (Valeton 1967).

Bauxites on karst

In karst bauxites, a lateral facies transition from a boehmite bauxite, via a diaspore bauxite, to a kaolinitic clay is known from the French Lower Cretaceous bauxites (Combes 1969; Valeton 1964).

A very marked lateral facies change from reworked laterites to bauxitic flint clays is known from the Lower Jurassic of the Negev, Israel (Goldbery 1979). In the east of the region, the reworked laterites (=Laterite Derivative Facies, LDF) rest on an undulatory surface of

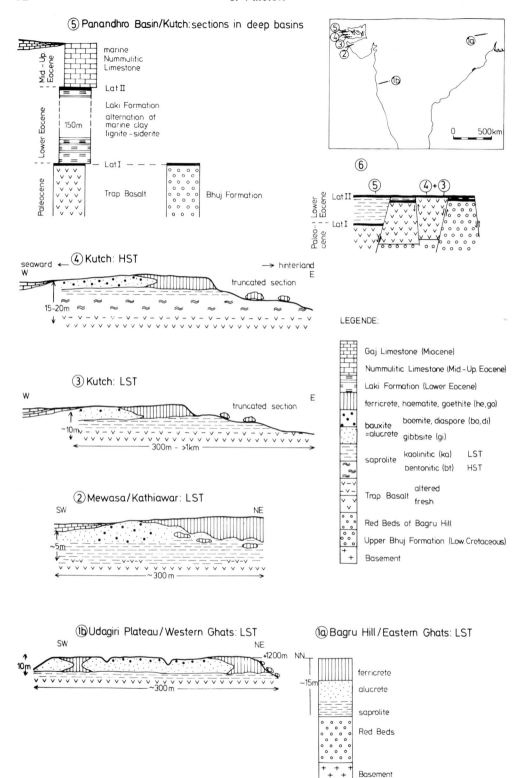

⑤ Panandhro Basin/Kutch: sections in deep basins

marine Nummulitic Limestone

Lat II

Laki Formation
alternation of marine clay lignite - siderite

Lat I

Trap Basalt   Bhuj Formation

Mid - Up. Eocene
Lower Eocene   150m
Paleocene

⑥

Lat II
Lat I

Paleo- / Lower cene / Eocene

⑤   ④+③

④ Kutch: HST

seaward ← W   → hinterland E
truncated section
15-20m

③ Kutch: LST

W   E
truncated section
~10m
300m - >1km

② Mewasa/Kathiawar: LST

SW   NE
~5m
~300 m

LEGENDE:

Gaj Limestone (Miocene)

Nummulitic Limestone (Mid-Up. Eocene)

Laki Formation (Lower Eocene)

ferricrete, haematite, goethite (he, go)

bauxite = alucrete   boemite, diaspore (bo, di)
                          gibbsite (gi)

saprolite   kaolinitic (ka)   LST
                 bentonitic (bt)   HST

Trap Basalt   altered
                 fresh

Red Beds of Bagru Hill

Upper Bhuj Formation (Low. Cretaceous)

Basement

① b Udagiri Plateau/Western Ghats: LST

SW   NE
                                    +1200m
10m
~300 m

① a Bagru Hill/Eastern Ghats: LST

NN
                ferricrete
~15m
                alucrete

                saprolite

                Red Beds

                Basement

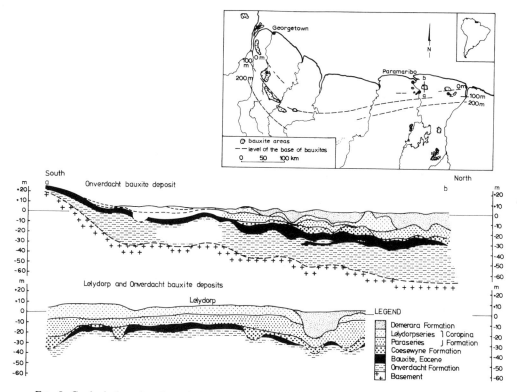

FIG. 5. Geological section through the Onverdacht area in the eastern Guyana foreland/Surinam with bauxite blankets dipping towards the north. The shallow channel system is formed by post-bauxitic erosion (after Valeton 1973a, b).

Triassic basement (Fig. 6), whilst in the west, flint clay facies was formed on karst from reworked laterites (LDF) which occur as infillings of solution cavities, sink holes and irregular depressions. The Laterite Derivative Facies comprises pisolite conglomerates, lateritic arenites and pedogenetically altered silt- and claystones. These units are composed mainly of kaolinite and haematite, but in their upper part occurs dolomicritic carbonate. In the eastern Nahal Ardon area the relatively weak *in situ* lateritisation, characterized by vertical mottling, has partially dissolved and removed the Fe^{3+} minerals and broken down the kaolinite (Valeton, Stütze & Goldbery 1983). In contrast, in the western flint clay facies of the Karstic zone the *in situ* transformation of the reworked laterite, as demonstrated by relic

FIG. 4. Vertical and lateral facies differentiation in selected sections of the Lower Tertiary alteration blanket around the Deccan peninsula. Saprolite subdivided in a lower bentonitic and upper kaolinitic zone low-silica type of section (LST): saprolite with kaolinite only; high-silica type of section (HST).
1a: Bagru Hill/Bihar Mtn.: Sequence of Precambrian basement—red beds—bauxite section (LST) on trap (modified from Valeton 1967). 1b: Udagiri plateau/Western Gaths: Laterite-bauxite facies pattern on trap (LST), intersected by younger rivers (Valeton 1967). 2: Mewasa/Kathiawar: Laterite-bauxite facies pattern on trap basalt (LST), after seaward erosion covered by marine Miocene Gaj Limestone (Valeton 1966). 3: Kutch: Laterite-bauxite facies pattern on trap basalt (LST), Upper Palaeocene, after seaward erosion, covered by marine Miocene. 5: Panandhoro Basin/Kutch: two bauxite horizons. Upper Palaeocene on trap basalt and Lower Eocene on top of the Laki Formation, covered by Middle Eocene marine Nummultic Limestone. 6: Relationship between contemporaneous tectonic movements, basin sediment fill and the formation of the first and second lateritic alteration blanket.

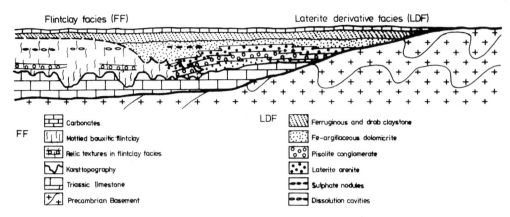

FIG. 6. Section through the Mishhor Formation, showing the laterite derivative facies in the east (resting on an undulating surface of Triassic limestone) passing into the flint clay facies towards the west (on karst topography) (redrawn from Goldbery 1979).

Formation of insitu laterites and bauxites in Kutch / Gujerat

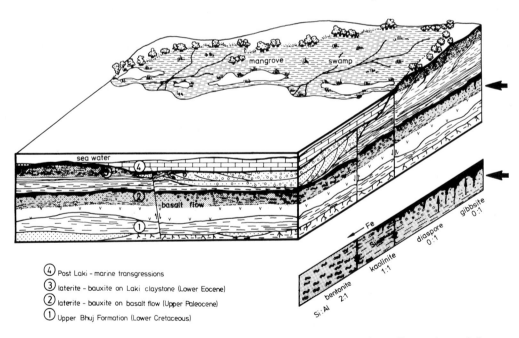

FIG. 7. Block diagram illustrating the interaction of basalt flows, tectonics, sedimentation and distribution of the alteration blankets I and II in Kutch/Gujerat. In the lower section, a schematic vertical and lateral distribution pattern of minerals in the alteration blanket on trap basalt is given.

structure, is much more pronounced. Deferrification and desilification have produced an iron-free boehmitic bauxite and high alumina flint clay (Fig. 6).

Sediments associated with *in situ* lateritic bauxites

This type of lateritic bauxites is frequently associated with the following sediment types, either as underlying or overlying deposits:

Red beds rich in detrital and dissolved material of reworked laterites including the LDF of Goldbery (1979);

lacustrine sediments or hypersaline precipitates containing concretions of gypsum or celestite;

lignites associated with dark clays, layers of siderite or concretionary pyrite and marcasite, or jarosite;

marine chemical sediments rich in newly formed iron materials (e.g. oolitic iron ores or glauconite).

Occasionally, some of these sediments are found as lateral facies variations of a bauxite formation (e.g. the Laki Formation in Kutch).

Deposition of the Laki Formation is controlled by synsedimentary tectonics (Fig. 4,6). Differential subsidence led to sedimentation in a fluvio-deltaic, lagoonal, or swampy marine environment. The fluvio-deltaic sediments are rich in reworked laterites and it is on this facies type that the second *in situ* bauxitisation occurs.

In most places, one or two of the above associations are found and a description of examples follows below.

Red beds

In India and in Surinam red beds are frequently associated with lateritic bauxites. These red beds are quite different from the well-known Palaeozoic or Permo-Triassic red beds of the northern hemisphere which are considered to have formed under a continental climate. Such red beds are rich in unstable minerals like feldspar and mica, have low contents of unstable heavy minerals, and are poor in any kind of Fe or Al precipitates. In contrast, the red under beds in Surinam are reworked laterite derivative sediments. Thus Surinam was taken as type locality and these red beds are called 'Surinam type red beds' (Valeton 1973a). They are an equivalent of Laterite Derivative Facies of Goldbery (1979). They are characterized by irregular stratification, slumping structures, angular or rounded boulders of red and white kaolinitic claystones and lateritic

iron crust, pisolites, very poor sorting of angular clasts, alternating clayey and sandy layers rich in quartz, high contents of stable heavy minerals, precipitates of colloidal material rich in haematite, kaolinite, gibbsite or in silica. In addition, they are often rich in burrows and root horizons, and maybe intersected by tidal channels. All these properties of low mechanical and high chemical maturity indicate a reworked laterite redeposited in an intertidal environment.

In India, in Kutch (Biswas 1977) and in the Bihar Mountains (Valeton 1967), very similar Cretaceous red beds are found immediately below the Trap Basalts. During the Lower Cretaceous, the Upper Bhuj Formation of northwestern Kutch developed as a fluviatile-deltaic sequence of grey and red sandstones. The sediments are characterized by poor sorting and high clast-angularity. Iron and silica were mobilized and reprecipitated separately in low-energy environments. Towards the north-west, near the village of Gadulin, the deltaic-clastic sequence contains lenses of coal and precipitates of jarosite. The coarse-grained sand layers are disturbed by dense systems of burrows, and the cement consists of dark-red precipitated iron. Typical grey fine-grained tidal flat sediments (in the sense of Reineck & Singh 1973) are intercalated in this sequence.

In Bagru Hill/Bihar, a facies containing reworked laterite is found between basement and bauxite indicating a near-coast environment which is highly significant for the interpretation of the formation of east coast bauxites on the high plateaus. From the base up, the section is divided into three units (Figs 4,1a & 8).

(1) Deeply weathered basement, interpreted as the preserved lower part of a pre-Upper Cretaceous lateritic soil section.

(2) A sedimentary sequence of about 126 m thickness similar to the red beds of the upper Bhuj Formation in Kutch. It is also characterized by a very irregular distribution of the following types of sediments: clay layers with slumps, soft white clay pebbles in a red haematitic matrix, alternation of coarse, sandy layers with horizons rich in stable heavy minerals, white, kaolinitic clay layers intersected by slickensides. The topmost sand layers are highly bioturbated and contain several root horizons that probably indicate terrestrial lowland surfaces.

(3) 20–30 m bauxite showing excellent macro- and microrelic textures of a former basalt flow. Near the contact to the underlying sediments there are

BAGRU HILL, BIHAR

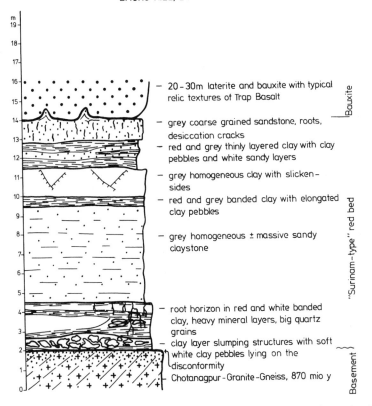

FIG. 8. The Bagru Hill section, Bihar Mountain, showing of the red beds intercalated between weathered basement and the bauxite on trap basalt.

typical flow textures, and occasionally inclusions of large sedimentary blocks occur in the former basalt.

The mineral association, textures and structures of the Bagru Hill section strongly resemble those of the Kutch and Surinam occurrences.

Lacustrine sediments and hypersaline precipitates

A variety of terrestrial, lacustrine and marine sediments were noted on top of karst bauxites (Bardossy 1982). The only case of hypersaline sediments occurring in such a position is known from the Lower Jurassic bauxitic flint clays of the Negev. Concretions of gypsum and celestite are form nodules in the Laterite Derivative Facies. Goldbery (1982) related the precipitation of these sulphates to soil-forming processes immediately before the bauxitisation. However, Valeton et al. (1981, 1983) suggested a marine source for the Sr- and Ca-sulphates and

attributed precipitation to soil formation in a coastal plain environment.

Lignites and clays with iron minerals

Sediments of this type are frequently found on the top of Mediterranean karst bauxites, but they are rarely found on top of lateritic bauxites.

The Panandhro Basin in Kutch/India, which is formed by subsidence after the bauxitisation of Trap Basalt is such an example (Balasubramaniam, Sabale & Namdas 1978). The Basin is filled with a sequence of lignites alternating with marine claystones which contain layers of concretionary marcasite-pyrite and layers of siderite. The Fe-rich precipitates are interpreted as a dissolution product of reworked laterite. A detailed investigation of this facies type is being undertaken by a German-Indian research team.

Marine chemical sediments rich in early diagenetic iron minerals

The relationship between lateritic bauxites and marine early diagenetic iron and silica minerals can be observed in Cretaceous to Lower Tertiary tropical shelf sediments and has been reported from a number of localities.

In Nigeria, Kogbe (1978) described an association of three rock types, reflecting an interfingering of marine and terrestrial environments of Palaeocene to post-Miocene age:

marine Fe-rich oolites of late Palaeocene age; crusts and concretionary laterites on the emerged surface of the oolite formation; Miocene and younger fluvial and lacustrine ferruginous sandstones with thin intercalated laterites, and a main laterite on top of the sequence.

An intensive neoformation of glauconite pellets characterizes Upper Cretaceous and Lower Tertiary sedimentation on the shelves around Equatorial Africa (Lambroy & Odin 1975; Odin 1978). The worldwide optimum of glauconite formation in shelf areas during this period is in many cases related to intensive chemical weathering on the continents. Good examples are described from Central Europe by Odin (1975, 1978) and Valeton, Abdul-Razzak & Klußmann (1982).

Model for the development of lateritic bauxite belts in space and time

As already stated the geographical distribution and the vertical and lateral facies variation within bauxites belts may be explained in terms of their formation in coastal plains. Various aspects of this model are discussed below.

Interfingering of marine fluvial and terrestrial sediments

Most of the lateritic bauxites of India, Equatorial Africa, Mississippi embayment, Surinam and Guiana are found on continental margins. The interfingering of marine, fluvial and terrestrial sediments in the bauxite belts of India, Surinam and North America indicate that this kind of alteration takes place in a near-shore environment.

The role of groundwater

For the solution, migration and precipitation of major and minor elements within the bauxitic alteration blanket, the following groundwater conditions must be fulfilled:

(1) net flow towards the sea;
(2) groundwater levels must be high and oscillatory in nature;
(3) E_h-conditions must be reducing.

It is suggested that the low-silica type (Fig. 4,3) and the high-silica type (Fig. 4,4) within the B_t horizon of the lateritic soil sections are produced by differences in drainage intensities. Poor drainage conditions lead to a high-silica type because silica is not efficiently removed and combines with Al to form new layer silicates with high silica content (e.g. smectite). Good drainage conditions produce the low-silica type and kaolinite is formed. Silica is the most mobile major element.

Fig. 9 shows schematically the relationship between groundwater level and the formation of the three lateritic bauxite facies within the B_{ox} horizon. The bauxites discussed in this paper may be interpreted in general terms using this figure. Fig. 9,2 shows the typical three horizons of the weathering profile which is produced by upward migration of all major elements. The absolute enrichment of Al and Fe is most pronounced in the B_{ox} horizon of the alteration blanket.

Lateral differentiation is especially well developed in the uppermost zone. Iron-rich ferricretes are exposed on the landward side and pass seawards into an Al-rich alucrete. It is suggested that the groundwater level intersects the palaeosurface at the ferricrete/alucrete transition zone (Fig. 9,2) and that Fe (which also migrates upward on the seaward side) is removed from this level by groundwater flow. Al, which is less mobile than Fe, remains and both zones are indurated to form crusts when the region finally emerges beyond the reach of the water table. In coastal plains, the groundwater level rises periodically to form superficial lakes (e.g. monsoon or seasonal floods). During high groundwater level, Fe migrates upward in the bivalent state, and during the dry period it is oxidized to Fe^{3+} and thus immobilized. Iron can only be dissolved and transported in the bivalent state under reducing conditions. As long as the alteration blanket is water-saturated, it has a soft colloidal consistency. Al as well as Ti, Fe, Si migrate in solution, but because of their varying mobility they are transported to different parts of the alteration blanket. Elements with the highest mobility are most easily transported by the groundwater to the sea.

E_h and pH conditions are affected by vegetation and groundwater circulation. As the latter is influenced by basin topography, it is suggested that the high-silica type bauxite in Kutch region (Fig. 4,4) formed in basins with poor drainage, whereas the low-silica type formed on swells with more effective drainage.

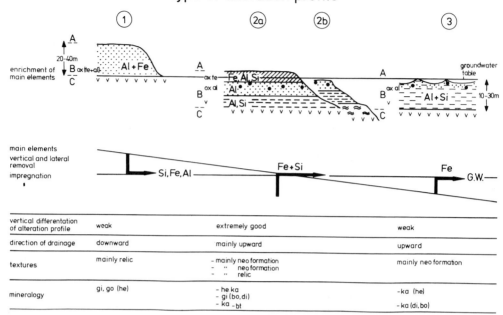

FIG. 9. Schematic sections of the three main types of bauxitic alteration blankets. 1: Formation of bauxites at various levels above the watertable, without separation of Al and Fe. 2a: Low-silica type, and 2b high-silica type; top of the section near the surface of the groundwater level; shows strong separation of Fe and Al in the B_{ox} horizon; the saprolite, $B_{t(=v)}$ horizon, is either kaolinitic or bentonitic below the kaolinite. 3: formation of flintclay below groundwater level by total extraction of Fe.

Characteristics of associated sediments

The associated sediments help to identify the palaeoenvironment of lateritic bauxite. The character of the sediments is influenced by source rock, type and rate of tectonic movements and local topography.

The source rocks are deeply weathered magmatic or metamorphic rocks, lateritic soils or clastic sequences containing a high percentage of reworked laterites.

Since the alteration blankets are formed in the marine-terrestrial transition zone, minor tectonic movements and eustatic sea-level changes profoundly affect the type of sediments deposited. Associated terrestrial sediments indicate uplift or regression immediately after bauxite formation (e.g. red beds associate with LDF). Slow subsidence or transgression leads to the formation of lacustrine and deltaic sediments, lignites and hypersaline precipitates in lagoons. This is demonstrated by the altered red beds, reworked laterites and precipitation of jarosite, e.g. in basins of the Eocene Laki Formation, and the alternation of lignites and marine clays in the Panandhro Basin in Kutch.

Rapid subsidence leads to a purely marine environment, which is considerably influenced by the geochemical activity in the hinterland. All the depleted elements from the laterites are supplied to the marine environment. Very little geochemical work has been done in this field. The geochemical investigations of flint clays in the Negev (Valeton et al. 1983), iron oolites and glauconites give some indication of the geochemical interaction between lateritic weathering and neoformation of marine iron minerals. During flint clay genesis, mobilized elements which contribute to the marine environment are, besides silica and iron, elements like Mn, V, Pb, lanthanides (Ce, La, Nd).

In Kutch, the whole facies association of the sediments pre-dating the Trap Basalt and during the time of laterite formation in the Lower Tertiary indicates a marine-terrestrial transition zone in a peneplained coastal area. In many other places all over the world, such as Eastern Guiana foreland, the Mississippi embayment, Nigeria and Equatorial Africa, the associated sediments indicate similar lowland environments. In addition, for the Mediterranean karst bauxites similar associations with near-coastal marine or terrestrial sediments are known in many places.

Duration of the bauxitisation process

The time which is needed for the formation of a 10–20 m thick lateritic alteration blanket is probably much shorter than believed by many scientists. In all cases where the underlying and overlying rocks can be dated, the duration of bauxitisation seems to be very short. In Hawaii, local bauxitisation occurs on 10 000 yr lava flows. Bushinsky (1967) described bauxitisation in 2–5 m y and Bardossy (1982) in 1–5 m y. In Kutch, for each of the two laterite horizons, the calculated time for bauxitisation was in the range of only a few million years.

High level bauxites

In many parts of the southern hemisphere, lateritic bauxites are situated on plateaus or on peneplains ranging in altitudes from a few hundred to 1000 or even 2000 m. In India, these bauxites are also situated in elongated belts. They often show the same vertical and lateral geochemical, textural and mineralogical differentiation as the equivalent weathering crusts in low-level positions. Only in rare cases, are they associated with coastal-plain sediments (e.g. Bagru Hill) (Figs 5,1a & 8).

Many bauxites on high levels on the Western Ghats, and also in the northern part of the Eastern Ghats in India, show a lateral facies change from sections with kaolinitic saprolite and iron crust, to an intermediate bauxite layer, to thick bauxite layers without capping iron crusts. The last named are, in addition, sometimes pisolitic, showing higher contents of boehmite. The pisolitic facies proper, with the highest economic quality, is often found in talus on the slopes. After induration in coastal plains, the massive or pisolitic bauxites are the hardest and most resistant to weathering and erosion, followed by some massive type of iron crust.

Valeton (1967) assumed that these bauxite-bearing laterites were formed at a low level and were subsequently uplifted to their present altitude after which the soil-hardening process began.

Bauxitisation might have started earlier, possibly in the Upper Cretaceous on the Precambrian basement in the southern part of the Eastern Ghats in the Shevaroy Hills/Salem District, Galiconda and related areas in Andhra Pradesh. These bauxites are slightly different. They are related to an undulating landscape where iron and aluminium are not separated during weathering and so do not show a vertical zonation. They belong to type 1 which was formed above the groundwater level (Fig. 9,1). The laterisation process of the whole bauxite belt is terminated by uplift and dissection of the peneplained landscape by young rivers. The lateritic bauxites are then transformed by subsequent pedogenetic processes producing to polygenetic soils.

ACKNOWLEDGMENTS: The stimulus to work on the problems discussed in this paper arose out of a discussion with Dr P. K. Bannerjee and other friends at the meeting of the IGCP 129 Working Group held in Trivandrum in 1979.

References

ALEVA, G. J. J. 1965. The buried bauxite deposits of Onverdacht, Surinam, South America. *Geol. Mijnb.* **44**, 45–58.
—— 1979. Bauxitic and other duricrusts in Surinam. A review. *Geol. Mijnb.* **58**, 321–36.
BALASUBRAMANIAM, K. S., SABALE, S. G. & NAMDAS, M. R. 1978. Geology and geochemistry and genesis of siderites from Panandhro basin, Madh series, Kutch district, Gujerat state. *J. Mines, Metals & Fuels*, November, 384–6.
BARDOSSY, G. 1982. *Karst Bauxites—Bauxite Deposits on Carbonate Rocks.* Elsevier, Amsterdam. 441 pp.
BISWAS, S. K. 1977. Mesozoic rock stratigraphy of Kutch, Gujerat. *Q. Jl geol. Min. metall. Soc. India*, **49**, 3+4, 1–51.
BONTE, A. 1965. Sur la formation en deux temps des bauxites. *C. r. Acad. Sci.*, Paris, **260**, 5076–7.

BUSHINSKY, G. J. 1967. Regularities in bauxite distribution in geosynclines. *Trudy SNJJGGJMS*, **66**, Novo Sibirsk, 9–25.
COMBES, S. J. 1969. *Recherches sur la génèse des bauxites dans le nord-est de l'Espagne, le Languedoc et l'Ariège (France).* Thèse Univ. de Montpellier Mém. CERGH III–IV, **1**, 375 pp.
—— 1978. Nouvelles données sur les relations entre la paléogéographie et la gîtologie de bauxites du troisième horizon dans la zone du Parnasse (Grèce). *4th int. Congr. ICSOBA, Athens*, Vol. 1, 92–100.
—— 1979. Observations Sédimentologiques, Paléogéographiques, minéralogiques et géochémiques sur les bauxites du deuxième horizon dans la zone du Parnasse (Grèce). *Bull. Soc. géol. Fr.* 7e serie, XXI, No. 4, 485–94.
FOX, C. S. 1923. The bauxite and aluminous laterite occurrence of India. *Mem. Surv. India*, **49**, 287 pp.

GOLDBERY, 1979. Sedimentology of the Lower Jurassic flint clay-bearing Mishhor Formation, Makhtesh Ramon, Israel. *Sedimentology*, **26**, 229–51.

—— 1982. Paleosols of the Lower Jurassic Mishhor and Ardon Formations ('Laterite Derivative Facies'), Makhtesh Ramon, Israel. *Sedimentology*, **29**, 669–90.

GORDON, M., TRACY, J. I. & ELLIS, M. W. 1958. Geology of the Arkansas bauxite region. *Prof. Pap. U.S. geol. Surv.* **299**, 268 pp.

GOUDIE, A. 1973. *Duricrusts in Tropical and Subtropical Landscapes.* Clarendon Press, Oxford. 174 pp.

GRUBIC, A. 1972. Répartition paléogéographique des bauxites dans les dinarides yougoslaves. *3rd Congr. int. ICSOBA, Nice 1972*, 145–50.

KELLER, W. D., WESTCOTT, J. F. & BLADSOE, A. O. 1954. The origin of Missouri fire clays. *Clay Clay Miner.* **2**, 7–46.

KLAMMER, G. 1976. Zur jungquartären Reliefgeschichte des Amazonastales. *Z. Geomorph., N. F.* **20**, 149–70.

KOGBE, C. A. 1978. Origin and composition of the ferruginous oolites and laterites of north-western Nigeria. *geol. Rundsclau, Afrikaheft*, **67**, 662–74.

LAMBROY, M. & ODIN, G. S. 1975. Nouveaux aspects concernant les glauconies du plateau continental nord ouest Espagnol. *Revue Géogr. phys. Géol. dyn.* **17**, 2, 99–120.

MILLOT, G. 1964. *Géologie des Argiles.* Masson, Paris.

—— & BONIFAS, M. 1955. Transformation isovolométriques dans les phénomènes de latérisation et de bauxitisation. *Bull. Serv. Carte géol. Als. Lorr.* **8** (1), 3–19.

NIA, R. 1968. *Geologische, petrographische, geochemische Untersuchungen zum Problem der Boehmit-Diaspor-Genese in griechischen Oberkreidebauxiten der Parnass-Khiona-Zone.* Thesis, Univ. Hamburg. 133 pp.

NICOLAS, J. 1968. Nouvelles données sur la génèse des bauxites ä mur karstique du sud-est de la France. *Miner. Deposita* **3**, 18–33.

—— & Esterle, M. 1965. Position et âge de la bauxite karstique d'Ollière (Var). *C. r. Acad. Sci., Paris*, **260**, 3722–3.

ODIN, G. S. 1975. *Les glauconies.* Thèse de Doktorat, Paris.

—— 1978. Nature, formation et signification des glauconies. *Int. Congr. Sedimentology, Jerusalem, 1978*, Abstracts, pp. 478–9.

OVERSTREET, E. C. 1964. Geology of the southeastern bauxite deposits. *Bull. U.S. geol. Surv.* **1199A**, 19 pp.

REINECK, H. E. & SINGH, J. B. 1973. *Depositional Sedimentary Environments.* Springer-Verlag, Berlin.

ROCH, E. & DEICHA, G. 1966. Sur les argillites de la région de Dragnignan (Var). *C. r. hebd. Séanc. Soc. géol. Fr.* **271**, 145–7.

SAHARASRABUDUE, Y. S. 1961. Unpublished report. Geological survey of India.

Soil Taxonomy 1965. U.S. Department of Agriculture, Washington.

TALATI, D. J. 1968. On the occurrences of clay deposits near Arsodia, Sabar Kantha district. *Miner. wealth, India*, **4** (1), 1–6.

—— 1970. On the origin, classification and review of Bharnagar bentonite, Gujerat state. *Miner. wealth, India*, **4** (2), 1–3.

VALETON, I. 1964. Facies problems of boehmite and diasporic bauxites. In: AMSTUTZ, G. C. (ed.) *Sedimentology and Ore Genesis. Developments in Sedimentology, Vol. 2*, 123–9. Elsevier, Amsterdam.

—— 1966. Laterale Faziesdifferenzierung Laterite-Bauxit und deren Beziehung zum Paläorelief in Gujerat, Indien. *Trav. ISCOBA, Zagreb 1966*, **(2)**, 50–82.

—— 1967. Bauxitführende Laterite auf dem Trappbasalt Indiens als fossile, polygenetisch veränderte Bodenbildung. *Sedim. Geol.* **1**, 7–56.

—— 1971. Tubular fossils in the bauxite and the underlying sediments of Surinam and Guiana. *Geol. Mijnb.* **50**, 733–41.

—— 1973a. Pre-bauxitic red sediments and sedimentary relics in Surinam bauxites. *Geol. Mijnb.* **52** (6), 317–32.

—— 1973b. Laterite als Leithorizonte zur Rekonstruktion tektonischer Vorgänge auf den Fesländern. *Geol. Rdsch.* **62**, 153–61.

—— 1980. Relationship between palaeoenvironment and bauxite formation within laterites. *1st European meet. Int. Ass. Sediment. Bochum 1980*, pp. 134–5.

——, STÜTZE, B. & GOLDBERY, R. 1981. Zur Petrographie und Geochemie der Flintclay-Formation in Makhtesh Ramon, Negev Israel. *DFG report*.

——, ——, & ——. 1983. Geochemical and mineralogical investigations of the Lower Jurassic flintclay bearing Mishhor and Ardon formations, Makhtesh Ramon, Israel. *J. sedim. Geol.* **32**, in press.

——, ABDUL-RAZZAK, & KLUßMANN, D. 1983. Mineralogy and geochemistry of glauconite pellets of Cretaceous sediments from Norst West Germany. *Geol. Jb.* **52**, 5–93.

VAN DER HAMMEN, T. & WYMSTRA, T. A. 1964. A palynological study on the Tertiary and Upper Cretaceous of British Guiana. *Leidse geol. Med.* **30**, 184–240.

WALTHER, J. 1915. Laterite in West Australien. *Z. dt. geol. Ges.* **67**, 113–140.

ZANONE, L. 1971. *La bauxite en Côte d'Ivoire.* Abidjan.

ZAPP, A. D. 1965. Bauxite deposits of the Andersonville district, Georgia. *Bull. U.S. geol. Surv.* **1199-G**, 37 pp.

IDA VALETON, Geologisch-Paläontologisches Institut, Universität Hamburg, Bundesstrasse 55, D 2000 Hamburg 13, W. Germany.

Geochemistry of a nickeliferous laterite profile, Liberdade, Brazil

J. Esson

SUMMARY: The 23 samples studied represent a complete vertical profile, 11 m thick, from topsoil to bedrock. Multiple bedrock samples were collected to obtain good estimates of bedrock variability and bulk composition. Nineteen of the 22 major and trace elements studied fall into five behavioural groups within each of which all element pairs show strong positive correlations: (1) Al, Ti, V, Nb and Zr are very highly enriched (factors 20–65 relative to bedrock) in horizon B; (2) Fe, Cr, Sc, Cu and Ce are also strongly enriched (factors 4–40) in horizon B, but show slightly greater enrichment (factors 10–50) in horizon C; (3) Ni, Y, La, Nd and Zn are characterized by concentrations (enrichment factors 8–90) in horizon C of 5–10 times their levels in horizon B; (4) Mn and Co are intermediate between (2) and (3); (5) Si and Mg are strongly depleted in all soil horizons. A mass-balance model, based on quantitative Ti retention, gives relative wt% losses over the profile as a whole of MgO 99.5, Na_2O 99.3, CaO 98.5, SiO_2 98.3, K_2O 96, NiO 91, CoO 89, MnO 88, Cr_2O_3 83, Fe_2O_3 79, Sc 78, Zn 75, La 69, Ce 65, Al 59, V 57, Nb 40, Nd 19, Y 12, TiO_2 0. Zr and Cu show *ca.* 10% relative gains, indicating stronger retention than Ti. *In situ* leaching of an estimated serpentinite thickness in excess of 200 m is required to produce 11 m of residual laterite.

At Liberdade a nickeliferous cap has developed *in situ* on an ultrabasic body consisting of serpentinite and serpentinized peridotite. The body has a surface area of 1.5 km² and forms a small hill, Morro de Corisco, at an altitude of 1200 m in a mature mountain range in southern Minas Gerais, close to its borders with the states of Rio de Janeiro and São Paulo. Drainage is complex and consists mostly of ephemeral streams. There is a strongly seasonal climate, with mean day-time temperatures of 15°C in winter, 30°C in summer, and winter minima rarely below 10°C. Monthly rainfall varies from below 50 mm in the dry season to 350 mm during summer, with an annual total of about 2000 mm. Relative humidity is typically 40–60%.

The thickness of laterite cover varies from 3 to more than 12 m, with an average of about 9 m. After a preliminary survey, an undisturbed profile with a thickness to bedrock of 11 m was selected for detailed study. Sample numbers 1–23 (Table 1) were collected by continuous channelling in the wall of an open pit and collecting a composite sample for each depth interval, usually 0.5 m. Samples of unweathered bedrock were collected from outcrops exposed in workings and shallow drill-cores. Various nickel silicate assemblages (garnierites) occur in thin, irregular bands and veins, mainly in the weathered rock at the base of the profile, but rarely constitute more than 5% of the bulk material at that level. There are serpentine-, talc- and chlorite-rich varieties of garnierite at Liberdade.

Variations in Munsell colour and texture were used as criteria for subdivision of the profile (Fig. 1 and Table 1). The Munsell colours of the horizons are: horizon A 5YR4/4; horizon B_{21} 2.5YR3/4; horizon B_{22} 2.5YR3/6; horizon C 5YR5/8. In the saprolitic horizon C, the division into C_1 and C_2 is based on the recognition of residual rock fabric in C_2. Horizon B_1 is transitional to A, B_3 is transitional between B_2 and C, and horizon D is bedrock.

The thickness and mineralogy (volume %) of each horizon can be summarized as follows:

(1) A 0.2 m, goethite (25), quartz (35), vermiculite (35), maghemite (5);
(2) B 7.25 m, goethite (40–50), quartz (25–35), kaolinite (10–20), gibbsite (5–20), maghemite (2–3);
(3) C_1 2.3 m, goethite (20–25), quartz (15–20), talc (10–15), chlorite (15–25), serpentine (5), maghemite (5–10);
(4) C_2 1.95 m, serpentine (50–85), chlorite (0–15), quartz (0–15), talc (0–5), goethite (0–5), magnetite (5), amphibole (few %);
(5) D bedrock, serpentine (90), magnetite (5), amphibole (few %), olivine, pyroxene (trace).

The full mineralogy and major element chemistry of the profile, including details of sample preparation and analytical methods, and the garnierites have been described by Esson & Surcan (1978a, b). This paper discusses the behaviour in the profile of 10 trace elements, determined by X-ray fluorescence, and gives

TABLE 1. *Trace element concentrations (ppm), including mean values for each horizon and depth intervals for each sample*

Sample no.	1	2	3	4	5	6	7	8	9	10	11	12	13
Depth, m													
Top	0.0	0.2	0.7	1.2	1.7	2.2	2.7	3.2	3.75	4.25	4.75	5.25	5.75
Base	0.2	0.7	1.2	1.7	2.2	2.7	3.2	3.75	4.25	4.75	5.25	5.75	6.25
Horizon	A	B_1	B_1	B_{21}	B_{21}	B_{21}	B_{21}	B_{22}	B_{22}	B_{22}	B_{22}	B_{22}	B_{22}
Sc	40	51	57	56	52	51	56	56	63	58	61	72	73
Y	48	17	28	25	18	21	26	22	24	18	17	13	14
La	71	25	18	17	12	10	19	12	19	17	9	6	11
Ce	147	170	188	165	157	160	183	208	185	156	114	95	147
Nd	72	30	35	32	42	29	29	17	20	24	27	26	39
Zr	140	304	321	355	308	286	354	400	350	344	305	272	253
Nb	9	31	32	29	31	29	31	37	39	35	35	26	26
V	230	384	390	386	388	372	388	396	405	410	421	416	405
Cu	80	122	127	123	115	103	119	110	130	118	77	86	102
Zn	363	169	183	171	159	153	148	146	152	152	135	154	179

Sample no.	14	15	16	17	18	19	20	21	22	23	24*	Range*
Depth, m												
Top	6.25	6.85	7.45	7.9	8.4	8.9	9.4	9.75	10.25	10.7	12.0	
Base	6.85	7.45	7.9	8.4	8.9	9.4	9.75	10.25	10.7	11.5	20.0	
Horizon	B_{22}	B_3	C_1	C_1	C_1	C_1	C_1	C_2	C_2	C_2	D(Bedrock)	
Sc	65	72	65	60	60	79	70	22	9	5	6.5	5-11
Y	16	23	39	105	121	150	159	161	153	96	1.5	1-3
La	12	22	47	105	114	109	155	165	137	79	5.3	2-8
Ce	197	415	291	109	89	67	59	32	18	11	12	6-15
Nd	30	53	83	156	197	222	226	191	151	106	2.5	1-8
Zr	250	134	46	20	18	15	12	10	8	5	4.5	3-6
Nb	27	18	9	7	4	4	6	2	5	5	1	0-·2
V	394	347	277	110	112	156	122	72	44	16	19	15-32
Cu	118	153	210	207	199	113	91	54	55	43	3.5	1-7
Zn	191	366	643	738	1050	1111	856	1108	801	82	46	36-53

Mean and standard deviation(s) for each horizon:

Horizon	B		C_1		C_2		D	
	mean	s	mean	s	mean	s	mean	s
Sc	60	8	67	8	12	9	6.5	2.4
Y	20	5	115	48	137	35	1.5	0.8
La	15	5	106	39	127	44	5.3	2.4
Ce	181	74	123	96	20	11	12	4.0
Nd	31	9	177	59	149	43	2.5	2.9
Zr	303	64	22	14	7.7	2.5	4.5	1.0
Nb	30	5	6	2	4	1.7	1	0.6
V	393	19	155	70	44	28	19	6.3
Cu	144	19	164	57	51	7	3.5	2.4
Zn	176	57	880	199	664	527	46	5.7

*Mean (column no.24) and range of 6 unweathered samples from outcrops and drill cores.

new major element results obtained from further work on shallow drill-core material for the serpentinite bedrock. Also presented are the significant major and trace element correlations within the profile and a revised mass-balance/elemental budget model, together with summaries of previously published major element analyses which were obtained mainly by atomic absorption. Elements, conventionally those contributing significantly to the analytical

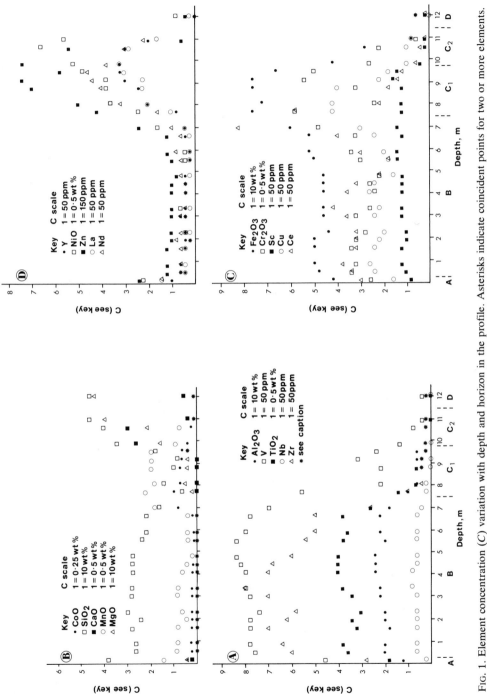

Fig. 1. Element concentration (C) variation with depth and horizon in the profile. Asterisks indicate coincident points for two or more elements. Other symbols and C scale factors are explained in the key to each diagram.

total, reported as oxides were retained as oxides in the mass balance model. The mathematical basis of the model is discussed in the Appendix.

Chemical variation with depth

Fig. 1 shows trace element variation with depth, from topsoil to bedrock. Previously published major element values (Esson & Surcan 1978a), recalculated on a volatile-free basis, are also plotted for comparison. The trace element results are presented in full in Table 1. Table 2 is a matrix of standard product-moment correlation coefficients (r) for all major and trace element pairs showing significant correlations, i.e. those at or above the 95% confidence level. Both the correlation matrix and the variation diagrams show that 15 of the 22 elements investigated fall into three distinct groups, each of which is characterized by the level in the profile at which the elements achieve their maximum concentrations.

The elements of the first group, Al, V, Ti, Nb and Zr, are mainly concentrated in horizon B (Fig. 1A), where they consistently reach levels of about 20, 20, 65, 30 and 65 times their respective concentrations in the bedrock (Tables 1 and 3). Of the elements in this Al-group, Al and V show the least and Zr the greatest variation within horizon B. Just above the base of horizon B the concentrations of all the Al-group elements plunge rapidly, by an order of magnitude for Ti, Zr and Nb, to level out briefly in horizon C_1 before their final decrease through horizon C_2 to the bedrock values. The slight increases in V and Al at a depth of 9.5–10 m are discussed below. Within this group all ten element pairs have correlation coefficients of 0.922–0.987 (Table 2), with a mean value of 0.961.

Fe, Cr, Sc, Cu and Ce constitute a group of elements which are also strongly enriched throughout horizon B (Fig. 1C), by factors of about 6.5, 4.5, 9, 40 and 15, respectively, relative to the bedrock. However these Fe-group

TABLE 2. *Element correlation matrix, showing only those correlation coefficients for which the confidence level exceeds 95%*

Main matrix (row element × column element):

	SiO_2	Al_2O_3	Fe_2O_3	MgO	NiO	Cr_2O_3	TiO_2	MnO	CoO	CaO	Na_2O	K_2O	Sc	Y	La
SiO_2															
Al_2O_3															
Fe_2O_3		-.969													
MgO	.657	-.729	-.733												
NiO		-.859		.499											
Cr_2O_3	-.793		.814	-.628											
TiO_2		.987			-.671	-.871									
MnO	-.703	-.421	.736		.453	.716	-.494								
CoO	-.695		.705		.409	.702	-.462	.982							
CaO	.428	-.554		.490	.674				-.534						
Na_2O	.620	-.563	-.687	.865					-.520	-.536					
K_2O		.599							-.518		.614				
Sc	-.836	.564	.826	-.861	-.436	.717	.498				-.656	-.719			
Y		-.766				.928		-.799	.527	.437	.613		-.439		
La		-.783				.928		-.816	.561	.484	.634		-.457	.979	
Ce	-.455	.600	.404	-.605	-.519	.521	.552			-.497	-.404		.557	-.544	-.512
Nd		-.741				.865		-.788	.672	.586	.435		-.503	.965	.949
Zr		.945			-.601	-.842	.957	-.545	-.538	-.477	-.474	.732		-.754	-.780
Nb		.960			-.619	-.817	.967	-.527	-.511	-.484	-.510	.727	.421	-.732	-.767
V		.986			-.774	-.859	.977	-.584	-.610			.551	.625	-.763	-.780
Cu	-.837		.828	-.621		.565			.641	.671	-.465	-.634	.645		
Zn	-.448	-.661	.520		.758	.430	-.704	.745	.682	.471		-.470		.887	.888

Upper-right block (trace-element sub-matrix):

	Zn	Cu	V	Nb	Zr	Nd	Ce
Zn		-.612	-.705	-.724	.935		
Cu						.587	
V				.929	.922	-.717	.646
Nb					.978	-.754	.473
Zr						-.781	.448
Nd							-.445

Correlation coefficient	Confidence level, %
± .611 to 1.0	> 99.9
± .515 to .610	99 – 99.9
± .404 to .514	95 – 98.9

TABLE 3. *Major element composition (wt%) of bedrock (no. 24) and slightly weathered bedrock (no. 23), together with means and standard deviations (s) for each soil horizon calculated from the results of Esson & Surcan (1978)*

Sample	SiO_2	Al_2O_3	Fe_2O_3	FeO	MgO	CaO	MnO	NiO	TiO_2	Cr_2O_3	CoO	Na_2O	K_2O	H_2O^*	Total
23	41.19	0.88	4.79	2.12	35.47	0.21	0.12	2.45	0.03	0.31	0.02	0.05	0.01	12.31	99.96
24+	40.92	0.91	4.35	1.49	38.44	0.25	0.10	0.32	0.03	0.23	0.02	0.10	0.01	12.88	100.05
L+	40.24	0.79	4.15	1.25	37.52	0.09	0.08	0.27	0.02	0.18	0.02	0.09	0.01	12.51	
H+	41.73	0.94	4.45	1.64	38.98	0.64	0.12	0.36	0.03	0.32	0.02	0.11	0.01	13.57	

Major element mean and standard deviation (anhydrous, total iron basis) for each horizon:

Horizon	SiO_2	Al_2O_3	Fe_2O_3	MgO	CaO	MnO	NiO	TiO_2	Cr_2O_3	CoO	Na_2O	K_2O
A(1 sample)	38.11	11.57	41.22	4.53	0.05	0.67	1.10	0.90	1.69	0.12	0.04	0.02
B(13 samples)	26.33	21.33	48.29	0.15	0.01	0.28	0.21	1.83	1.49	0.04	0.02	0.02
s		2.56	1.35	2.83	0.04	0.00	0.10	0.04	0.16	0.20	0.02	0.01	0.01
B_3(1 sample)	15.83	18.56	59.21	0.78	0.01	0.90	0.76	1.29	2.41	0.20	0.03	0.01
C_1(5 samples)	11.22	6.39	72.84	3.91	0.06	1.00	1.85	0.32	2.18	0.19	0.02	0.01
s		4.51	2.07	4.62	1.12	0.09	0.08	0.42	0.25	0.71	0.04	0.01	0.00
C_2(3 samples)	40.50	2.07	26.00	26.13	1.00	0.33	2.88	0.06	0.92	0.06	0.04	0.01
s		6.26	1.00	16.83	12.90	0.66	0.18	0.34	0.03	0.51	0.04	0.02	0.00
D (6 samples)	46.85	1.04	6.88	44.01	0.29	0.11	0.37	0.03	0.26	0.02	0.11	0.01
s		0.44	0.08	0.30	0.76	0.24	0.02	0.04	0.00	0.06	0.00	0.01	0.00

* H_2O is total water determined on air-dried samples.

+ Sample 24 is mean of 6 specimens, compositional range L - H.

elements reach even higher levels of enrichment in horizon C_1 before the rapid decrease in horizon C_2 towards the concentrations found in the bedrock. Ce reaches its maximum value, near the transition between horizons B and C_1, and begins to decline slightly higher in the profile than the other elements in the group. Cr has a strong second peak close to 9.5 m, coinciding with well-defined small peaks for V, Al and Sc. Within the Fe-group elements the correlation coefficients for the element pairs not including Ce are in the range 0.565–0.828 (mean 0.732), whereas those involving Ce are 0.404–0.587 (mean 0.517). These values reflect the slightly different distribution of Ce in the profile (Fig. 1C), In fact, Ce is somewhat more strongly correlated with the Al-group elements (r = 0.448–0.646, mean 0.544) but differs markedly from them by both being less strongly enriched throughout horizon B as a whole and having a strong, sharply defined peak in its distribution pattern. Thus Ce shows some affinities with both the Al- and Fe-groups and its behaviour in this environment is clearly intermediate between the two

The third group of elements, Ni, Y, Zn La and Nd, is notable for the strong maxima in horizon C (Fig. 1D), where the elements typically reach 8, 90, 20, 25 and 70 times their respective concentrations in the bedrock. In horizon B also Y, Zn, La and Nd are moderately enriched, by factors of about 13, 4, 3 and

12, respectively, relative to bedrock, whereas Ni is depleted by about 50%. The range of correlation coefficients for all element pairs in this group is 0.758–0.979 (mean 0.822) and, as expected (Fig. 1A,D), there are strong negative correlations (r = -0.601 to -0.859, mean -0.767) between the elements of the Ni-group and those of the Al-group. The negative correlations of Ce with Y, La and Nd emphasize the departure of Ce from typical rare earth behaviour; possible reasons for this are discussed below.

Five other elements, not belonging to any of the three main groups but showing significant variation with depth, are plotted in Fig. 1(B). The behaviour of Mn and Co, with almost identical distribution patterns (r = 0.952), resembles that of the Ni-group (Fig. 1D) in some respects but Mn and Co are, by comparison, only moderately enriched in horizon C where they reach 5–10 times the bedrock values. In this respect Mn and Co have some affinity with the Fe-group (Fig. 1C), but the latter show very much greater enrichment in horizon B. The correlation coefficients of Mn and Co with the five Ni-group elements are in the range 0.409–0.745 (mean 0.556). Of the Fe-group, only Fe, Cr and Cu show significant correlations with Mn and Co (r = 0.641–0.736, mean 0.695).

Ca (Fig. 1B) is strongly depleted, relative to bedrock, in all horizons except C_2 where it is enriched by a factor of about 5. Consequently, it shows significant positive correlations with the Ni-group (r = 0.435–0.674, mean 0.565). Ca also has significant relationships with Mg (0.490) and Si (0.428), both of which are depleted, relative to their bedrock values, in horizons A, B and C. The most striking negative correlation is that between Si and Fe (r = −0.969) and the very marked inverse relationship is apparent in Fig. 1(B,C).

Mass-balance model results

Within the upper 90% of horizon B there is relatively little variation in the major and trace element composition and mineralogy (Fig. 1, Table 1 and Esson & Surcan 1978a), indicating that in this part of the profile: (a) there is an almost steady-state condition in which a negligible amount of change is taking place, (b) all elements are being removed in the same proportions as the concentrations in the soil at this level or (c) the relative enrichment or depletion rates of the elements are constant. Ti and Zr

(Tables 1 & 3) are notable for their very high degree of enrichment in horizon B, where they reach about 65 times their concentrations in the bedrock. Because of this resistant behaviour, Ti was used as the reference element in previous discussion of the major-element budget for the profile (Esson & Surcan 1978a, 1979). In the revised mass-balance model presented here Ti is retained as the reference element in preference to Zr because, although Zr is rather less mobile (see below) in the profile than Ti, the analyses of Ti in the bedrock are more accurate than those for Zr. Apart from the inclusion of trace elements, the principal changes in the revised model involve the use of new data for the bedrock major-element composition (Table 3). The basis of the model (see Appendix for details) is that the soil profile was formed by differential chemical weathering of bedrock by descending groundwater and that all Ti was retained in the residue. Table 4 summarizes the results obtained from this model when using the maximum estimate (50%) for the soil porosity. Changes in porosity do not affect the element balance relative to Ti, but do

TABLE 4. *Wt% element losses and gains (marked +), from the Ti-retained mass-balance model, for sub-horizons, horizons and the full soil profile*

Horizon	A	B$_1$	B$_{21}$	B$_{22}$	B$_3$	C$_1$	C$_2$	No. 20	B	C	A+B+C
Thickness, m											
Soil	0.20	1.00	2.00	3.65	0.60	2.30	1.95	0.35	7.25	4.25	11.70
Parent rock	2.9	30.9	54.7	118.4	14.6	13.7	2.1	0.57	218.6	15.7	237.3
All elements	96.2	98.1	97.9	98.2	97.3	89.4	32.3	61.6	98.1	81.9	97.0
SiO$_2$	96.9	99.0	98.7	99.0	99.1	97.5	37.2	85.2	98.9	89.6	98.3
Al$_2$O$_3$	57.6	63.4	59.0	62.4	52.7	36.1	+1.8	+150	61.0	31.1	59.0
Fe$_2$O$_3$*	77.1	88.7	85.5	87.5	77.2	+12.7	+73.1	+263	86.1	+20.7	79.0
MgO	99.6	100	100	100	100	99.0	49.5	96.8	100	92.6	99.5
CaO	99.4	99.9	99.9	99.9	99.9	97.9	+49.9	70.0	99.9	78.6	98.5
MnO	77.8	93.5	94.0	96.5	79.2	7.5	+41.4	+223	94.1	1.1	87.9
NiO	88.6	98.7	98.9	99.0	94.5	47.2	+422	+153	98.5	+14.4	91.0
TiO$_2$	0.0	0.0	0.0	0.0	0.0	0.0	0.0	0.0	0.0	0.0	0.0
Cr$_2$O$_3$	75.4	88.3	87.5	90.6	75.7	13.8	+68.0	+265	88.3	3.1	82.7
CoO	80.5	94.9	95.2	97.3	77.0	15.5	+23.4	+163	94.9	9.6	89.2
Na$_2$O	99.8	99.8	99.6	99.7	99.2	98.0	70.9	88.7	99.6	94.5	99.3
K$_2$O	92.2	95.9	95.5	96.5	97.3	94.9	44.2	81.3	96.2	88.2	95.6
Sc	78.2	85.6	83.9	83.8	72.7	+0.8	15.1	+284	83.3	1.3	77.9
Y	8.3	79.1	76.4	84.3	69.5	+501	+3870	+2949	79.6	+943	11.8
La	45.7	92.0	93.9	95.7	88.3	+120	+1312	+1089	93.6	+277	69.0
Ce	53.2	72.2	70.9	76.5	8.2	+9.0	9.7	+88.6	69.7	+6.5	64.7
Nd	8.3	79.8	76.9	84.6	53.1	+520	+2810	+2789	79.0	+820	19.3
Zr	+7.0	+16.4	+36.9	+9.8	28.9	52.7	13.2	8.0	+14.8	47.5	+10.7
Nb	65.6	41.4	37.0	43.2	52.2	37.2	+201	+130	42.3	5.9	39.9
V	53.7	62.1	57.0	62.1	51.5	13.7	+10.1	+146	60.2	10.6	56.9
Cu	+1.9	22.7	19.5	37.3	+35.4	+491	+960	+1063	25.6	+553	+12.8
Zn	69.2	92.7	92.6	93.7	78.4	+108	+485	+630	92.0	+158	75.4

* Total iron as Fe$_2$O$_3$

influence the thickness of serpentinite bedrock involved in producing the soil, each 10% (absolute) reduction in porosity increasing by about 35 m the thickness of bedrock required to produce the full soil profile (see Appendix).

In horizon C gains relative to Ti are shown by a number of elements, the Ni-group and Cu being most notable with gains of 400–3800%. Such features are interpreted as the result of sufficient amounts of the elements, leached from higher levels in the profile, being redeposited in horizon C to achieve their highest concentrations in the profile (Fig. 1D). It was noted earlier that the elements V, Al, Cr and Sc, although strongly associated with horizon B, have sharp abundance peaks at 9.5–10 m, close to the C_1–C_2 transition. Results from the mass-balance model (Table 4) show that these peaks are due to local enrichment by redeposition near the position of sample no. 20. High gains relative to Ti are also observed at a depth of 9–10 m for Fe, Mn, Co and Ce, resulting in a slight peak in the Fe-distribution and shoulders in the distributions of Mn, Co and Ce (Fig. 1B,C). Concentration peaks for Fe, Cr, Cu, Mn and Co at 7–8 m do not coincide with gains in these elements relative to Ti. A possible explanation of this apparent anomaly is that there has been some movement of Ti within the profile. Indeed the mass-balance results for Zr indicate that throughout horizon B there has been a net gain of Zr relative to Ti (Table 4), which can be interpreted equally well as a net loss of Ti relative to Zr. In the same way the relative Zr-loss in horizon C could be the result of a relative Ti-gain. The distribution of Ti with depth does in fact show a distinct shoulder, rather than a smooth downward curve, at a depth of 7–8 m, where the Ti-retained model produces the largest relative Zr-loss. Thus, perhaps sufficient Ti has been redeposited at this level to obscure the true pattern of gains and losses for this part of the profile.

Despite the reservations expressed above about the accuracy of the low Zr values obtained for the bedrock, the mass-balance model was also calculated using Zr instead of Ti as the reference element. The main effects of using the Zr-based model are, in order of decreasing importance:

(a) in the depth range 7.5–9.5 m there is a Ti-gain relative to Zr of about 100%;

(b) the depths at which Ce and Cu achieve their greatest gains are reduced by about 2 to 7.5–8.5 m, i.e. close to the depths of their maximum concentrations;

(c) the depths of maximum gains for other elements partly redeposited in horizon C (Table 4) remain unchanged but the gains are slightly increased;

(d) the overall gain in Cu (Table 4) is reduced to almost zero;

(e) element losses in horizon C are slightly reduced but those in horizon B, and for the profile as a whole, are increased by small amounts;

(f) the thickness of parent rock transformed to produce the present thickness of soil is increased by about 10% to 262 m.

All of these changes are compatible with some Ti having been mobilized, some redeposited and some lost from the profile. It is highly likely that Zr too has been subject to some movement. The effects of other departures from the basic assumptions made in the mass-balance model have been discussed elsewhere (Esson & Surcan 1978a, 1979) and are referred to below.

Discussion and conclusions

As mentioned earlier, Ce is more strongly retained in horizon B than Y, La and Nd. Under the oxidizing conditions prevailing in most of the profile, a considerable proportion of the Ce present would be Ce^{4+} which behaves more like Th^{4+} than the normal +3 states of Y, La and Nd (Felsche & Herrman 1970). Consequently, Ce^{4+} tends to be enriched in the residual horizon B, while some Ce^{3+} enrichment will occur, as for Y, La and Nd, by redeposition in horizon C. The outcome of this duality is that Ce behaves like those elements, the Al- and Fe-groups discussed above, enriched in both horizons B and C rather than those, including Y, La and Nd, enriched mainly in horizon C.

The analytical values of Y, Nd and Nb, and perhaps Cu, for the bedrock samples may not be sufficiently accurate to obtain good estimates of their profile budgets. Indeed, it would be expected that the Y and Nd budgets should be closer to La and Zn, the Nb closer to V, than they are. However, these reservations do not affect the distribution patterns of these elements or the profile budgets for other elements.

There are few reports of detailed major and trace element studies of complete nickeliferous laterite profiles (Schellman 1964; Zeissink 1969, 1971), but it is well established that most of the trace elements studied and all except a few major elements are more concentrated in the laterite than in the parental ultramafic rock. However, the element distribution within profiles from different deposits is subject to considerable variation due to differences in the maturity of the profiles (Schellman 1971; Zeissink 1971; Ahmad & Morris 1978). A greater degree of leaching results in increased maturity, increased element separation and a greater tendency for relatively mobile elements, such as Ni, Zn, rare earths (+3 states), Mn and Co, and even moderately immobile elements, e.g. Fe, Cr, Sc and Cu, to reach their maximum concentrations at lower levels in the profile. By both geochemical and mineralogical criteria, the Liberdade profile is very mature.

Although subject to the conditions mentioned above, in the Appendix and discussed by Esson & Surcan (1978a, 1979), the mass-balance model presented here shows that the profile has not been formed simply by leaching the most mobile elements. Mg, Si, Ca and Na, from the bedrock. Even most of the elements with much enhanced concentrations in the laterite have suffered considerable losses by leaching (Table 4). The model also illustrates the importance of redeposition, at lower levels, of elements leached from higher parts of the profile and provides an estimate of the very considerable thickness, over 200 m, of bedrock which has been processed to produce the present 11 m thick residual deposit.

References

AHMAD, S. & MORRIS, D. F. C. 1978. Geochemistry of some lateritic nickel-ores with particular reference to the distribution of noble metals. *Mineralog. Mag.* **42**, 143 & M4–8.

ESSON, J. & SURCAN, L. C. 1978a. Chemistry and mineralogy of section through lateritic nickel deposit at Liberdade, Brazil. *Trans. Instn Min. metall.* **87**, B53–60.

—— & —— 1978b. The occurrence, mineralogy and chemistry of some garnierites from Brazil. *Bull. Bur. Rech. geol. min. Orléans.* 1978 sect. II, 263–74.

—— & —— 1979. Discussion of chemistry and mineralogy of section through lateritic nickel deposit at Liberdade, Brazil. *Trans. Instn. Min. metall.* **88**, B74–5.

FELSCHE, J. & HERRMAN, A. G. 1970. Yttrium and lanthanides. *In:* WEDEPOHL, K H (ed.) *Handbook of Geochemistry, II.* Springer-Verlag, Berlin.

NORRISH, K. & CHAPPELL, B. W. 1977. X-ray fluorescence spectrometry. *In:* ZUSSMAN, J. (ed.) *Physical Methods in Determinative Mineralogy, 2nd edn.* Academic Press, London.

SCHELLMAN, W. 1964. Zur lateritischen Verwitterung von Serpentinit. *Geol. Jb.* **81**, 645–78.

—— 1971. Über Beziehungen lateritscher Eisen-, Nickel-, Aluminium- und Manganerze zu ihren Ausgangsgesteinen. *Miner. Deposita*, **6**, 275–91.

ZEISSINK, H. E. 1969. The mineralogy and geochemistry of a nickeliferous laterite profile (Greenvale, Queensland, Australia). *Miner. Deposita*, **4**, 132–52.

—— 1971. Trace element behaviour in two nickeliferous laterite profiles. *Chem. Geol.* **7**, 25–36.

J. ESSON, Department of Geology, The University, Manchester M13 9PL.

Appendix

The results presented in Table 4 are derived from a simple model based on the following conditions:

(1) the present-day profile results from differential leaching of bedrock by percolating groundwater;

(2) no significant extraneous material has been added to the profile;

(3) differential losses from the profile by mechanical means have been negligible compared with those by chemical processes;

(4) an index constituent has been quantitatively retained in the residue;

(5) movement of the reactive solutions has been essentially vertical, with a net downward motion (profile is well-drained and annual precipitation exceeds evaporation;

(6) lateral movement has had a relatively small net effect, i.e. any loss/gain by such processes has been balanced by a corresponding gain/loss.

The purpose of the model is to estimate (a) the thickness of bedrock consumed in forming the soil profile and (b) and elemental balance for each sampling interval in the profile. Aggregate bedrock thicknesses and elemental balances for individual horizons and the full profile are obtained by summing the results from (b) over the measured thicknesses. All calculations are based on unit area of profile.

Consider a sampling interval of thickness T formed, according to the model, by differential leaching of bedrock. The mass of index constituent in this interval is given by $IDT/100$ where I is wt% of index constituent and D the bulk density of the dry sample. Let T' be the thickness of bedrock containing an equal mass of index constituent. Then $IDT/100 = I'D'T'/100$, where I' and D' are the wt% index constituent and density for the bedrock. Thus the model bedrock thickness consumed to produce thickness T of the soil profile is given by

$$T' = (IDT)/(I'D'). \tag{1}$$

For any other constituent, the mass in dry soil of thickness T is given by $EDT/100$, where E is the constituent wt%. Similarly, a thickness T' of bedrock contains $E'D'T'/100$, where E' is the constituent wt% in bedrock. These two expressions can be used to evaluate the wt% of the constituent lost during the conversion of thickness T' of bedrock into a thickness T of soil. The result is $100[1 - (EDT)/(E'D'T')]$.

Using equation (1) to eliminate T', this reduces to

wt% of constituent lost

$$= 100[1 - (EI')/(E'I)]. \tag{2}$$

A positive result for expression (2) indicates a gain, rather than a loss, relative to the index constituent. The index constituent could be a resistant mineral or a chemical constituent. In this case none of the bedrock minerals were sufficiently resistant and Ti, as discussed in the main text, was used. From expression (2) it can be seen that only wt% analyses are required to estimate the losses and gains of constituents relative to Ti. In order to estimate T', however, the bulk density of the dry soil is required and is difficult to measure because variable amounts of shrinkage and crumbling occur on drying.

Values of the mean particle density for dried, powdered samples from horizon B are the range $3.05-3.50$ g cm^{-3}. Crude measurements of dry bulk density indicate a maximum porosity of about 50%. For the sake of uniformity the D values used were taken as 50% of the mean particle densities, i.e. D values in the range $1.52-1.75$. Bedrock density measurements gave an average value of $2.75(D')$. If D values are too low, then T' results are minimum estimates.

Effects of departures from model conditions (2) and (4) can be tested directly. If 50% of the Ti in the soil profile has been introduced from another source, the 237 m of parent rock required by the model (Table 4, line 2) is reduced by 50% and the wt% of all elements lost falls from 97 to 94. If Al_2O_3, less likely to be selectively introduced from an extraneous source, is used as the index constituent the model requires 97 m of bedrock and the aggregate loss of all elements is 96.2 wt%. However, factors affecting the results in the opposite direction are (a) some index constituent lost from profile and (b) losses by mechanical processes (condition 3), e.g. surface erosion and removal of solids in suspension. Conditions (5) and (6) are more difficult to test but, as it is the least mobile minor/major constituent in the system, Ti is not likely to have been preferentially enriched in the profile by upward or lateral migration. Enrichment of other elements, relative to Ti, by such processes would reduce their apparent losses from the profile and, in the case of some elements in horizon C, increase their apparent gains at particular levels in the profile.

RED BEDS

Reddening of tropical coastal dune sands

R. Gardner

SUMMARY: This paper describes the pedological processes, including that of reddening, in latosol formation in dune sands on humid tropical coastlines. A time-scale is established for these processes in South-east India, Sri Lanka and the Natal Province of the Republic of South Africa. Weathering *in situ* of the dune sands has occurred since their deposition in all three locations. Evidence from thin section and TEM analyses shows that detrital addition of clay minerals and iron oxides has been negligible. The major pedogenic processes (as shown by thin section, SEM, TEM, XRD and chemical analyses) in the formation of the latosols in the field areas are: weathering of garnet, feldspar, hornblende and opaque minerals; formation of kaolinite and illite clay minerals; oxidation of iron to haematite, and loss of carbonate and partial removal of silica from weathered horizons, followed by their deposition lower down the profile in some cases. Dating of the red latosols (2.5 YR 4/6–10R 4/8), radiometrically and using archaeological evidence, indicates that formation to depths of 10 m has taken place in under 20,000 years in the coastal dune sands in the field areas. It is suggested that reddening of desert sands by mineral breakdown could take place on time-scales that are much shorter than has previously been assumed.

Reddening in desert sands as a result of mineral breakdown *in situ* is thought to take at least 100,000 years, and typical estimates are of the order of one million years for the brick red colours (Walker 1967; Folk 1976; Turner 1980). Pedogenic reddening of coastal dunes in and very near to the tropics takes two major forms, latosolic and podsolic, of which the former is by far the more prevalent. It is generally assumed that the effect of humidity decreases the time required for reddening, as it speeds up the processes of mineral breakdown, although there is a shortage of radiometric dates for latosolic reddening of dune sands.

Reddening associated with the formation of tropical podsols in beach ridge and dune sands seems to occur very rapidly (Andriesse 1969/70; Pye 1981), possibly due to the very different pedogenic processes involved. Outside the tropics reddening takes place during the formation of terra rossa soils, latosols and some podsols, although reddened soils are rare in cool areas since goethite rather than haematite is the prevalent iron oxide.

This paper focusses on the formation of red latosols in coastal sand dunes. The characteristics of the soils are described and the major processes and rates of formation are discussed. Results are presented from studies in three field areas, namely South-east India, Sri Lanka and Natal (Republic of South Africa). A common origin is suggested for the deposits of all three areas.

Pedogenic reddening of dune sands

The distribution of reddened dune sand within the tropics is shown in Fig. 1. The latosols of the coastal dunes and the red desert soils of the interior are formed as a result of broadly similar processes. In both cases weathering of minerals leads to the liberation of iron which is oxidized to ferrihydrite and then converted to haematite following internal dehydration, as described by Fischer & Schwertmann (1975).

More intense leaching in latosols, characteristic of more humid areas, leads to the removal of alkalies and alkaline earths and the formation, typically, of kaolinite. The combination of kaolinite and haematite has been shown to be very stable (Fordham 1970). In desert soils less intense leaching results in retention of soluble cations within the profile and the formation of montmorillonite, palygorskite and illite clay minerals (Birkeland 1974; Walker, Waugh & Crone 1978). Calcium carbonate and gypsum may also be important components of red desert soils. In both soil types illuviation and translocation of clays and iron oxides down the profile are limited, giving rise to poor horizon development within the soils and reddening throughout the soil profile.

Tropical podsolisation, involving the accumulation of a reddened organic rich B horizon, is a relatively rare phenomenon in well-drained dune sands (Fig. 1), and results from very different pedological processes from those form-

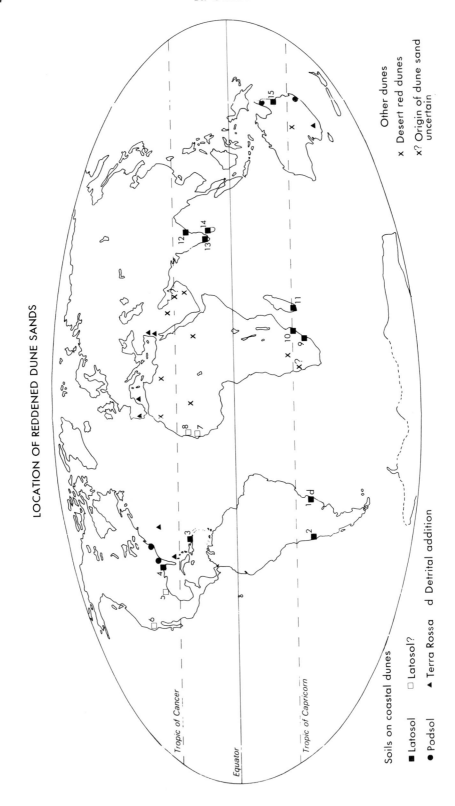

LOCATION OF REDDENED DUNE SANDS

Soils on coastal dunes

■ Latosol   □ Latosol?

● Podsol   ▲ Terra Rossa   d Detrital addition

Other dunes

x   Desert red dunes

x? Origin of dune sand
     uncertain

Fig. 1. Known locations of reddened dune sands (adapted from Gardner & Pye 1981). Red latosols formed on coastal dunes are numbered and referenced.

1. BIGARELLA, J. J. 1975. Lagoa dunefield, state of Santa Catarina, Brazil; a model of aeolian and pluvial activity. *Bolm parana. Geoci.* **33**, 133–67.
2. PASCOFF, R. 1962. Note Preliminaire sur certain analogie du Quaternaire chilien avec Quaternaire marocain. *C. r. Séanc. mens. Soc. Sci. nat. phys. Maroc,* **28** (7), 129–37.
3. KAYE, C. A. 1959. Shoreline features and Quaternary shoreline changes, Puerto Rico. *Prof. Pap. U.S. geol. Surv.* **317-B**, 49–140.
4. SETLOW, L. W. 1978. Age determination of reddened coastal dunes in northwest Florida, U.S.A., by the use of scanning electron microscopy. *In:* WHALLEY, W. B. (ed.) *Scanning Electron Microscopy in the Study of Sediments,* 283–306. Geoabstracts, Norwich.
5. PRICE, W. A. 1962. Stages of oxidation coloration in dune and barrier sands with age. *Bull. geol. Soc. Am.* **73**, 1281–3.
6. NORRIS, R. M. & NORRIS, K. S. 1961. Algodones dunes of southeastern California. *Bull. geol. Soc. Am.* **72**, 605–20.
   VEDDER, J. G. & NORRIS, R. M. 1963. Geology of San Nicolas Island, California. *Prof. Pap. U.S. geol. Surv.* **369**, 65 pp.
7. DAVEAU, S. 1965. Dunes ravinees et depots du Quaternaire Recent dans le Sahel Mauritanien. *Revue Geogr. Afrique Occidentale,* **1**, 7–47.
   GROVE, A. T. & WARREN, A. 1968. Quaternary landforms and climate on the south side of the Sahara. *Geogr. J.* **134**, 194–208.
   BREED, C. S., FRYBERGER, S. G., ANDREWS, S., McCAULEY, J. LENNARTZ, F., GEBEL, D. & HORSTMAN, F. 1979. Regional studies of sand seas using Landsat (ERTS) imagery. *In:* McKEE, D. (ed.) *A Study of Global Sand Seas. Prof. Pap. U.S. geol. Surv.* **1052**, 307–97.
8. TRICART, J. & BROCHU, M. 1955. Le grand erg ancien du Trarza et du Cayor (Sud-ouest de la Mauritanie et Nord du Senegal). *Revue Geomorph. dyn.* **4**, 145–76.
   WORRALL, G. A. 1969. The red sands of the southern Sahara. *Bull. Liason, Ass. Senegalaise Etude Quat. l'Ouest Africain,* **21**, 36–9.
   COUREL, M. F. 1975. Nouvelles observations sur les systemes dunaires du Cap Vert (Senegal) *Travaux et documents de geographie tropicale* 22, Univ. Bordeaux, 217–41.
9. McCARTHY, M. J. 1967. Stratigraphical and sedimentological evidence from the Durban region of major sea level movements since the Late Tertiary. *Trans. geol. Soc. S. Afr.* **70**, 135–65.
   MAUD, R. R. 1968. See general references.
   DAVIES, O. 1976. See general references.
10. HOBDAY, D. K. 1977. Late Quaternary sedimentary history of Inhaca Island, Mozambique. *Trans. geol. Soc. S. Afr.* **80**, 183–91.
11. BATTISTINI, R. 1964. *L'extreme sud de Madagascar, etude geomorphologique.* Thesis, University of Paris. 636 pp.
12. DURGAPRASADA RAO, N. V. N. & SRIHARI, Y. 1980. Clay mineralogy of the late Pleistocene red sediments of the Visakhapatnam region, east coast of India. *Sedim. Geol.* **27**, 213–27.
13. FOOTE, R. B. 1883. See general references.
    AHMAD, E. 1972. See general references.
    GARDNER, R. A. M. 1981. See general references.
14. COORAY, P. G. 1968. See general references.
    DE ALWIS, K. A. & PLUTH, D. J. 1976. See general references.
15. HOPLEY, D. 1970. *Geomorphology of the Burdekin delta, north Queensland.* Department of Geography, James Cook University, Monograph series 1.

ing latosols and red desert soils. In particular, the role of organic acids in promoting mineral breakdown, complexing of iron, and aiding in the movement of clays down the profile distinguishes this type of soil (Mohr, Van Baren & Van Schuylenborgh 1972; De Coninck 1980).

Coastal dune sands in South-east India, Sri Lanka and Natal

South-east India

The reddened coastal dune sands (locally termed teris) are found on the extreme south-east coast of the Peninsula, between Rameswaram and Kanyakumari (a distance of 200 km) in Tamilnadu State. This area is now one of the driest in India, receiving between 500 and 900 mm of rainfall annually, largely between October and December during the north-easterly Monsoon (Cadet 1979). Annual mean temperature is 28°C and for most months of the year evaporation losses are greater than precipitation.

The teri sands are found up to 16 km inland from the coast, although more commonly they form a belt approximately 5 km wide roughly parallel to the present shoreline, as described by Foote (1883) and Gardner (1981b). The reddened sands reach a maximum height of 60 m above sea-level and overlie calcrete developed on calcareous aeolianite. The latter is underlain by carbonate-cemented nearshore marine sands at 6–8 m above sea-level. The teri sands are generally reddened throughout their depth, that is, up to 10 m in places, and little profile differentiation can be observed in the field, the sands appearing as a dense, structureless deposit. In a few profiles carbonates removed from the reddened sands have been redeposited at depth within the profile. However, in general the carbonates have been totally removed.

The deep red colours (2.54 YR 4/6–10R 4/8) of the sands are produced by free ferric oxide (largely haematite), the average content in the sands being 1.6%. The amount of clay-sized material varies from 2.8% in aeolian reworked sands to 29% in the undisturbed teri sands. In all cases kaolinite is the predominant clay mineral, with lesser amounts of illite, as seen in Table 1. Gibbsite and montmorillonite are absent, and easily weathered silicate minerals (as classified by Goldich 1938) are rare, the majority of the sand grains comprising quartz and opaque minerals. Mineralogical and chemical characteristics do not appear to differ systematically throughout any of the profiles.

Sri Lanka

The fossil reddened sand dunes occupy approximately one-third of the coastline of the island, between Mundal on the north-west coast and Mullaitivu on the north-east coast. This area currently receives slightly more rainfall (900–1400 mm y^{-1}) than the neighbouring coastline in South-east India. The rain falls at similar periods of the year, the higher rainfall resulting in denser vegetation on the dune sands.

The sands are found as a series of ridges, numbering between one and six and extending up to 25 km inland. Earlier workers were concerned about their origin, and suggested aeolian deposition of red earths from inland (Wayland 1919), coastal alluvial deposits (Moormann & Panabokke 1961), and marine, littoral and, possibly, dune complexes (Cooray 1967; Deraniyagala 1976). Subsequent research has shown that the sands are largely coastal dunes (Gardner 1981b).

These dunes are similar to the teris in terms of colour, lack of sedimentary structures and observable horizonation in the soil, depth of reddened profile and clay mineralogy, as shown in Table 1. However, the sands differ slightly in terms of sand-grain mineralogy, reflecting a facies change in the basement rock (Wadia 1957; Cooray 1967). Thus, hornblende is more common and garnet less abundant in the sands in Sri Lanka, a difference observed more in modern dune sands than in the reddened sands from which the majority of the less stable minerals have been removed (as described by De Alwis & Pluth 1976a,b). The final weathered products in the two locations are very similar, but the red sands in Sri Lanka overlie Tertiary and older rocks often veneered by lateritic gravel and pebble deposits of colluvial or marine origin (Wayland 1919; Cooray 1967; Deraniyagala 1976).

Natal, Republic of South Africa

Reddened dune sands are found along the coastline of Natal and Zululand from Durban northwards into Mozambique. The inland extent and width of the belt increases in the north to over 20 km. The area receives summer rainfall of approximately 1000 mm y^{-1}, and the mean annual temperature is 20°C. The climatic and soil conditions result in widespread cultivation of the dunes, primarily for sugar cane.

The reddened sands overlie, and are in places weathered from, calcareous aeolianite which rests, near Durban, on an 8 m raised shoreline (Maud 1968). Farther north, the reddened sands overlie the Port Durnford Beds, thought

TABLE 1. *Summary of mineralogical characteristics of samples from reddened dune sands in South-east India, Sri Lanka and Natal, Republic of South Africa. Results from analyses of 100 samples*

	Quartz	Feldspar	Hornblende	Biotite	Garnet	Opaques	Ferric oxide (free)	Clay minerals	Clay size fraction
India									
Red dune sand	90.0%	1.5%	0.0	0.0	<0.5%	4.5%	1.6%	Kaolinite 80% Illite 20%	3.0–29.0% (by weight) Less than 1%
Unaltered dune sand	82.0	10.0	0.0	0.0	3.0	4.0	0.0		
Sri Lanka									
Red dune sand	90.0	2.0	<0.5	<0.5	0.5	5.0	1.2	Kaolinite 85% Illite 15%	8.0–36.0% (by weight) Less than 1%
Unaltered dune sand	80.0	9.0	3.0	0.5	1.0	5.0	0.0		
Natal									
Red dune sand	86.0	6.0	1.5	0.0	0.5	4.0	1.4	[a]Kaolinite 75% Illite 25%	10.0–29.0% (by weight) Less than 2%
Unaltered dune sand	74.0	11.0	5.0	0.0	2.0	7.0	0.0		

[a] Additional small amounts of montmorillonite are present in three samples.

The free ferric oxide was extracted by means of the sodium–citrate–dithionite method of Mehra & Jackson (1960). This method extracts both amorphous and crystalline oxides from the sample. The iron content of the solute was measured by AAS, and values in ppm were recalculated as % Fe_2O_3 in the sample.

Clay minerals were identified using X-ray diffraction after removal of the ferric oxide. Untreated, magnesium-saturated and pre-heated samples were analysed using a copper anode. Relative percentages were calculated using the areas under the XRD peaks, and by measuring the ratios of the main peaks. The different size fractions of the clays were not analysed separately. Grain size was determined by sieving (sand grades) and pipette and hydrophotometer (Rendell 1974) techniques (silt and clay grades) after dispersion of the sample in calgon. Results were recalculated as percentages by weight.

Thin section analyses were undertaken on samples removed undisturbed from the field. The partially indurated nature of the sediments made this possible, with the aid of metal containers. Samples were then impregnated by clear glue ('UHU' was found to be best) prior to sectioning to a standard thickness of 0.03 mm. Percentage (by volume) of different minerals was calculated from point counting 300 grains.

to be estuarine/lagoonal in origin (Hobday 1975). Here a maximum of six fossil ridges have been identified (Davies 1976). Reddening to depths of 10 m is common, the sands attaining colours of 2.5 YR 4/6 as a result of liberation of free ferric oxide (1.2% on average), and appearing as dense, structureless deposits. Kaolinite and illite make up the majority of the clay-size fraction, which varies between 12 and 28% by weight. The red sands show a depletion in feldspar, hornblende and opaque minerals compared to the unweathered sands (Table 1). Thus, again, the south-east African dune sands are similar to those of South-east India.

Soil forming processes in the field areas

The reddened sands in all three areas exhibit deep and homogeneous profiles, characterized by matrices enriched in haematite and kaolinite, with lesser amounts of illite. In contrast to earlier suggestions of detrital accumulations of the reddened sand in India (Foote 1883; Ahmad 1972) and Sri Lanka (Wayland 1919), this work shows that the matrix has formed in the three locations as a result of the precipitation of ions liberated during the breakdown of the less stable minerals within the sands. The work of Maud (1968) in Natal and De Alwis & Pluth

(1976a,b) in Sri Lanka support these conclusions.

Examination by SEM (Cambridge S150) shows evidence of dissolution of many of the remaining less stable minerals within the reddened sands. This is particularly well shown on feldspar grains in the form of etching along cleavage planes (Fig. 2a) and, less spectacularly, on garnet grains (Fig. 2b). The evidence is supported by thin section studies, which show that clay replacement of feldspar has resulted in the formation of clay pseudomorphs in a few instances, or, more usually, in silt-sized relic cores of the mineral. Hornblende is similarly affected, whereas the surfaces of garnet and some opaque minerals are distinctly haematised (Fig. 3). Clay replacement and haematisation can be observed throughout the reddened profiles. Evidence of mineral dissolution in thin section, as reported by Walker et al. (1978), is less common.

Support for the hypothesis of intrastatal alteration of the sands is provided by determinations of the relative abundance of heavy minerals and feldspar grains (Table 1). Comparison with underlying aeolianites in the cases of India and Natal, and with modern dunes in the cases of India and Sri Lanka, suggests, on assuming that the mineralogy of the reddened sands was

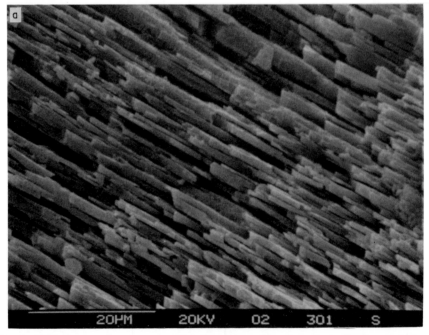

FIG. 2. SEM photographs of sand grains from the reddened sands in South-east India. (a) Dissolution of feldspar along cleavage planes. (b) Chemical etching on the surface of a garnet grain. (c) Precipitation of amorphous silica on a quartz grain.

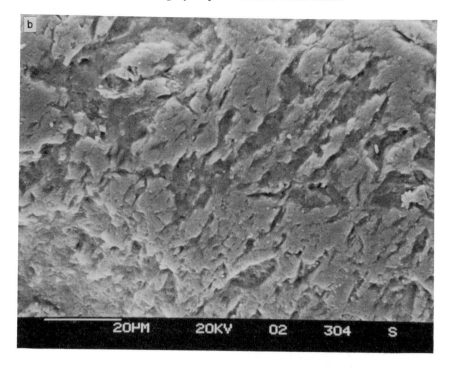

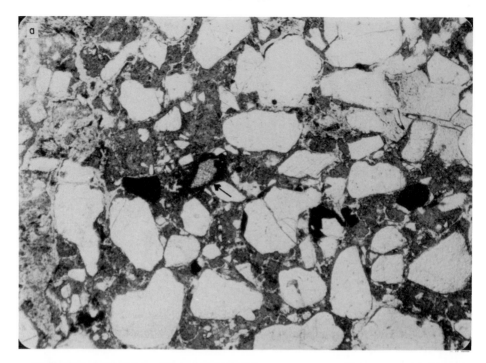

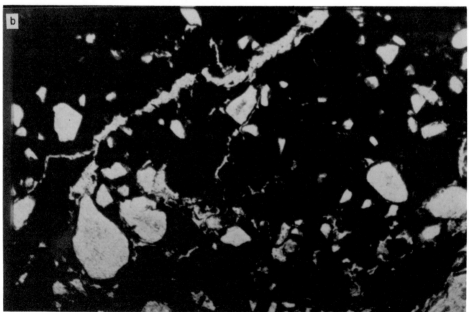

FIG. 3. Thin section micrographs of the reddened sands in South-east India. (a) Note the clotted texture of the kaolinite/haematite matrix, the abundance of craze and skew planes, and the dark haematite rim around a disintegrating garnet grain (×40). (b) A particularly dense sample viewed under crossed nicols, showing the absence of well-defined, oriented cutans and the lack of easily weathered minerals (×30).

originally broadly similar to that of the unweathered sands in each area, that depletion of the following minerals has taken place. All carbonates have been lost, and there is marked depletion of garnet and feldspar in India (Fig. 3b), and of hornblende, garnet and feldspar in Natal and Sri Lanka. Significant differences in absolute numbers of opaque grains were also recorded in all areas. The data suggest that garnet (almandite) may be a major source of iron in the teri sands of India (Gardner 1981a), even though it is traditionally regarded as a relatively stable mineral (Pettijohn 1941). It would seem that its disintegration is favoured by the dune micro-environment, namely warm, moist but well drained, and oxidizing conditions with intrastral solutions having slightly to significantly acid pH values (6.9–4.4 recorded in the Indian sands). The relative instability of garnet under conditions of chemical weathering, was noted many years ago by Dryden & Dryden (1946) and Graham (cited in Jackson & Sherman 1953).

Further evidence from SEM studies indicates that desilicification occurs in association with the weathering, some of the silica being deposited in the lower levels of the profiles in India as coatings of amorphous silica on quartz grains (Fig. 2c). Studies of the surfaces of quartz grains show that little of the silica is supplied from dissolution of the quartz particles; breakdown of heavy minerals and feldspars, probably, provides the source.

Studies by X-ray diffraction of the clay mineralogy indicate (Table 1) that in nearly all cases, kaolinite and illite are the only clay minerals present. Gibbsite is absent from all samples. A few analyses show the proportion of illite, although relatively small, to increase slightly with depth in the profile. This is not a general trend, and it may reflect localized differences in the degree of leaching and groundwater geochemistry; the illite probably formed as a precursor to kaolinite in locations with less leaching. Chemical analyses of the clay matrix using AAS and flame photometry (after leaching the soil by IM ammonium acetate) showed the near absence of exchangeable cations of Na, Mg, Ca and K, a result in accordance with the mineralogical studies. Amounts of free aluminium oxide were also very low, less than 0.5%.

Examination of the kaolinite by TEM showed the flakes to be small (<0.5 mm diameter) hexagons, with little evidence of the physical destruction that would result from aeolian reworking of the flakes (Wilson & Pittman 1977), as seen in Fig. 4.

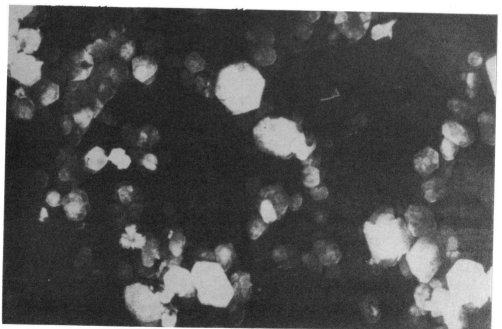

FIG. 4. TEM photograph showing hexagonal flakes of kaolinite present in the reddened sands (×40,000).

The formation of the clays gives rise to the highly bi-modal grain-size distribution characteristic of the red sands. Although the amount of clay differs within and between profiles, no systematic trends were discerned from the samples of red 'sand' analysed in each area. Differences in clay content probably relate to variations in the original mineralogy and degree of alteration of the sands.

The absence in thin section of *oriented* clay and/or iron oxide cutans on mineral grains (as classified by Brewer 1964) indicates that large-scale translocation of clays has not taken place within any of the profiles. Although the mineral grains are coated by the clay/iron oxide matrix, this is easily dispersed by gentle shaking in calgon. This coating does not constitute a well-defined cutan (Brewer 1964) but is rather a poorly defined, if present at all, covering of the matrix on the grains and is optically no different from the matrix. This suggests that infiltration of detrital clays has not been an important process in the formation of the red sands. The stability of the matrix, once formed, is enhanced by the attachment of haematite to the kaolinite (Follett 1965; Fordham 1970), the absence of undecomposed matter or other reducing agents (Bromfield 1954; Hem & Cropper 1959; Turner 1980) and the decrease in permeability in the sands.

The evidence presented above indicates that the pedogenic reddening of the sands is a result of weathering *in situ*, giving rise to deep latosols. In summary, the main mineralogical and textural evidence is the homogeneity of the profiles, the lack of evidence for large-scale illuviation, the shape of the clay particles, the absence of sand-sized silt and clay aggregates often found in detritally derived (aeolian) red sediments, and the evidence for mineral weathering. The absence of gibbsite, currently forming in the wetter zones (>1500 mm rainfall) of India (Gardner 1981a,b) and Sri Lanka (De Alwis 1971), suggests that weathering has taken place under climatic conditions little wetter, if at all, than at present.

Dating the dune sands

Radiometric dates could only be obtained for South-east India, because of the paucity of datable material (Fig. 5). The samples from India consisted of aragonite shells from calcareous aeolianite/calcrete and raised marine/lagoonal sediments. The shells were checked for contamination by XRD and in thin section, and in all cases were found to comprise aragonite. The presence of aragonite landsnails in calcareous aeolianite is somewhat surprising. However, similar occurrences have been reported from Mexico (Ward 1975), where they were ascribed to protective organic coatings around the aragonite shells. The ^{13}C values for the marine shells support the evidence for the lack of contamination (see Gross 1964), which indicates that the date of 38,000 BP is not unreliable, even though it is near the limit for radiocarbon dating.

The dates provide a consistent framework, and imply that the formation of the coastal dunes, now reddened, took place at the height of the last glaciation (approximately 21,000–17,000 BP), during which time there is considerable evidence for greater aridity and aeolian activity elsewhere in the tropics (Rognon & Williams 1977; Sarnthein 1978; Bowler 1978; Street & Grove 1979). Reddening to depths of 10 m has thus occurred within less than 20,000 years, possibly considerably less given the stability of the latosols once formed. This evidence for the rapid rate of reddening is consistent with the preservation of the morphology of the sands, and with the archaeological artefacts (microliths) on the surface of the sands. Microliths have been dated elsewhere in the subcontinent to less than 10,000 years, and more probably 4000–8000 BP (Allchin 1963). Foote (1883) reportedly found microliths within the sands, although subsequent searches by Zeuner & Allchin (1956) and the author failed to reveal artefacts *in situ* within undisturbed deposits.

Evidence from artefacts in Sri Lanka indicates a similar age for the red dunes there. Microlithic artefacts, the same as those in India, are found on the surface of the dune sands, and Middle Stone Age material was recovered from the lateritic gravels that underlie the dune sands near Aruakalu. Similar Middle Stone Age artefacts are dated to less than 40,000 years in India (Agrawal, Guzder & Kusumgar 1972). Deraniyagala (1976) has also reported finding microliths within the red sands. In addition, the comparability of form, pedogenic weathering, preservation and location of the sands in India and neighbouring Sri Lanka, and the fact that an earlier red weathering phase is not documented in India, all suggest that the two reddened dune formations are of similar age.

Dating of the sands in Natal is more difficult (Maud 1968). Recent archaeological work on artefacts from radiometrically dated sites in Swaziland suggests that the Middle Stone Age tools found on top of the Port Durnford Beds and in the aeolianite, beneath reddened sands,

STRATIGRAPHY & AGES OF REDDENED DUNE SANDS

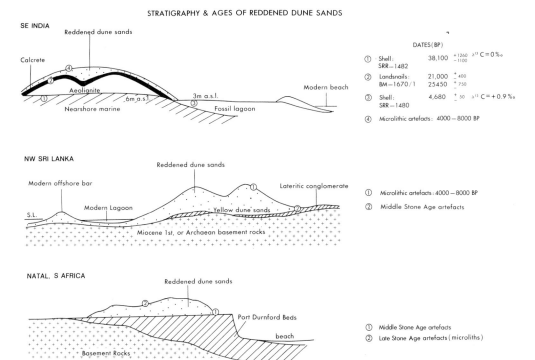

FIG. 5. Stratigraphy and ages of the reddened dune sands in South-east India, Sri Lanka and Natal. Generalized sections shown.

are of the order of 25,000–19,000 years old (Price-Williams, pers. comm.). It is hoped to obtain radiometric dates from the coastal sequences in the near future. Microliths are found on the surface of the reddened sands.

Conclusion

In brief, it has been shown that dark red latosols formed *in situ* to depths of 10 m on the coastal dune sands in South-east India, Sri Lanka and Natal. Breakdown of garnet (almandite), hornblende, opaques and feldspars led to the formation, after leaching of the soluble cations and small amounts of silica, of a dense matrix enriched in ferric oxide (haematite), kaolinite and illite. The sand grains remaining after weathering are mainly quartz, with some opaques. Mineralogical, chemical and textural characteristics appear homogeneous throughout most of the profiles. Radiocarbon dating and evidence from archaeological material has shown the time-scale for this process to be less than 20,000 years.

Reddening and weathering of desert sands *in situ*, from breakdown of less stable minerals containing iron, may either follow the above processes, or, if the cations are not leached from the profile, montmorillonite and palygorskite clays may be formed (Walker *et al.* 1978). In either case reddening occurs so long as oxidizing conditions exist within the dune sands. In some instances mineral breakdown may be aided by salt weathering (Goudie, Cooke & Doornkamp 1979).

Typically, desert dunes contain significant quantities of pore water (Glennie 1970), and quite frequently are subjected to rain falls, and occasionally are subjected to flooding. (For example, parts of the Kalahari currently receive over 400 mm of rain per year, falling in 20–30 events.) Thus certain dune sands, where minerals are breaking down *in situ*, may not require 100,000 years, or more, to redden in this way. In fact, much shorter time-scales seem possible, and each area of reddened desert dune sand should be examined on its own merits.

References

AGRAWAL, D. P., GUZDER, S. & KUSUMGAR, S. 1972. *Radiocarbon chronology in Indian prehistoric archaeology*. Proceedings, Tata Institute for Fundamental Research, Bombay.

AHMAD, E. 1972. *Coastal Geomorphology of India*. New Delhi, Orient Longman.

ALLCHIN, B. 1963. The Indian Stone Age Sequence. *Jl R. anthrop. Inst.* **XCIII**, 210–34.

ANDRIESSE, J. P. 1969/70. The development of the podsol morphology in the tropical lowlands of Sarawak (Malaysia). *Geoderma*, **3**, 261–79.

BIRKELAND, P. W. 1974. *Pedology, Weathering and Geomorphological Research*. Oxford University Press.

BOWLER, J. M. 1978. Glacial age aeolian events at high and low latitudes: a southern hemisphere perspective. *In:* VAN ZINDEREN BAKKER, E. M. (ed.) *Antarctic Glacial History and World Palaeoenvironments*, 149–72. A. A. Balkema, Rotterdam.

BREWER, R. 1964. *Fabric and Mineral Analysis of Soils*. Wiley, New York.

BROMFIELD, S. M. 1954. The reduction of iron oxide by bacteria. *J. Soil Sci.* **5**, 129–39.

CADET, D. 1979. Meteorology of the Indian Summer Monsoon. *Nature*, **279**, 761–7.

COORAY, P. G. 1967. *An Introduction to the Geology of Ceylon*. National Museum of Ceylon Publication, Colombo. 62 pp.

DAVIES, O. 1976. The older coastal dunes in Natal and Zululand and their relation to former shore-lines. *Ann. S. Afr. Mus.* **71**, 19–31.

DE ALWIS, K. A. 1971. *The pedology of the red latosols of Ceylon*. Ph.D. thesis, University of Alberta.

DE ALWIS, K. A. & PLUTH, D. J. 1976a. The red latosols of Sri Lanka: I. Macromorphological, physical and chemical properties, genesis and classification. *J. Soil Sci. Soc. Am.* **40**, 912–9.

—— 1976b. The red latosols of Sri Lanka: II. Mineralogy and weathering. *J. Soil Sci. Soc. Am.* **40**, 920–28.

DE CONINCK, F. 1980. Major mechanisms in the formation of spodic horizons. *Geoderma*, **24**, 101–28.

DERANIYAGALA, S. U. 1976. The geomorphology and pedology of three sedimentary formations containing a mesolithic industry in the lowlands of the dry zone of Sri Lanka. *In:* KENNEDY, K. A. R. & POSSEHL, G. L. (eds) *Ecological Backgrounds of South Asian Prehistory*. South Asia Occasional Papers, Cornell University.

DRYDEN, A. L. (JR) & DRYDEN, C. 1946. Comparative rates of weathering of some common heavy minerals. *J. sedim. Petrol.* **16**, 89–96.

FISCHER, W. R. & SCHWERTMANN, U. 1975. The formation of haematite from amorphous iron (III) hydroxide. *Clays Clay Miner.* **23**, 33–7.

FOLK, R. L. 1976. Reddening of desert sands: Simpson Desert, Northern Territory, Australia. *J. sedim. Petrol.* **46**, 604–15.

FOLLETT, E. A. C. 1965. The retention of amorphous,

colloidal 'ferric hydroxide' by kaolinites. *J. Soil. Sci.* **16**, 334–41.

FOOTE, R. B. 1883. On the geology of Madura and Tinnevelly Districts, *Mem. geol. Surv. India*, **20**, 1–103.

FORDHAM, A. W. 1970. Sorption and precipitation of iron on kaolinite. III. The solubility of iron III hydroxides precipitated in the presence of kaolinite. *Aust. J. Soil Res.* **8**, 107–122.

GARDNER, R. A. M. 1981a. Reddening of dune sands —evidence from South-east India. *Earth Surf. Processes*, **6**, 459–68.

—— 1981b. *Geomorphology and environmental change in South-east India and Sri Lanka*. D.Phil. thesis, Oxford University. 643 pp.

GARDNER, R. A. M. & PYE, K. 1981. Nature, origin and palaeoenvironmental significance of red coastal and desert dune sands. *Progr. Phys. Geogr.* **5**, 514–34.

GLENNIE, K. W. 1970. *Desert Sedimentary Environments*. Developments in Sedimentology 14. Elsevier, Amsterdam.

GOLDICH, S. S. 1938. A study in rock weathering. *J. Geol.* **46**, 17–58.

GOUDIE, A. S., COOKE, R. U. & DOORNKAMP, J. C. 1979. The formation of silt from quartz dune sand by salt-weathering processes in deserts. *J. Arid Environ.* **2**, 105–12.

GROSS, M. G. 1964. Variations in the O^{18}/O^{16} and C^{13}/C^{12} ratios of diagenetically altered limestones in the Bermuda Islands. *J. Geol.* **72**, 170–94.

HEM, J. D. & CROPPER, W. H. 1959. Survey of ferrous-ferric chemical equilibria and redox potentials. *Wat. Supply Pap. US geol. Surv. 1459A.* 31 pp.

HOBDAY, D. K. 1975. The Port Durnford formation. *Trans. geol. Soc. S. Afr.* **77**, 141–9.

JACKSON, M. L. & SHERMAN, G. D. 1953. Chemical weathering of minerals in soils. *Adv. Agron.* **5**, 219–318.

MAUD, R. R. 1968. Quaternary geomorphology and soil development in coastal Natal. *Z. Geomorf., Suppl. Band* **7**, 155–99.

MEHRA, O. P. & JACKSON, M. L. 1960. Iron oxide removal from soils and clays by a dithionite-citrate system buffered with sodium bicarbonate. *Proc. 7th Conf. Clays and Clay Minerals*, 317–27.

MOHR, E. C. J., VAN BAREN, F. A. & VAN SCHUYLENBORGH, J. 1972. *Tropical Soils: a comprehensive study of their genesis*. Mouton-Ichtiar Bar-Van Hoeve, The Hague, Paris, Djakarta. 481 pp.

MOORMANN, F. R. & PANABOKKE, C. R. 1961. Soils of Ceylon. *Trop. Agric. Mag. Ceylon agric. Soc.* **117**, 1–70.

PETTIJOHN, F. J. 1941. Persistence of heavy minerals and geologic age. *J. Geol.* **49**, 610–25.

PYE, K. 1981. Rate of dune reddening in a humid tropical climate. *Nature*, **290**, 582–4.

RENDELL, H. M. 1974. A note on the photo-extinction method of sediment size analysis, *Bloomsbury Geographer*, **3**.

ROGNON, P. & WILLIAMS, M. A. J. 1977. Late Quaternary climatic changes in Australia and North Africa: a preliminary interpretation. *Palaeogeogr. Palaeoclim. Palaeoecol.* **21**, 285–327.

SARNTHEIN, M. 1978. Sand deserts during glacial maximum and climatic optimum. *Nature*, **272**, 43–6.

STREET, F. A. & GROVE, A. T. 1979. Global maps of lake-level fluctuations. *Quat. Res.* **12**, 83–118.

TURNER, P. 1980. *Continental Red Beds*. Developments in Sedimentology, 29. Elsevier, Amsterdam. 562 pp.

WADIA, D. N. 1957. *Geology of India and Burma*. Tata McGraw-Hill, London. 508 pp.

WALKER, T. R. 1967. Formation of red beds in modern and ancient deserts. *Bull. geol. Soc. Am.* **78**, 353–68.

WALKER, T. R., WAUGH, B. & CRONE, A. J. 1978. Diagenesis in first cycle desert alluvium of Cenozoic age, south-western United States and northwestern Mexico. *Bull. geol. Soc. Am.* **89**, 19–32.

WARD, W. C. 1975. Petrology and diagenesis of carbonate aeolianites of northeastern Yucatan Peninsula, Mexico. *In:* WANTLAND, K. F. & PUSEY, W. C. (eds) *Belize Shelf: carbonate sediments, clastic sediments and ecology*. AAPG Studies in Geology, vol. 2, 500–71.

WAYLAND, E. J. 1919. Outlines of the Stone Age of Ceylon. *Spolia zeylan.* **11**, 41–58.

WILSON, M. D. & PITTMAN, E. D. 1977. Authigenic clays in sandstones: recognition and influence on reservoir properties and palaeoenvironmental analysis. *J. sedim. Petrol.* **47**, 3–31.

ZEUNER, F. E. & ALLCHIN, B. 1956. The microlithic sites of Tinnevelly District, Madras State. *Ancient India*, **12**, 4–20.

RITA GARDNER, Department of Geography, Kings College, Strand, London WC2R 2LS.

Post-depositional reddening of late Quaternary coastal dune sands, north-eastern Australia

K. Pye

SUMMARY: Deeply weathered quartz sand dunes of late Quaternary age occur at Cape Bedford and Cape Flattery on the seasonally humid tropical east coast of North Queensland, Australia. Many dunes have deep podzolic profiles with reddened B horizons up to 8 m thick. The red colour is produced by iron oxides and hydroxides which, together with kaolinite and minor gibbsite, form interstitial matrix and cutans on quartz grains. The red sands are generally uncemented, but nodular ironstone layers are present in some sections. Radiocarbon dating has shown that one red podzol profile near Cape Flattery has formed in the last 7500 C–14 years. Progressive podzolization has removed much of the iron pigment from the oldest dunes in the area, which are >48 000 C–14 years old. These dunes are degraded and contain subsurface horizons which are cemented by humate.

Recent years have seen the development of a sophisticated model to account for reddening and diagenesis of sediments in arid tropical environments (Walker 1967; Walker, Waugh & Crone 1978), but less is known about post-depositional alteration of sediments in humid tropical and subtropical climates (Gardner & Pye 1981). Latosolic reddening of coastal sediments in subhumid Sri Lanka, south-east India and Natal has recently been investigated by De Alwis & Pluth (1976) and by Gardner (1981, 1983), but no studies of podzolic reddening in wetter parts of the tropics have been published. This paper examines the nature and development of podzolic weathering profiles which have formed in late Quaternary quartz dune sands at Cape Bedford and Cape Flattery, North Queensland. The study was made as part of a wider regional investigation of the formation and Quaternary history of coastal dunes in North Queensland (Pye 1980).

The Cape Bedford–Cape Flattery dunefield, covering approximately 600 km², lies at latitude 14°50′ south on the north-east coast of Queensland (Fig. 1). The climate is seasonally humid tropical, with about 75% of the mean annual rainfall total (1800 mm y⁻¹) falling between January and April. Quartz sand has been supplied to the area from Palaeozoic granites and metamorphic rocks to the south and south-west of Cooktown, and from Mesozoic sedimentary rocks in the Endeavour River basin to the west. The dunefield consists of elongate parabolic dunes which have advanced from south-east to north-west over fluvial deposits and low bedrock hills (Pye 1982a). Most of the dunes are vegetated and podzolized to varying degrees, but in many places blowout formation has led to reworking of weathered sands by active dunes.

Investigative methods

Stratigraphic data were obtained from more than 70 power drill and hand auger holes and from coastal cliff and gully sections. Morphological information was provided by field mapping and air photograph interpretation. Sand samples were collected at various levels in drill holes and from exposures for laboratory analysis. Grain size analysis was performed by wet sieving to remove fines, and by mechanical dry sieving using a Ro-tap shaker. Some fine fractions were retained for Coulter Counter analysis. Mineralogy of the light and heavy sand fractions after bromoform separation was determined by X-ray diffraction, electron microprobe analysis, and optical microscopy. Clay fractions were also identified by X-ray diffraction. Extractable iron, aluminium and magnesium contents were determined by atomic absorption spectrophotometry on dilute HCl extracts. Sand grain surface textures were examined by scanning electron microscopy in conjunction with energy dispersive analysis (EDAX). Further attempts were made to identify iron oxides and hydroxides by transmission electron microscopy, differential thermal analysis, and Mössbauer spectroscopy.

Weathering profile characteristics

Morphological, stratigraphic and C-14 data have allowed recognition of three main dune sand bodies for which the names *Mitchell Dunesand* (youngest), *Bedford Dunesand*, and *Flattery Dunesand* (oldest) have been suggested (Pye & Switsur 1981). It is possible, however, that future work will allow subdivision of these units. Dunes of the youngest generation consist of undifferentiated or very weakly podzolized

117

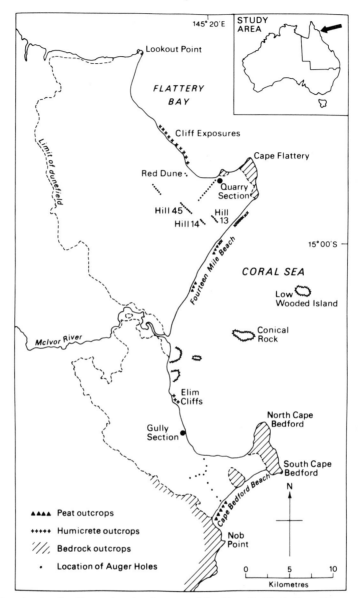

Fig. 1. Map showing location and major features of the Cape Bedford–Cape Flattery dunefield, North Queensland.

white quartz sand, representing material which has been reworked from older dunes in the past few hundred years. The Bedford Dunesands are deeply weathered and podzolized with reddening in the B horizons. Leached A2 horizons in these sands are up to 6 m deep. Flattery Dunesands are also deeply podzolized with thick humate-impregnated B horizons. They have generally lost their original ferric oxide colouration, although residual reddening is locally visible. Complete sections showing superimposition of all three units are rare, since successive aeolian phases have partially stripped earlier weathering profiles. Mitchell sands overlie both Bedford sands and Flattery sands, while Bedford sands are seen to overlie Flattery sands in cliffs south of the McIvor River mouth and to the north-west of Cape Flattery.

The profile characteristics and composition of the humate-bearing Flattery Dunesands have

Fig. 2. A 20 m high quarry section showing major horizon differentiation in a weathered dune profile near Cape Flattery.

already been discussed in detail (Pye 1982b), and therefore the main concern of this paper is the reddened Bedford sands. These are well exposed in a 20 m high quarry section located about 2 km north-west of Cape Flattery (Fig. 2). In this section 3 m of leached white quartz sands, stained grey by organic matter in the uppermost 50 cm, overlie 6–8 m of orange and red (2.5YR4/8) B horizon sands. The latter in turn overlie at least 8 m of yellow, pale orange and white C horizon sands which display banding and mottling. A concentration of humic acids imparts a brownish colour to the top of the B horizon and base of the A2 horizon, and nodular ironstone fragments occur near the top of the reddened zone. The mineralogical and chemical properties of this profile are discussed below.

A second exposure of red podzolized dune sands has been created by recent marine erosion at Elim Cliffs, located about 4 km northwest of Cape Bedford (Fig. 1). At the top of this section 1–3 m of leached white A horizon sands overlie 5–8 m of orange and red B horizon sands, which in turn overlie more than 40 m of yellow, pale orange and white banded and mottled sands. Erosional unconformities are visible in the lower part of the section in some places, but sands above and below have a similar weathered appearance. A major ironstone layer again occurs near the top of main reddened B horizon, and thinner cemented layers are found at lower levels. The nature and origin of the ironstones are discussed by Pye (1983a).

Textural and mineralogical composition of the weathered sands

The Cape Bedford–Cape Flattery dunesands as a whole are fine-grained, well-sorted and slightly negatively skewed, although important within dunefield variations exist (Pye 1982c). In the weathered Bedford Dunesand profile

shown in Fig. 2 mean grain size increases with depth, despite higher silt and clay content in the B and C horizons (Fig. 3). Upwards fining is a primary sedimentary feature also found in the modern active dunes, but evidence suggests that some size reduction by *in situ* weathering of quartz does occur in the A horizons (Pye 1983b). The fines so produced are mechanically translocated through the sand column and are deposited in the B and C horizons. Some podzol B horizons contain 10–15% quartz silt, compared with 0–0.2% in active dunes.

Fig. 3 shows that extractable Fe and Al content is very low in the A horizon sands, but that the B horizon contains concentrations of both iron and aluminium. Levels of Ca, Mg, Mn, K and Ti are very low at all levels in the profile. The range of Fe and Al contents in the red sands is comparable with that in other ferruginized dune sand bodies in North Queensland (Table 1).

Under the SEM, reddened sand grains are seen to possess cutans composed of aggregates of platy particles, mostly less than 2 μm in diameter, which are orientated with their flat surfaces parallel to that of the quartz grains (Fig. 4a,b). X-ray diffraction has shown the fine material principally to consist of kaolinite and fine quartz, with subsidiary gibbsite and ferric oxides and hydroxides (Fig. 5). In addition to

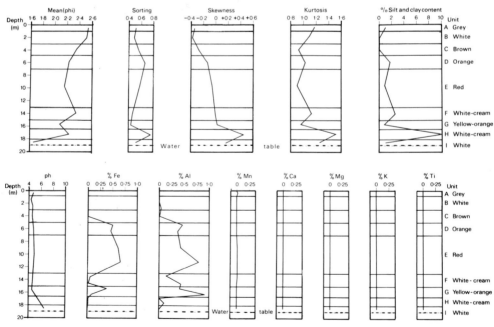

FIG. 3. Summary diagram showing variation of textural and chemical properties with depth in the Cape Flattery quarry section.

TABLE 1. *Percentage extractable Fe and Al in North Queensland weathered dune sands (determined by atomic absorption spectrophotometry on dilute HCl extracts)*

		% Fe	% Al
Cape Bedford			
Red sands		0.91–1.66	0.06–0.41
Orange sands	B horizon	0.20–0.38	0.06–5.18
Ironstone layer		0.95–3.20	0.17–2.60
Pale yellow sands	C horizon	0.02–0.62	0.07–0.90
Cape Flattery			
Red sands		0.62–5.20	0.25–1.70
Orange sands	B horizon	0.35–2.55	0.07–2.70
Pale yellow sands	C horizon	0.05–0.39	0.02–1.52
Leached white sands	A horizon	0–0.02	0–0.04
Mount Inkerman (Burdekin Delta)			
Red sands	Upper	1.04–1.38	0.61–4.26
Red-yellow sands	Lower	0.55–1.11	0.24–3.11
Ramsay Bay (Hinchinbrook Is.)			
Red sands	B horizon	0.14–0.25	0.05–0.18
Brown sands	B horizon	0.42–0.67	0.15–0.35
Leached white sands	A horizon	0.01–0.02	0–0.01
Temple Bay			
Red sands	B horizon	0.22–1.31	0.16–4.20
Leached white sands	A horizon	0–0.07	0–0.06
Cowley Beach (Innisfail) (beach ridges)			
Yellowish–red sands		1.42–3.20	0.90–2.48

goethite and haematite, there is some evidence from XRD and Mössbauer spectroscopy that a very finely crystalline ferric hydrate is also present.

Electron microprobe, EDAX and XRD results show that quartz constitutes more than 99% of sand grains in all three dune units at Cape Flattery. Feldspars are entirely absent. Heavy mineral content ranges from 0.05 to 0.7%, averaging about 0.2%. The main heavy minerals present are zircon, ilmenite, rutile, tourmaline, andalusite and anatase.

The humate-stained B horizon sands of the Flattery Dunesand unit, described more fully in Pye (1982b), are depleted in Fe (range 0.005–0.280%) but retain high levels of Al (range 0.168–0.900%), chiefly in the form of gibbsite. Organic carbon content is higher than in the Bedford Dunesand B horizons, averaging 5–6% and occurring as organans on quartz grain surfaces.

Processes of weathering and podzolization

Three groups of processes are involved in post-depositional modification of dune sands by podzolization: (a) weathering, (b) eluviation (leaching), and (c) illuviation.

Weathering

There can be little doubt that most of the iron in the Cape Flattery red sands owes its origin to *in situ* weathering of iron–bearing detrital mineral grains. In thin section clay-iron cutans are sometimes seen to be absent at points of grain contact, suggesting they are not detrital, while the ironstone layers are clearly post-depositional. Only a few per cent of the total iron present could be accounted for by brown ferric hydrate coatings on quartz grains inherited from the beach environment, while the morphology and physiographic setting of the reddened sand bodies make it unlikely that iron has been introduced by infiltrating groundwater or surface wash. Secondary addition of iron-bearing dust is also unlikely, since 99% of effective wind energy in this area is directed from the sea, and potential iron-rich source rocks are of restricted occurrence on the adjacent part of Cape York Peninsula.

However, difficulty arises in identifying the minerals which may have released iron on weathering, since the sands now contain only

K. Pye

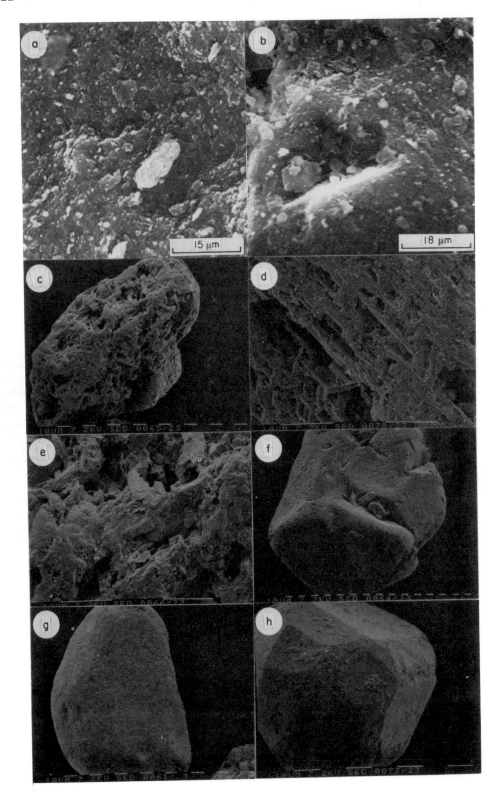

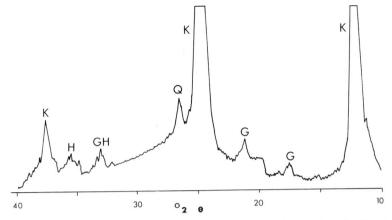

FIG. 5. X-ray diffraction trace of the <2 μm fraction of B horizon red sands: K = kaolinite; G = goethite; H = haematite; Q = quartz (CuKα radiation). Gibbsite is present in some other samples.

quartz and an assemblage of resistant heavy minerals. Two main possibilities exist: either the sands originally contained some feldspars and other weatherable heavy minerals which have since been completely destroyed, or weathering of the heavy mineral species presently found in the sands is responsible for the formation of all the iron and aluminium hydroxides present. The abundance of kaolinite and

gibbsite in the sands suggest that the second possibility is unlikely, since ilmenite, zircon and rutile contain only very low levels of aluminium. Unfortunately, the problem of the original composition of the sands cannot be resolved by reference to the modern beach and foredune sands, since the latter consist only of sediment which has been reworked from the weathered dunes by marine action (Table 2).

TABLE 2. *Composition of heavy mineral fractions from Cape Bedford–Cape Flattery dune and beach sands (%)*

Mineral	Cape Flattery quarry section horizons			Elim Cliffs dune C horizon	Elim Beach	C Bedford Beach	Fourteen Mile Beach
	A	B	C				
Ilmenite	tr	26	28	11	22	3	20
Zircon	7	33	22	30	67	3	25
Rutile	3	14	22	10	3	25	18
Tourmaline	28	6	5	13	6	23	11
Andalusite	34	3	5	15	tr	25	6
Staurolite	7	2	3	3	tr	5	5
Brookite	tr	6	3	5	tr	3	5
Anatase	21	10	11	14	tr	13	10

FIG. 4. Scanning electron micrographs of sand grains from the Cape Flattery quarry section. (a) and (b): surface textures of reddened quartz grains from the B horizon, showing adhering particles of kaolinite and ferric hydroxides; (c) residual anatase grain; (d) & (e) details of the surface texture of the grain shown in (c); (f) zircon grain showing signs of solution of edges and along lines of weakness; (g) rounded tourmaline grain with smooth, unetched surface texture; (h) staurolite grain showing crystal faces unaltered by chemical weathering (all heavy minerals from A1 horizon sands).

TABLE 3. *Heavy mineral content of the A, B and C horizon sands in the Cape Flattery quarry section (% weight of bulk sample after bromoform separation)*

Horizon	Sample	Depth below surface (m)	Sand colour	Munsell notation	% heavy minerals
A1	CF232	2.0	white	2.5 Y/–	0.159
A2	CF233	3.2	white	2.5 Y/–	0.134
B1	CF234	5.5	red-brown	5 YR 5/8	0.245
B2	CF238	11.5	red	2.5 YR 4/8	0.655
C1	CF241	15.0	pale yellow	2.5 Y 8/4	0.111
C2	CF245	19.0	white	2.5 Y/–	0.686

The ultimate origin of the sand is probably the granites and metamorphic rocks south of Cooktown and the Jurassic sandstones to the west, but it is unknown to what extent the sediment was deposited, weathered and reworked on the continental shelf during the Pleistocene. It is also possible that in the past, olivine, feldspar and other minerals were supplied from the Cenozoic basalt flows in the McIvor and Endeavour River valleys (Morgan 1968), although there is no evidence of such supply to modern day beaches.

Although the presence of kaolinite and gibbsite suggest the sands once contained feldspar, evidence does suggest that some of the heavy minerals now found in the sands are being weathered under conditions of extreme leaching. In the profile shown in Fig. 2, for example, total heavy mineral contents of the A1 and A2 horizons are much lower than those of the B and C horizons (Table 3). Whereas ilmenite is a major component of the latter, it is virtually absent in the A horizons where its main alteration product, anatase, is correspondingly more abundant. The abundance of zircon and rutile is also reduced in the A horizons, while tourmaline, staurolite and andalusite are relatively more common than in the B and C horizon sands. SEM examination further suggests that ilmenite, zircon and rutile grains have suffered solution (Fig. 4c,d,e,f), whereas tourmaline, andalusite and staurolite have been less affected (Fig. 4g,h). EDAX spectra confirm that the latter minerals are compositionally unaltered while ilmenite grains are either depleted in iron or completely altered to anatase (Fig. 6).

There is less evidence of complete alteration of ilmenite to anatase in the B horizon sands, but microprobe analyses indicate that many grains have lost up to one-third of their original iron content (Table 4). Grey & Reid (1975) have suggested that alteration of ilmenite ($Fe^{2+}Ti^{4+}O_3$) under oxidizing conditions is a two-

stage process. In the first stage ferrous iron is oxidized to ferric iron and up to one-third of the ferric ions may be removed through an essentially unaltered oxygen lattice by hydroxyl ions, leaving a residual product which Grey & Reid term 'pseudorutile' ($Fe_2^{3+}Ti_3O_9$). In the second stage both iron and oxygen need to be removed to convert 'pseudorutile' to pure TiO_2, resulting in disruption of the crystal lattice. Such a two-stage process is supported by the results obtained in this study.

Alteration of ilmenite in the upper oxidized few metres of sediments has previously been reported by Temple (1966), Pirkle & Yoho (1970), Force (1976) and Minard, Force & Hayes (1976). Experimental work has indicated that solution of ilmenite is enhanced by a range of mineral acids (Dumon 1976), and it is likely that organic acids released by the acidophyllous vegetation at Cape Flattery have a similar effect. It is therefore concluded that weathering of ilmenite has contributed, in a significant degree, to the reddening of the subsurface dune sands.

Eluviation

Eluviation of weathering products from podzolic A horizons may be accomplished by simple rainwater flushing (Wright & Foss, 1968), but it usually also involves the formation of soluble organo-metallic complexes (Schnitzer & Skinner 1963; Schnitzer & Khan 1972; Mohr, van Baren & van Schuylenborgh 1972). Leaching is especially favoured on deep, well-drained parent materials such as dune sands which are often initially low in bases and which support acidophyllous vegetation cover.

Illuviation

The factors which promote deposition of translocated material may be either physical or chemical (Mohr *et al.* 1972; De Coninck 1980).

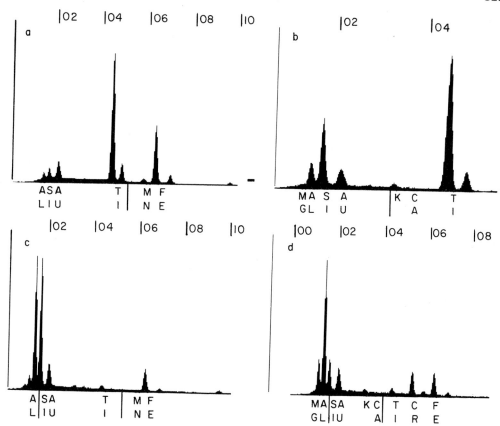

FIG. 6. EDAX spectra of heavy mineral grains from A2 horizon of the Cape Flattery section: (a) altered ilmenite grain showing iron depletion; (b) residual anatase grain with Al and Si peaks probably representing kaolinite trapped in surface etch pits; (c) unaltered andalusite grain; (d) unaltered staurolite grain.

TABLE 4. *Chemical composition of selected altered and unaltered ilmenite grains from the Cape Flattery B horizon sands, as indicated by microprobe spot analyses*

	Unaltered ilmenite			Altered ilmenite			
	1	2	3	1	2	3	4
SiO_2	0.260	0.321	0.312	0.205	0.303	0.294	0.325
Al_2O_3		2.000		0.207			
FeO	45.124	46.688	47.243	34.840	30.323	36.115	30.329
MgO		1.433		0.228			
CaO					0.082	0.111	0.017
MnO	5.752	0.711	1.559	5.709	4.059	1.042	3.415
Cr_2O_3			0.140				
TiO_2	48.716	47.467	50.259	55.204	58.944	58.000	60.890
Total	99.851	98.620	99.512	96.394	93.408	95.562	94.975
TiO_2/FeO ratio	1.08	1.02	1.06	1.58	1.94	1.61	2.01

Downward movement may be impeded by drying out of the profile, by a reduction in sediment permeability, or interception by the water table. Soluble organo-metallic complexes may be precipitated if a critical ratio of metal to fulvic acid is exceeded (Schnitzer 1969), or if a rise in subsoil pH causes hydrolysis of the organic matter.

Progressive podzolization

Podzolization is a progressive process in which the depth of the leached A2 horizon increases with time (e.g. Thompson 1981). Sesquioxides and clay minerals in the upper part of the B horizon are continually being remobilized by organic acids and redeposited lower in the profile. Andriesse (1969/70) inferred from podzol chronosequences in Sarawak that a preliminary red weathering phase is followed by progressive bleaching and finally build up of dark brown organic colloids (humate) in the subsoil. A similar sequence is evident at Cape Flattery, where colour differences between the reddened Bedford Dunesands and the much older Flattery Dunesands are marked. An even later stage, in which the dark brown colouration is lost by oxidation of humate to leave grey-white residual gibbsitic sands, has been suggested in southern Queensland dune sands (Ward, Little & Thompson 1979). A schematic model, showing the stages in progressive podzolization, is presented in Fig. 7.

Rates of podzolization

Data regarding rates of podzolization in different parts of the world are incomplete. Paton *et al*. (1976) showed that dune sand replaced after mining disturbance in southern Queensland developed a miniature podzol profile in 4.5 years, with the formation of bleached A2 soil 'pipes' after 9 years. Chandler (1942) estimated that mature podzol development on glacial deposits in Alaska had taken 1000–1500 years, while Andriesse (1969/70) found podzols with reddened B horizons in Sarawak beach ridge sands considered to be less than 5000 years old. On the other hand, Ward & Little (1975) and Ward (1977) envisaged a much longer timescale for podzol development in southern Queensland dune sands, suggesting it would take 5000 years to form an A horizon 30 cm thick and over 100 000 years to form one 4 m thick.

Radiocarbon dates obtained from the Cape Flattery deposits suggest that weathering and podzolization has been much more rapid than Ward and Little might expect (Pye 1981). Three samples of charcoal collected from depths of 2, 4 and 8 m in the profile shown in Fig. 2 yielded respective ages of 7480 ± 75 (Q–2081), 8200 ± 85 (Q–2082), and 7560 ± 90 (B–2494) C-14 years BP. The charcoal samples were repeatedly treated with NaOH and HCl to remove organic acid contaminants before the assays were carried out. Since the charcoal must have been incorporated in the sands while the dune was active, the dates indicate that stabilization and weathering have occurred within the last 7500 C-14 years. A further date, obtained from an *in situ* tree root embedded in the humate-impregnated B horizon of a Flattery Dunesand profile near Cape Flattery, indicated an age greater than 48,000 C-14 years BP.

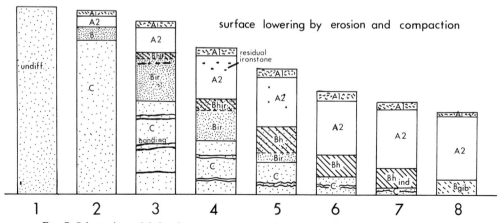

FIG. 7. Schematic model showing stages in progressive podzolization of a quartz sand body.

Podzol profiles elsewhere in eastern Australia

Weathered dune complexes occur in several places on the east coast of Australia between Cape York and Bass Strait (Coaldrake 1962; McElroy 1953; Jennings 1959; Thom, Bowman & Roy 1981; Pye 1982c). In southern Queensland podzolized dunes with coloured subsurface sands are found at Teewah, Rainbow Beach and on Fraser Island and North Stradbroke Island. In general, however, reddening of podzolic B horizons is less pronounced than at Cape Bedford and Cape Flattery. This may partly be because the cooler climate in the south is less favourable for the formation and preservation of haematite (cf. Schmalz 1968), but it is also possible that at the time of deposition the southern Queensland dune sands contained fewer iron-bearing minerals. Ilmenite forms a smaller percentage of the beach and dune heavy mineral assemblages in New South Wales and southern Queensland compared with North Queensland (Whitworth 1931; Connah 1948; Beazley 1948; Gardner 1955), and sands in the former areas *at the present time* have a lower potential for reddening. The differences in ilmenite abundance may reflect the fact that much of the sand in New South Wales and Queensland coastal embayments has been supplied from Mesozoic sandstones in the Clarence, Richmond and Maryborough Basins, whereas in North Queensland a higher proportion of coastal sand is derived directly from granitic and metamorphic rocks. Some ilmenite may also have been supplied from numerous Cenozoic basalt outcrops in North Queensland.

Degraded dune sands with humate-impregnated podzolic B horizons are found at several places in southern Queensland and New South Wales (McGarity 1956; Ward *et al.* 1979; Thom *et al.* 1981). Similar weathering profiles are found in beach ridge sands which have been shown by uranium series dating to be last interglacial in age (Marshall & Thom 1976), but as yet it is unknown to what extent dune deposits in different areas are correlative.

Conclusions

Under humid tropical and subtropical conditions such as experienced in eastern Australia quartz dune sands experience progressive podzolization. At Cape Bedford and Cape Flattery in tropical North Queensland podzol profiles formed in dunes less than 7500 C-14 years old have developed reddened B horizons up to 8 m thick. Older dune deposits in the same area, C-14 dated as more than 48 000 years old, now contain humate-cemented B horizons and are interpreted as having lost much of their original ferric iron by progressive podzolization.

At the time of deposition the Cape Flattery dune sands probably contained some feldspar and other easily weathered minerals which have since been completely altered to kaolinite and gibbsite. Weathering of ilmenite is considered to have provided a major source of iron. Zircon and rutile grains also appear to be slowly weathered under present acid leaching conditions in podzolic A horizons, but tourmaline and andalusite are less affected.

The modern active dunes at Cape Bedford and Cape Flattery consist entirely of sand which has been reworked from older weathered dunes. They consist of more than 99.7% quartz with minor amounts of resistant heavy minerals, providing a high quality source of glass sands. It is suggested that reworking of quartz sands which have experienced advanced humid tropical podzolization is one way in which mature quartz arenites may be formed. The existence of an ABC profile, framework grain mineralogy consisting almost exclusively of quartz and resistant heavy minerals, and clay mineralogy dominated by kaolinite and gibbsite with goethite (and sometimes haematite), are diagnostic properties of 'mature' tropical podzol profiles which may aid their recognition in the fossil record. Profiles with B horizons containing abundant organic carbon (humate) and gibbsite are found in very old, 'supermature' podzols which, in the later stages, developed under conditions of imperfect drainage. It should be noted, however, that not all humate-rich sediments represent a late stage phenomenon of podzolization; humate may also accumulate in swamps, lakes, coastal embayments and poorly drained sediments within only a relatively few years of sedimentation (Coaldrake 1955; Swanson & Palacas 1965; Pye 1982b).

ACKNOWLEDGMENTS: This paper is based on research undertaken while the author held a NERC Research Studentship at the University of Cambridge. Logistic support provided by Cape Flattery Silica Mines Pty Ltd and the Hopevale Aboriginal Community is gratefully acknowledged.

References

ANDRIESSE, J. P. 1969/70. The development of the podzol morphology in the tropical lowlands of Sarawak (Malaysia). *Geoderma*, **3**, 261–79.

BEAZLEY, A. W. 1948. Heavy mineral beach sands of Southern Queensland. 1. The nature, distribution and extent, and the manner of formation of the deposits. *Proc. R. Soc. Qd* **59**, 109–40.

CHANDLER, R. F. 1942. The time required for podsol profile development as evidenced by the Menderhall glacial deposits near Juneau, Alaska. *Proc. Soil Sci. Soc. Am.* **7**, 454–9.

COALDRAKE, J. E. 1955. Fossil soil hardpans and coastal sandrock in Southern Queensland. *Aust. J. Sci.* **17**, 132–3.

—— 1962. The coastal dunes of Southern Queensland. *Proc. R. Soc. Qd* **72**, 101–16.

CONNAH, T. H. 1948. Reconnaissance survey report of beach sand deposits, southeast Queensland. *Qd Govt Min. J.* **49**, 223–45.

DE ALWIS, K. A. & PLUTH, D. J. 1976. The red latosols of Sri Lanka. *J. Soil Sci. Soc. Am.* **40**, 912–28.

DE CONINCK, F. 1980. Major mechanisms in the formation of spodic horizons. *Geoderma*, **24**, 101–28.

DUMON, J-C. 1976. Action d'acides organiques divers sur des minéraux titanés (ilmenite et rutile). Comparison de leur pouvoir d'extraction du titane avec celui d'acides minéraux. *Bull. Soc. géol. Fr.* **18**, 75–9.

FORCE, E. K. 1976. Metamorphic source rocks of titanium placer deposits—a geochemical cycle. *Prof. Pap. U.S. geol. Surv.* **959-B**, 13 pp.

GARDNER, D. C. 1955. Beach sand heavy mineral deposits of eastern Australia. *Bull. Bur. Miner. Resour. Geol. Geophys. Aust.* **28**, 1–103.

GARDNER, R. A. M. 1981. Reddening of dune sands—evidence from Southeast India. *Earth Surf. Processes*, **6**, 459–68.

—— 1983. Reddening of tropical coastal dune sands. *In:* WILSON, C. (ed.) *Residual Deposits. Spec. Publ. geol. Soc. Lond.* **11**, 103–15. Blackwell Scientific Publications, Oxford.

—— & PYE, K. 1981. Nature, origin and palaeoenvironmental significance of red coastal and desert dune sands. *Prog. Phys. Geog.* **5**, 514–34.

GREY, I. E. & REID, A. F. 1975. The structure of pseudorutile and its role in the natural alteration of ilmenite. *Am. Miner.* **60**, 898–906.

JENNINGS, J. N. 1959. The coastal geomorphology of King Island, Bass Strait, in relation to changes in the relative level of land and sea. *Rec. Queen Vict. Mus.* NS **11**, 39 pp.

MARSHALL, J. F. & THOM, B. G. 1976. The sea level in the last interglacial. *Nature*, **263**, 120–1.

MCELROY, C. T. 1953. Successive profile development in sand dunes at Port Kembla, New South Wales. *Aust. J. Sci.* **16**, 112–5.

MCGARITY, J. W. 1956. Coastal sandrock formation at Evans Head. *Proc. Linn. Soc. N.S.W.* **81**, 52–8.

MINARD, J. P., FORCE, E. R. & HAYES, G. W. 1976. Alluvial ilmenite placer deposits. Central Virginia. *Prof. Pap. U.S. geol. Surv.* **959-H**, 15 pp.

MOHR, E. C. J., VAN BAREN, F. A. & VAN SCHUYLENBORGH, J. 1972. *Tropical Soils*. Mouton-Ichtiar-Van Hoeve, The Hague. 481 pp.

MORGAN, W. R. 1968. The geology and petrology of Cainozoic basalt rocks in the Cooktown area, North Queensland. *J. geol. Soc. Aust.* **15**, 65–78.

PATON, T. R., MITCHELL, P. B., ADAMSON, D., BUCHANAN, R. A., FOX, M. D. & BOWMAN, G. 1976. Speed of podzolization. *Nature*, **260**, 601–2.

PIRKLE, E. & YOHO, W. H. 1970. The heavy mineral ore body of Trail Ridge, Florida. *Econ. Geol.* **65**, 17–30.

PYE, K. 1980. *Geomorphic evolution of coastal sand dunes in a humid tropical environment, North Queensland*. Unpublished Ph.D. Thesis, Cambridge University. 413 pp.

—— 1981. Rate of dune reddening in a humid tropical climate. *Nature*, **290**, 582–4.

—— 1982a. Morphological development of coastal dunes in a humid tropical environment, Cape Bedford and Cape Flattery, North Queensland, Australia. *Geogr. Annlr.* **64A**, 213–27.

—— 1982b. Characteristics and significance of some humate cemented sands (humicretes) at Cape Flattery, Queensland, Australia. *Geol. Mag.* **119**, 229–42.

—— 1982c. Negatively skewed aeolian sands from a humid tropical coastal dunefield, Northern Australia. *Sedim. Geol.* **31**, 249–66.

—— 1983c. The coastal dune formations of Northern Cape York Peninsula, Queensland. *Proc. R. Soc. Qd*, **94**, 33–47.

—— 1983a. SEM and microprobe study of iron oxide-cemented duric layers (petroferric horizons) in late Quaternary coastal dune sands of northeast Australia. *In:* WHALLEY, W. B. & KRINSLEY, D. H. (eds) *Scanning Electron Microscopy in Geology*. GeoAbstracts, Norwich. In press.

—— 1983b. Formation of quartz silt by humid tropical weathering of dune sands. *Sedim. Geol.* (in press).

—— & SWITSUR, V. R. 1981. Radiocarbon dates from the Cape Bedford and Cape Flattery dunefields, North Queensland. *Search*, **12**, 225–6.

SCHMALZ, R. F. 1968. Formation of red beds in modern and ancient deserts. Discussion. *Bull. geol. Soc. Am.* **79**, 277–80.

SCHNITZER, M. 1969. Reactions between fulvic acid, a soil humic compound, and inorganic soil constituents. *Proc. Soil Sci. Soc. Am.* **33**, 75–81.

—— & SKINNER, S. I. M. 1963. Organo-metallic interactions in soils. 1. Reactions between number of metal ions and the organic matter of a podzol Bh horizon. *Soil Sci.* **96**, 181–7.

—— & KHAN, S. U. 1972. *Humic Substances in the Environment*. Dekker, New York.

SWANSON, V. E. & PALACAS, J. G. 1965. Humate in coastal sands of Northwest Florida. *Bull. U.S. geol. Surv.* **1214-B**, B1-B29.

TEMPLE, A. K. 1966. Alteration of ilmenite. *Econ. Geol.* **61**, 695–714.

THOM, B. G., BOWMAN, G. M. & ROY, P. S. 1981. Late Quaternary evolution of coastal sand barriers, Port Stephens–Myall Lakes area, Central New South Wales, Australia. *Quat. Res.* **15**, 345–64.

THOMPSON, C. H. 1981. Podzol chronosequences on coastal dunes in eastern Australia. *Nature*, **291**, 59–61.

WALKER, T. R. 1967. Formation of red beds in modern and ancient deserts. *Bull. geol. Soc. Am.* **78**, 353–68.

——, WAUGH, B. & CRONE, A. J. 1978. Diagenesis in first cycle desert alluvium of Cenozoic age, southwestern United States and northwestern Mexico. *Bull. geol. Soc. Am.* **89**, 19–32.

WARD, W. T. 1977. Sand movement on Fraser Island. *Anthropology Mus., Univ. Qd Occ. Papers in Anthropol.* **8**, 113–26.

—— & LITTLE, I. P. 1975. Times of coastal sand accumulation in southeast Queensland. *Proc. ecol. Soc. Aust.* **9**, 313–7.

——, LITTLE, I. P. & THOMPSON, C. H. 1979. Stratigraphy of two sandrocks at Rainbow Beach, Queensland, and some notes on their composition. *Palaeogeogr. Palaeoclim. Palaeoecol.* **26**, 305–16.

WHITWORTH, H. F. 1931. The mineralogy and origin of the natural beach concentrates of New South Wales. *J. Proc. R. Soc. N.S.W.* **6**, 59–74.

WRIGHT, W. R. & FOSS, J. E. 1968. Movement of silt-sized particles in sand columns. *Proc. Soil Sci. Soc. Am.* **32**, 446–8.

K. PYE, Department of Earth Sciences, Downing Street, Cambridge CB2 3EQ.

Origin of red beds in a moist tropical climate (Etruria Formation, Upper Carboniferous, UK)

B. M. Besly & P. Turner

SUMMARY: The Etruria Formation is a Westphalian mudstone-dominated red bed succession, which is interdigitated with, and overlies coal-bearing paralic sediments in Central England. Sedimentological studies indicate that the formation mainly comprises swamp and muddy alluvial plain deposits. The distribution of red beds shows that the red coloration formed in areas of improved drainage soon after deposition, and was associated with the formation of complex palaeosols. The occurrence of coals and humid climate floras within the red bed succession suggests that the onset of oxidizing conditions was not the result of climatic change. There are no fundamental differences in the mineralogy of associated grey and red beds: both have distinctive residual characteristics, and consist mostly of kaolinite and quartz. The alluvium was derived from intensely weathered source areas, and was reddened during and soon after deposition by the dehydration of detrital ferric hydroxides and the oxidation of ferrous iron associated with organic material. Palaeomagnetic results support the view that reddening took place shortly after deposition, when the area lay in close proximity to the Carboniferous equator. The Etruria Formation thus provides an excellent example of the genesis of red beds as muddy alluvium under moist tropical climatic conditions.

In recent years documentation of the diagenetic processes occurring in Cenozoic and modern arid zone alluvium has led to the erection of a sophisticated model for the origin of red beds (Walker 1967; Walker, Waugh & Crone 1978). This model involves the intrastratal dissolution of unstable ferromagnesian minerals and the consequent precipitation of a suite of authigenic minerals including clay minerals, quartz, K-feldspar and haematite. The formation of red beds in this way requires a long period of time. Although there have been no detailed studies of moist tropical alluvium comparable to those made on arid-zone alluvium, the diagenetic model has been extended to include the deposits of such areas (Walker 1974). However, the lack of knowledge concerning the transport and diagenesis of iron in humid tropical climates and the lack of detailed studies of ancient examples means that red beds associated with humid palaeoclimate indicators (coal seams, humid-zone floras) are still poorly understood.

Several processes have been invoked for the origin of red beds in humid climates other than the diagenetic model of Walker (1974). These include the erosion and redeposition of red lateritic soils as red alluvium (Krynine 1949, 1950), the dehydration of detrital ferric hydroxides transported in close association with clay minerals as 'brown alluvium' (Van Houten 1964, 1968, 1972), and the post-depositional oxidation of ferrous iron during fluctuation of the ground water table (Czyscinski, Byrnes &

Pedlow 1978). None of these processes, nor Walker's diagenetic model, can alone account for the origin of red beds in the moist tropical conditions which prevailed during the deposition of the Carboniferous coal measures in north-western Europe. Previous studies (Archer 1965; Downing & Squirrel 1965) have, however, provided evidence that early post-depositional oxidation of grey, organic rich sediments occurred in the formation of such red beds in the South Wales coalfield.

This paper presents a description of the sedimentology, mineralogy, geochemistry and palaeomagnetism of the Etruria Formation, a Westphalian alluvial red bed sequence in Central England which passes laterally into a grey, paralic coal-bearing sequence. The results indicate that prolonged diagenetic changes were not important in the formation of this reddened, humid climate alluvium. It is suggested that red beds were formed as a result of sedimentologically and climatically induced fluctuations of the groundwater table during and shortly after deposition.

Stratigraphy

The Etruria Formation, usually known by the lithologically incorrect name 'Etruria Marl Formation' (Ramsbottom et al. 1978) is a mudstone-dominated red bed sequence which outcrops in the Westphalian coalfields of Central England (Fig. 1). The stratigraphy of the

131

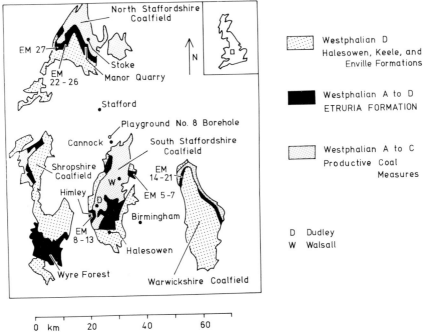

FIG. 1. Location map of Westphalian outcrops in central England. Numbers preceded by EM indicate localities of samples subjected to palaeomagnetic analysis.

Formation is well documented as a result of bore-holes drilled by the National Coal Board (NCB).

In the type area of North Staffordshire (Gibson 1905) the Etruria Formation overlies the Productive Coal Measures, its base occurring just above the boundary between the *Torispora securis* and *Thymospora obscura* miospore zones (Westphalian C/D boundary: Butterworth & Smith 1976). To the south and west,

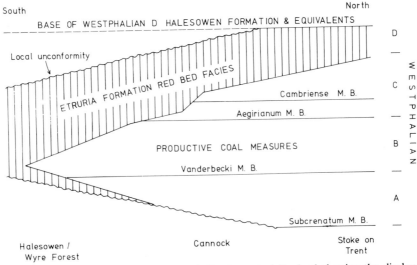

FIG. 2. Regional stratigraphy of the Westphalian in central England showing the diachronous distribution of the Etruria Formation red beds. M.B. indicates stratigraphically important marine bands.

towards the southern margins of the depositional basin (Calver 1969), the base of the formation is strongly diachronous (Fig. 2). It occurs near the horizon of the *Cambriense* Marine Band (mid-Westphalian C) in the Cannock area (Ramsbottom *et al.* 1978; unpublished NCB data) and near the horizon of the *Aegerianum* Marine Band (Westphalian B/C boundary) in the Dudley area (Poole 1970; M. A. Butterworth, pers. comm.). In the extreme south of the basin, around Halesowen and in the Wyre Forest, red beds occur at all horizons in the Westphalian A to C sequence (Poole 1966, 1970; unpublished NCB data).

In all areas there is a marked inverse correlation between the total thickness of the productive coal-bearing sequence, and the age at which red beds first occur in the sequence. Near the margins of the basin the coal bearing sequence is thin and condensed (Ramsbottom *et al.* 1978) and red beds appear earlier than in the thick sequences in the basin centre.

Within the formation diagnostic flora and fauna are largely absent, and the age of the bulk of the formation is not known. It is everywhere overlain by the Halesowen Formation and its equivalents, which contain *Anthraconauta tenuis*, and which are thus of lower Westphalian D age (Ramsbottom *et al.* 1978).

The detailed stratigraphic relationship between the red and the grey, coal bearing sediments has been documented in areas where the NCB has drilled close space boreholes (2–5 km apart). A typical example of the lateral transition from grey to red beds is found in North Staffordshire (Fig. 3). The lateral transition takes place by the interdigitation of grey and red beds over a thickness of up to 100 m, the red beds occurring in tongues between coal seam marker horizons. These tongues have tops which are consistent with the coals, and discordant bases, the red intercalations thickening towards the margins of the basin and amalgamating to form a continuous red bed sequence. The amalgamation of red bed intercalations at the expense of coal-bearing sediment always occurs in the direction in which the whole sequence thins.

Sedimentology

The Etruria Formation red beds occur in three facies associations, which are described below. Their stratigraphic relationship is illustrated schematically in Fig. 4.

Transitional, swamp dominated alluvial association

The upward change from the grey coal-bearing sediments into red beds takes place in a transitional sequence in which grey and red beds are interdigitated. The grey beds consist of

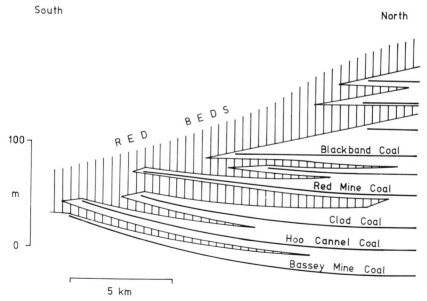

FIG. 3. Interdigitation of coal-bearing grey beds and red beds in the southern part of the North Staffordshire coalfield. Data from NCB boreholes and shafts, collated by E. L. Boardman.

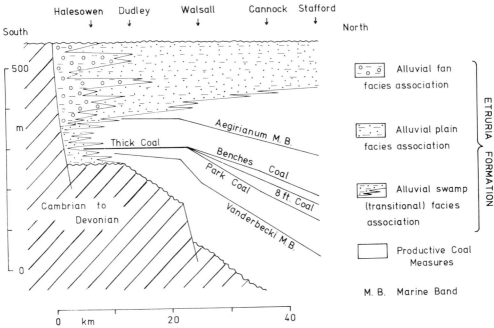

FIG. 4. Schematic illustration of facies relationships within coal measure and Etruria Formation sediments in the South Staffordshire coalfield. For location of section, see Fig. 1.

repeated units in which laminated grey mudstone containing plant debris passes upwards successively into grey mudstone homogenised by plant root action, seat-earth type palaeosols, and coals. The laminated mudstone is often absent, in which case much of the sequence consists of rooted mudstone, palaeosol, and coal. Occasional erosionally based, upward fining channel sandstones are present. In North Staffordshire some of the coal seams are overlain by laminated sapropel rich siderite ironstones—the Blackband ironstones (Gibson 1905; Boardman 1978)—and varved mudstones. The palaeosols consist of a carbonaceous or coaly layer, overlying grey slickensided mudstone containing siderite concretions and *Stigmaria* roots. They are thus comparable with coal measure seat-earths described by Huddle & Patterson (1961) and Roeschmann (1971), and with palaeosols in the Triassic South Head Palaeosol Series of S.E. Australia (Retallack 1977). The upper coaly layer indicates that the original soil had a peat surface layer. Other than this little soil horizon development is present. This type of palaeosol may be compared with immature alluvial soils forming at the present day in backswamp areas of tropical river systems, such as the Fly River in Papua New Guinea (Paijmans *et al.* 1971). Such soils form under a permanently high water table, and this

suggests that the grey beds in the transitional association are predominantly swamp deposits. The existence of near permanent waterlogging is supported by the extensive occurrence of lycopod roots (Scott 1979). The laminated ironstones and varved mudstones are interpreted by Boardman (1978) as the deposits of lakes which formed at the end of peat forming episodes.

The appearance of red beds in a vertical section through the transitional facies association usually takes the form of repeated intercalations of red mottled horizons, which become thicker upwards until an entirely red sequence is obtained (Fig. 5). The first appearance of red pigment occurs in the lower horizon of seat-earth palaeosols. These palaeosols are similar to seat-earths found in the grey beds, in that they have a coaly upper horizon and contain carbonaceous roots and siderite. Within their lower horizon they contain a zone of red or brown pigmentation, either as a uniform coloration, or concentrated in mottles associated with root traces and slickensided surfaces. Occasionally siderite or goethite root fill concretions are present.

Where red or brown pigmentation first appears in the succession it forms a recognizable contemporaneously oxidized horizon in a swamp palaeosol. The presence of carbonace-

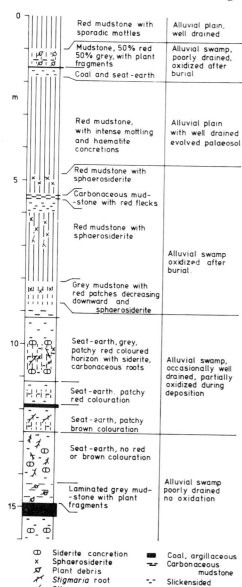

Legend:

Symbol	Description	Symbol	Description
⊕	Siderite concretion	■	Coal, argillaceous
×	Sphaerosiderite	⊏	Carbonaceous mudstone
⊿	Plant debris		
⋏	*Stigmaria* root	⌐	Slickensided texture
⋏	Other root		

|||| Red pigment

FIG. 5. Vertical section of a mudstone sequence showing the transition between Coal Measures and Etruria Formation red beds in NCB Playground No. 8 borehole, Cannock.

Coastal Plain of Surinam (Slager & Van Schuylenborgh 1970). In this case gleying results from seasonal water table drop in well drained freshwater swamps. Similar contemporaneous oxidation is also recorded in the well-drained swamp depositional environment described by Coleman (1966) in the Atchafalaya basin.

Higher in the sequence illustrated in Fig. 5 the red intercalations become thicker. They consist, in part, of superimposed oxidized seat-earth horizons, but also include the sediment which underlies the palaeosols, which is not rooted and which contains abundant plant material. Here oxidation of the sediment has been more extensive, and is no longer confined to horizons within the palaeosol.

The depositional environments of the transitional association sediments are closely comparable with the alluvial floodplain association described from the Westphalian in northern England by Scott (1979). The present authors, however, prefer to refer to the association as an alluvial swamp association, to emphasize its predominantly waterlogged nature. There appears to be no difference in depositional environment between sequences which contain red intercalations, and sequences which lack them, except that in the former better drainage conditions prevailed in the backswamp areas. The lateral continuity of the red intercalations (Fig. 3) shows that this improvement in drainage occurred over wide areas. The tendency of the intercalations to thicken and amalgamate where there is a net thinning of the entire sequence (Fig. 3) suggests that the improved drainage was partly controlled by subsidence.

Alluvial plain facies association

Sediments of this association make up the majority of the Etruria Formation red beds. They consist of fining upward units of massive, red, silty mudstone, whose tops usually consist of a palaeosol. The bases of these units usually contain thin, cross-laminated, sand and siltstone layers, which are interpreted as crevasse splay or levée deposits. Some contain thicker sand bodies, characterized by internal upward fining, erosive bases, and epsilon cross-bedding (Allen 1963). These are interpreted as deposits of meandering fluvial channels. The sandstones are usually green, although red pigmented sandstones occur sporadically. Successive upward fining units represent episodes of floodplain sedimentation, which were separated by periods of pedogenesis during which sediment supply was low or non-existent. Both

ous roots and siderite suggests that the environment was dominantly waterlogged. The red mottles around roots and planes of weakness suggests that they were formed by gleying (Bloomfield 1964). Possibly analogous gleyed soils are forming at present in the Younger

channel and floodplain sediments contain plant fragments, which are found as very poorly preserved impressions in the red lithologies.

Two types of palaeosol are found in the floodplain association. The first type is similar to the palaeosols found in the swamp association, except that it has undergone partial or complete oxidation after burial. Profiles are of similar thickness to those of the swamp association. Where partial oxidation has taken place the organic content of the coaly layer is greatly reduced, and a characteristic shaley appearance is developed. At this stage of oxidation siderite concretions and bedded ironstones are occasionally preserved in an unoxidized or partially oxidized state even though surrounded by red mudstone. Where complete oxidation has taken place, siderite concretions and bedded ironstones are oxidized to haematite. Oxidized ironstones retain their bedded relationship to the underlying palaeosol, and sometimes show lamination and traces of bivalve shells. The original nature of the fully oxidized seat-earths is apparent in their texture (cf. Archer 1965). The presence of oxidized concretionary fills of *Stigmaria* roots indicates that deposition and original plant growth occurred in waterlogged conditions (Scott 1979). Where any organic

material remains in the coaly layer symmetrical zones of unoxidized sediment remain above and beneath it.

The second type of palaeosol in the alluvial plain facies association is of a more evolved nature. Where a complete profile is observed it is usually much thicker than the seat-earth profiles, reaching 3 m in well-developed examples. Three horizons are recognizable within profiles. The upper horizon consists of very pale grey mudstone (Fig. 6) with a leached appearance. This is usually overlain by red floodplain sediments, the boundary between the two being sharp. Where the palaeosol is developed on a silty substrate, the silt content is usually depleted in this horizon. The upper horizon passes gradationally downwards into a grey and red mottled horizon, which sometimes contains diffuse, irregularly bounded haematite concretions, varying in size between 1 and 10 cm, which enclose quartz grains and aggregates of mudstone. These concretions resemble the nodular variety of lateritic ironstone defined by Pullan (1967 and pers. comm.). Like them, the palaeosol concretions are associated with red mottles. In rare cases, a crude pisolitic structure is visible in thin-section. Occasionally the concretions are sufficiently abundant to coalesce

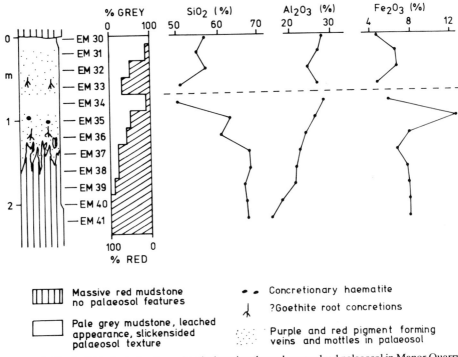

FIG. 6. Vertical lithological and geochemical section through an evolved palaeosol in Manor Quarry, Fenton, Stoke-on-Trent.

into a semi-continuous layer. Where macroscopic haematite concretions are absent, thin sections of the mottled horizon show abundant concentration of haematite into veins and nodules, indicating the mobilization of ferric iron during soil formation (cf. Brewer 1964). Beneath the mottled and concretionary horizon is a horizon of less pronounced mottling and colour veining passing gradationally downwards into unmodified red floodplain sediment.

Features such as those seen in this type of palaeosol are known in several types of present day soil, notably in a variety of podzolic and ferrallitic tropical soils (Mohr, Van Baren & Van Schuylenborgh 1972). Interpretation of the palaeosols is rendered difficult, partly by the variety of soils in which such features may have formed, and partly as a result of stacking of palaeosol profiles, which causes horizons of separate phases of soil formation to be superimposed in a manner which makes it difficult to recognize individual profiles in the field (Fig. 6: cf. Freytet 1971). In many cases the upper horizon of the palaeosol contains root concretions similar to those found in the oxidized seat-earths. These indicate that the original soil profile has been modified by polyphase soil evolution (cf. Freytet 1971), a seat-earth type soil having developed on the top of the more evolved soil profile during subsidence and waterlogging, prior to the resumption of sedimentation.

Where the palaeosol profile contains a horizon showing recognizable macroscopic haematite concretions and a thick leached upper horizon, the palaeosol may be regarded as 'lateritic', although the lack of descriptions of laterites developed on Quaternary alluvium prevents refinement of the interpretation. The geochemical results given below are consistent with this view. In other cases, the original soil type cannot be identified. Soils described by Mohr *et al.* (1977) containing comparable features all develop under good drainage conditions.

Where the red mudstones of this facies association do not show palaeosol features, they often contain small (1 mm) spherulitic calcite concretions. These are also found in reworked horizons in the alluvium, and are thus of very early post-depositional origin. Comparison with the well-drained swamp environment of Coleman (1966) suggests that these probably formed at a time when oxidizing conditions prevailed in the alluvium.

Alluvial fan association

This association is similar to the alluvial plain association, except that meandering alluvial channel deposits are largely absent, being replaced by sheets of texturally immature conglomerate. These sheets do not usually exceed 1 m in thickness, and occur in stacked sequences up to 10 m thick. Individual sheets consist of normally graded, clast supported conglomerates. They closely resemble sheet conglomerates described by Nagetgaal (1966), which were interpreted as sheet flood deposits. Channelized, cross-bedded conglomerates and pebbly sands are occasionally found, as are boulder beds of intra- and extra-formational blocks supported in a mud matrix. The sequences of sheet conglomerates are separated by mudstones and palaeosols similar to those in the alluvial plain association.

The alluvial fan association is concentrated along fault bounded margins to the depositional basin. It was deposited by flash floods, conglomerate bedload rivers, and mudflows, which built an apron of coalescing low angle fans at the edges of the basin.

Mineralogy and geochemistry

A preliminary study of the mineralogy and geochemistry of the Etruria Formation has been made using the samples collected for palaeomagnetic study, a sequence of samples from interbedded organic rich grey and red beds in Himley Quarry near Dudley, and sequences of samples from evolved palaeosols and red mudstones in Manor Quarry, Stoke-on-Trent. A variety of red, drab, leached grey, and organic rich grey mudstones have been examined to determine the mineralogy of the sediment and the distribution of iron, and to identify chemical processes which occurred during formation of the evolved palaeosols.

Detrital mineralogy and distribution of iron

X-ray diffraction study of the mudstones shows that they consist of kaolinite, quartz, and haematite, with minor amounts of illite, siderite, and goethite. Muscovite occurs as a detrital component, and is abundant only in a limited area to the south of Dudley. Illite is more abundant in silty mudstones and siltstones, indicating that it too is of detrital origin.

The principal constituents of the Etruria Formation mudstones are kaolinite and quartz. There are no fundamental differences in the occurrence of these minerals between the red alluvial plain mudstones and the intercalated palaeosols. Furthermore, the samples from an interbedded sequence of grey coal bearing

TABLE 1. *Comparison of the chemical composition of interbedded grey and red mudstone at Himley Quarry, Dudley*

Sample	Colour	SiO$_2$	Al$_2$O$_3$	Fe$_2$O$_3$	CaO	Na$_2$O	K$_2$O	MgO	TiO
EM52	Grey	48.7	25.4	4.6	0.15	0.40	3.07	1.21	0.74
EM53	Red	50.6	24.8	7.0	0.12	1.99	3.22	1.29	0.84
EM54	Grey	64.1	15.6	4.2	0.11	0.34	2.24	0.80	0.57
EM55	Red	57.3	14.6	10.1	0.17	0.79	1.86	1.11	0.55

mudstones and red Etruria-type mudstones show no difference in the occurrence of quartz and kaolinite. We therefore conclude that these minerals are largely of detrital origin.

Representative lithologies have been analysed for Si, Al, Fe, Mn, Ti, Ca, Mg, K and Na using standard atomic absorption techniques. The mudstones are characterized by high contents of iron (up to 13.2% Fe$_2$O$_3$) and alumina (up to 27.9% Al$_2$O$_3$) and generally low contents (<1%) of calcium and magnesium (as CaO and MgO). The sandstones, on the other hand, have less iron and alumina and are generally richer in calcium and magnesium.

The iron content of the mudstones varies according to their colour. In Himley Quarry, grey unoxidized organic rich mudstones interbedded with red beds contain 2.7–4.7% total iron, while the red beds contain 7.0–10.1%. There are no other marked differences in the chemical composition of grey, unoxidized mudstones and interbedded red beds. This is illustrated in Table 1 which compares the composition of two grey mudstones and two red beds from the Himley Quarry section. We take this to indicate that iron was reduced and mobilized in the waterlogged, organic rich grey mudstones (cf. Bloomfield 1964) but became fixed in the red mudstones as a result of better drainage, organic matter oxidation and formation of haematite.

The main diagenetic minerals in the mudstone are siderite and haematite. Siderite occurs mainly as small concretions in association with unoxidized organic rich, or oxidized seat-earths. It also occurs as a disseminated component and as small concretions in a number of red mudstones, which usually underlie seat-earths, and which may or may not have been post-depositionally oxidized. The persistence of siderite in the red mudstones indicates that this mineral formed at an early pre-reddening stage of post-depositional diagenetic alteration, at which time the sediment probably contained a considerable quantity of organic matter. Its subsequent oxidation may have contributed fine grained pigmentary haematite to the sedi-

ment. In most of the red mudstone sequence siderite is absent, and the only concretionary species are haematite (in palaeosols) and occasionally calcite. The latter is of very early diagenetic origin, as it is subject to sedimentary reworking. Such early concretionary calcite probably formed in oxidizing conditions (Coleman 1966) to the exclusion of siderite.

The marked difference of iron content and the apparently mutually exclusive occurrence of siderite and calcite in the red mudstones suggests that two distinct processes of reddening were involved in the early diagenesis of these rocks. Mudstones deposited under initially reducing conditions are characterised by the presence of siderite and organic matter and by low initial iron contents. During subsequent penetrative oxidation, the organic material may have been destroyed, in which case the sediment may have been reddened. In contrast, red mudstones characterized by high iron contents and by early calcite concretions were probably deposited under initially oxidizing conditions. The iron in these parts of the sequence was probably deposited as a detrital hydrated ferric hydroxide precursor, which subsequently inverted to haematite.

The principal constituents of the Etruria Formation make up a distinct assemblage, characteristic of derivation from an intensely weathered source with residual mineralogy. The similarity between the mineralogy of organic rich sediments and red beds, and the detrital nature of some of the haematite precursor pigment, suggests that these characteristics of the Formation are truly inherited. In this respect, it resembles other Westphalian coal-bearing mudstone sequences, which generally contain abundant kaolinite, which is regarded as having been derived from intensely weathered source areas (Ashley & Pearson 1978; Pearson 1979).

Evolved palaeosols

The principal geochemical variations in the red mudstones are associated with the palaeosols, and so provide information concerning the nature of the original soil types which became palaeosols. Fig. 6 shows the geochemical variation in major elements through a complex palaeosol profile sampled at Manor Quarry, Stoke-on-Trent. Two geochemically distinct superimposed soil profiles can be recognised: a lower profile (samples EM 34-41); and an upper profile (samples EM 30-33).

The uppermost horizon in the lower profile consists of very pale grey mudstone, is strongly leached and shows considerable loss of both iron and silica (Fig. 6). Below this is a mottled horizon with small haematite concretions which is reflected by a sharp increase in the ferric oxide content. This mottled, concretionary horizon passes gradationally downwards through veined and mottled sediment into unmodified red floodplain sediments. This lower horizon shows little geochemical change apart from a progressive decrease in the amount of alumina, presumably reflecting a greater abundance of kaolinite within the palaeosol profile. Similar geochemical changes have been described in the Etruria Formation by Holdridge (1959).

The overlying profile shows a number of differences from the lower profile and appears to be less well evolved. An upper leached horizon is present; this is depleted in iron but not in silica (Fig. 6). It passes gradationally downwards into a horizon which is slightly iron-enriched but which does not show concretionary development. As this profile may, in part, have been formed on material which constituted the upper horizon of the underlying soil profile, it is not possible to assess the significance of these results.

These chemical analyses illustrate the complexity of the Etruria Formation palaeosols. The significant loss of iron and silica from the upper horizon of the lower profile does, however, tend to confirm the suggestion already made that some of the soils had a 'lateritic' character. Similar results were obtained by Holdridge (1959), who reached a similar conclusion. Where chemical analyses are not available, the palaeosols can only be described as being of a well drained and highly evolved type.

Palaeomagnetism

Detailed palaeomagnetic studies of the Etruria Formation have been made in order to determine the palaeolatitude at which reddening took place and thus provide some evidence which might contribute towards the interpretation of the prevailing climatic conditions.

The unstable nature of the mudrocks precluded the drilling of specimens in the field. Orientated hand samples of mudrocks and sandstones were collected from various localities (Fig. 1). Specimen cores were drilled from the sandstones in the laboratory. From the mudrocks orientated specimen cubes of approximately 8 cm^3 were prepared by dry sawing, using a tungsten blade. These were trimmed to fit a cylindrical Perspex sample holder, and magnetic measurement was carried out using a Digico spinner magnetometer. An optimum number of six specimens was prepared from each sample in this way. Initial measurement of all the specimens showed considerable variation in the intensity and direction of the natural remanent magnetization (NRM). In general the mudstones are more strongly magnetized than the sandstones, by a factor of 2-3.

Representative specimens were selected for partial thermal demagnetization in order to examine the structure of the NRM. This method was deemed most appropriate in view of the mineralogical analysis, which showed that fine grained pigmentary haematite is the principal oxide phase present. The coercivity of fine particle haematite is so high that it is unresponsive to practical alternating field (AF) demagnetization (Dunlop 1971; Tauxe, Kent & Opdyke 1980). The results of the partial thermal demagnetization are presented as a stereographic projection of successive directional changes, a total intensity decay curve, and a Zijderveld diagram (Zijderveld 1967; Dunlop 1979).

The thermal demagnetization behaviour of the red mudstones and palaeosols falls into two groups. In the first, illustrated by EM9 (Fig. 7), the initial direction of magnetization is very shallow, with virtually zero inclination and southerly declination. At successively higher temperatures there is little directional change and a steady decrease in total intensity from an initial value of about 20 μG to nearly zero at 700°C. The Zijderveld diagram consists essentially of a single straight line segment indicating a single component of magnetization (the initial break in the line between 20 and 100°C is due to the removal of a weak, viscous component). All the specimens from this particular site were subsequently subjected to 'thermal cleaning' at 300°C to remove the effect of the viscous components. This resulted in a close grouping of the specimen directions in this sample (mean: Dec. = 186°, Inc. = −9° and α_{95} = 12). Similar behaviour during partial thermal demagnetization was observed for a number of the red

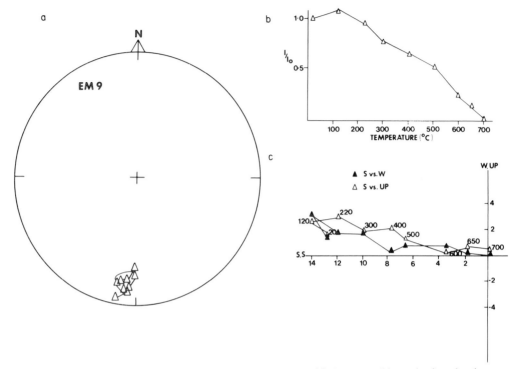

FIG. 7. Partial thermal demagnetization of sample EM9. (a) Stereographic projection showing directional changes at successive temperature increments; open symbols = upward inclination, closed symbols = downward inclination. (b) Normalized curve showing the decay of total intensity. (c) Zijderveld diagram showing the successive end-points during thermal demagnetization. Intensity in μG.

mudrocks (e.g. EM5, EM10, EM14, EM22 and EM27).

In the second group the initial intensity is generally much stronger and the directional behaviour, illustrated by EM20 (Fig. 8) rather more complicated. The initial direction is steep with positive inclination and with increments up to 400°C moving to progressively shallower inclination as the total intensity decays very rapidly (Fig. 8b) so that at 200°C only about 50% of the initial remanence remains. The orientation of this soft magnetic component is best seen on the Zijderveld diagram in Fig. 8(c). A steeply dipping component with easterly declination is indicated by the segment between 20 and 300°C and the removal of this reveals a southerly directed component with shallow inclination between 300 and 600°C. In order to remove the steeply dipping component, which is probably a viscous component acquired in the present Earth's field all the specimens of this sample were heated to 300°C and a site mean direction of Dec. = 194°, Inc. = +10°, α_{95} = 16.5 was defined. Thermal

cleaning at this temperature was found to be effective in removing viscous components of magnetization in all cases. Other samples showing similar behaviour include EM19, EM21 and EM26.

A number of channel sandstones from the alluvial floodplain association have also been studied palaeomagnetically (Table 2). The one red sandstone examined shows properties which are closely comparable to those of the red mudstones, with a similar blocking temperature spectrum, and only one non-viscous component of magnetization. This indicates that fine particle pigmentary haematite is the most important remanence carrier in red sandstones as well as red mudstones. Broadly similar results are shown by the green sandstones (Table 2) as illustrated by EM25 (Fig. 9). In this case there is a minor viscous component which is removed on heating above 150°C. A stable, southerly directed component with shallow upward inclination is then isolated between 150 and 400°C. Above this temperature there is a slight rise in total intensity and the direction

TABLE 2. *Palaeomagnetic results from the Etruria Formation after partial thermal demagnetization at 300°C and correction for tectonic dip. N = number of specimens per site, $\bar{D}$ = mean declination, $\bar{I}$ = mean inclination, α_{95} = 95% cone of confidence, and K = precision parameter. (WC = Westphalian C, WD = Westphalian D, L = Lower, U = Upper) Locations refer to British National Grid*

Sample no.	Location	Lithology	Colour	Age	N	$\bar{D}$	$\bar{I}$	α_{95}	K	Intensity (µG)
EM5	SK 047 026	Mudstone	Red	WC?	4	196	−1	12.6	54.0	8.0
EM6	SK 047 026	Mudstone	Yellow	WC?	5	218	−16	27.8	8.5	1.0
EM7	SK 047 026	Sandstone	Red	WC?	5	201	−26	10.5	53.0	6.5
EM8	SO 898 890	Mudstone	Red	WC(L)	4	179	−8	21.6	18.9	10.0
EM9	SO 898 890	Mudstone	Red	WC(L)	5	186	−9	12.3	39.4	18.0
EM10	SO 898 890	Mudstone	Red	WC(L)	4	200	+1	13.1	50.0	20.0
EM12	SO 898 890	Sandstone	Green	WC(L)	5	214	−13	19.4	16.4	10.3
EM14	SP 220 999	Mudstone	Red	WC(L)	6	189	+1	15.2	20.2	6.0
EM15	SP 220 999	Sandstone	Green	WC(L)	5	221	+14	26.0	9.5	4.7
EM17	SP 220 999	Mudstone	Red	WC(L)	6	180	−3	12.3	30.5	6.0
EM19	SP 222 999	Mudstone	Red	WC(L)	6	183	+15	11.9	32.5	20.0
EM20	SP 222 999	Mudstone	Red	WC(L)	6	194	+10	16.9	16.5	30.0
EM21	SP 222 999	Mudstone	Red	WC(L)	5	203	+15	11.4	45.4	20.0
EM22	SJ 830 460	Mudstone	Red	WC(U)/WD(L)	7	200	−8	11.4	28.9	6.0
EM23	SJ 830 460	Mudstone	Red	WC(U)/WD(L)	6	198	−7	13.2	26.6	20.0
EM24	SJ 830 460	Sandstone	Green	WC(U)/WD(L)	7	199	−13	12.8	33.9	5.5
EM25	SJ 830 460	Sandstone	Green	WC(U)/WD(L)	5	198	+2	5.9	164.8	5.2
EM26	SJ 828 467	Mudstone	Red	WD(L)	6	197	+9	22.9	9.4	15.0
EM27	SJ 788 451	Mudstone	Red	WD(L)	3	191	+12	12.1	103.4	3.0
				Mean direction	19	197°	−1°	6.9	24.5	

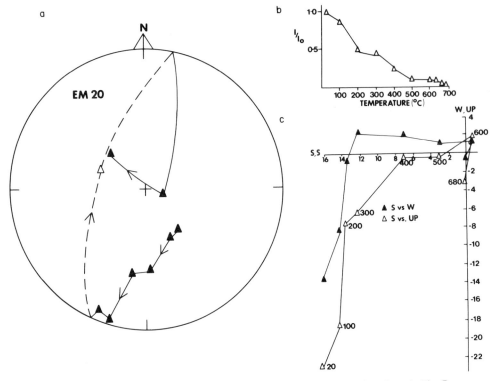

FIG. 8. Partial thermal demagnetization of sample EM20. Other details as in Fig. 7.

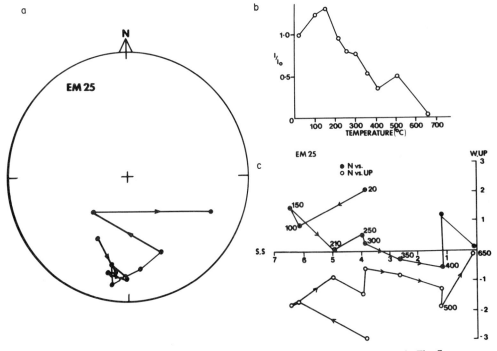

FIG. 9. Partial thermal demagnetization of sample EM25. Other details as in Fig. 7.

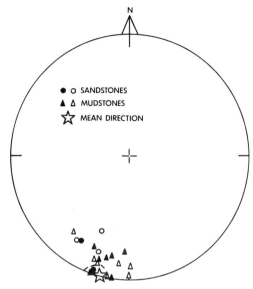

● ○ SANDSTONES
▲ △ MUDSTONES
☆ MEAN DIRECTION

FIG. 10. Stereographic projection of site mean directions for the Etruria Formation; open symbols = upward inclination, closed symbols = downward inclination. Circles denote sandstones; triangles denote mudstones. Mean direction: Dec. = 197°, Inc. = −1°.

changes suggesting the presence of a superimposed 'normal' component. This component is very clearly seen in EM12 where individual specimens with directly antiparallel 'normal' and 'reversed' directions are present.

Both red and green sandstones have been thermally cleaned at 300°C and the site mean directions (Table 2) calculated on the basis of all specimens for EM7, EM24 and EM25, and only on the 'reversed' specimens for EM12.

The thermal demagnetization of the Etruria Formation indicates that the characteristic remanent magnetization is reversed and directed toward the SSW with a very shallow inclination (Fig. 10 and Table 2).

A mean distribution of magnetization for the Etruria Formation has been calculated by giving equal weight for all site means shown in Table 2. This mean direction of Dec. = 197°, Inc. = −1°, α_{95} = 6.9 is consistent with other Upper Carboniferous results, and yields a pole position of 35°N, 190°E. This agrees closely with the late Carboniferous pole for northern Eurasia based on the compilation of Irving (1977) and provides good corroborative evidence that the magnetization and reddening of the Etruria Formation took place during the Westphalian.

More precise palaeomagnetic information regarding the age of magnetization can be inferred by consideration of the Upper Carboniferous magnetic reversal stratigraphy. The geomagnetic field was mostly reversed throughout the Carboniferous Period but there were two normal events which provide palaeomagnetic marker horizons. These events occurred in Westphalian A (Roy 1977) and Westphalian C (Noltimier & Ellwood 1977). The fact that most of the mudstones are probably lower Westphalian C age and carry a single reversed component of magnetization is consistent with the view that they were magnetized in the time interval between the Westphalian A normal event and before the Westphalian C normal event. The presence of normally magnetized specimens in EM12 indicates that the magnetization history of the green sandstones may have been more prolonged, extending to the Westphalian C normal event. The samples EM21 to 27 of upper Westphalian C to Westphalian D age must have been magnetized after this event.

The palaeolatitude of the Etruria Formation is readily determined from its mean direction of magnetization, assuming the geocentric dipole model for the Carboniferous field, from the relationship $2 \tan L = \tan I$. The mean inclination of −1° gives a palaeolatitude virtually on the Carboniferous equator. This, and the associated palaeoclimatic implications, are consistent with our model for the origin of red beds and palaeosols in the Etruria Formation.

Sedimentological model for the origin of the red beds

A consistent pattern emerges from the consideration of both the vertical and lateral transition from grey coal-bearing sediments into red beds. In a vertical section (Fig. 5) red pigment first appears in the lower part of a seat-earth, and occupies an increasing proportion of the sediment in the successive units of sediment between coal horizons or major palaeosols. Where the colour becomes dominantly red, the sediment beneath the soil horizons is oxidized, as are coal seams and seat-earths. At the same time more evolved, well-drained palaeosols appear in the sequence. A similar relationship between red pigment and lithology is to be expected in the lateral transition from grey to red beds beneath one coal marker horizon.

Sedimentological observations indicate that the increasing importance of red pigment in the sequence reflects a transition from deposition in a waterlogged swamp to deposition in nearly permanently well-drained conditions. The changes in sedimentation that caused this must have acted in a way that resulted in a lowering of the water table in the alluvial pile (cf. Friend 1966). Three models are possible:

(1) The improved drainage resulted from progradation of topographically higher and thus better drained swamp and floodplain environments into the water-logged swamp area. This may account for the presence of oxidized horizons in the well-drained swamp palaeosols, but is unlikely to explain the penetrative, post-burial, oxidation of coals and seat-earths.

(2) Improved drainage may have been caused by channel incision in the deposi-tional alluvial system. Red beds are being formed in this manner at the pres-ent in the Pleistocene of south-western Papua New Guinea (Paijmans et al. 1971), where grey alluvial plain mud-stones, dated as less than 27 000 yr BP, have been dissected as a result of gentle folding and lowering of sea-level. These sediments now have red podzolic and lateritic soils developed on them, and are deeply weathered and extensively reddened.

While this example confirms the potential for very rapid formation of red beds in the humid tropids, it is unlikely that it can be applied as a model for the Etruria Formation. While this process might have operated at the basin margins it is unlikely to have occurred at the centre of the basin, where the thickest sequence of red beds accumulated. Friend (1978) has expressed strong doubts as to the possibility of alluvial incision being a common occurrence in depositional landforms.

(3) The water table in the alluvial environ-ment marginal to the coal forming swamps may have dropped as a result of evaporative loss during the dry season of a strongly seasonal climate. This phenomenon has been described in the Upper Nile swamps of Sudan (Rzóska 1974). In this area large expanses of topographically poorly differentiated floodplain are flooded during a rainy season lasting for six months of the year. During the rest of the year rainfall is slight, and, in areas away from the main fluvial channels, the water table drops sharply, being largely inaccessible in

boreholes. The soils developed in these seasonally well drained swamps are not described, but the area is bounded by an area of basement on which laterites are forming (Morison, Hoyle & Hope-Simpson 1948).

A similar situation can be envisaged for the Etruria Formation red beds. In this case the drop in water table may have resulted partly from differential subsidence rates between areas of grey and red bed formation, and partly from evaporation. The existence of a seasonal monsoonal climate in England during the Westphalian has been suggested on sedimento-logical grounds by Broadhurst, Simpson & Hardy (1980).

Discussion and conclusions

In our study of the Etruria Formation we have found no evidence that intrastratal alteration of metastable ferromagnesian silicates contributed pigment to the red alluvium (cf. Walker 1974). The genesis of haematite pigment took place soon after deposition either by the dehydration of detrital ferric oxyhydroxides, or by the oxi-dation of ferrous iron associated with organic material.

The available evidence may be summarized as follows:

(1) Where intercalated organic-rich grey beds and red beds occur, the boundaries between the two correspond approxi-mately to depositional surfaces. The geometry and overall distribution of red beds and interdigitated coal-bearing sedi-ments was controlled by the position and behaviour of the water table during, and soon after, deposition. This, in turn, was controlled by three factors: (a) the progradation of areas of well-drained alluvial plain into waterlogged swamp areas; (b) the marked lowering of the water table away from topographically low areas of permanent waterlogging as a result of seasonal evaporative loss of water; (c) the maintenance of well-drained and poorly drained alluvial plain and backswamp areas by the action of differential subsidence rates between the centre of the basin and its margins.

The sedimentary processes involved in steps (a) and (b) are summarized in Fig. 11.

(2) The marked concentration of pigmentary haematite in the mudrocks (up to 20% in some cases) implies that haematite or

1) PERSISTENT ALLUVIAL SWAMP SEDIMENTATION

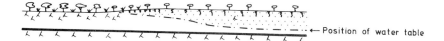

Water table permanently at or above sediment surface

2) PROGRADATION OF WELL DRAINED SWAMP AND ALLUVIAL PLAIN - WATER TABLE DROP IN WELL DRAINED AREAS INDUCED BY EVAPORATION

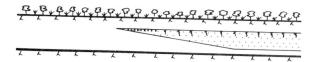

← Position of water table

3) RESUMPTION OF COAL FORMING ALLUVIAL SWAMP CONDITIONS AFTER DECREASE IN SEDIMENT SUPPLY AND/OR INCREASE IN SUBSIDENCE RATE

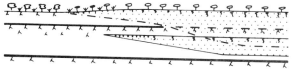

4) RENEWED PROGRADATION OF WELL DRAINED SWAMP AND FLOODPLAIN

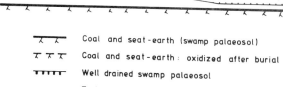

With more extensive development of well drained facies greater and more prolonged drop in water table allows oxidation of earlier organic rich sediment

⌁⌁	Coal and seat-earth (swamp palaeosol)
⊤⊤⊤	Coal and seat-earth: oxidized after burial
⊤⊤⊤⊤	Well drained swamp palaeosol
⊤⊤⊤	Evolved palaeosol
⍩⍩	Swamp flora
↲⌇⌇	Floodplain flora

— · — Water table

[⋯⋯] Reddened sediment

FIG. 11. Suggested sedimentological model for the formation of interbedded and diachronous red beds marginal to a coal-forming swamp area in the Westphalian of central England.

precursor hydroxides were present before permeability was lost as a result of compaction and cementation. Studies of modern overbank muddy alluvium indicates that lithification takes place comparatively rapidly, often within 10 000 years of deposition (Ho & Coleman 1968), and we therefore believe that the thick mudstone sequences in the Etruria Formation could not have maintained sufficient permeability to allow the introduction of large amounts of ferric oxyhydroxides by intrastatal solutions. By contrast in sequences where reddening postdates lithification, haematite is concentrated in porous and permeable sandstones rather than in mudrocks (Mykura 1960).

(3) There is evidence of significant post-depositional oxidation in the form of oxidized coal seams, ironstones, and plant-bearing mudstones. However, there is no evidence that this phase of oxidation involved intrastratal alteration of ferromagnesian minerals and the formation of haematite pigment as described by Walker (1974), although it may have promoted the inversion of detrital hydrated ferric oxides to haematite. The main argument in favour of this is the distinctive residual mineralogy of the Etruria mudstones, which was clearly inherited from an intensely weathered source, and could not have provided the unstable ferromagnesian minerals needed for this process.

(4) Thin section and macroscopic textural study of the evolved palaeosols indicates that haematite (and/or its precursor hydroxide) concentrations formed during pedogenesis. The pedogenic mobilization of ferric oxides must have occurred before the sediment was consolidated and in close proximity to surface exposure.

(5) The palaeomagnetic studies indicate that the reddening (and magnetization) must have taken place in Westphalian times. Many of the red mudstones contain a single reversed component of magnetization which must have been acquired prior to or after the Westphalian C normal event.

We believe, therefore, that the red beds of the Etruria Formation were formed during and shortly after deposition. This process probably involved the accumulation of fine grained alluvium rich in ferric oxyhydroxides and containing organic matter. Reddening resulted from the development of good drainage, organic matter oxidation, and the consequent dehydration of ferric oxyhydroxides (cf. Friend 1966; Van Houten, 1968, 1972). Pedogenesis, involving the subsequent remobilization of iron oxides, was a natural extension of this process and culminated in the formation of evolved palaeosols with 'lateritic' character.

ACKNOWLEDGMENTS: We would like to thank R. H. Hoare and E. L. Boardman, of the National Coal Board (Western Area), for allowing access to borehole cores and unpublished stratigraphic records. Permission to enter brick quarry exposures was granted by G. H. Downing Ltd, Hinton Perry and Davenhill Ltd, Ibstock Brick Co. Ltd, and Wilnecote Brick Co. Ltd. The study was undertaken while BMB was in receipt of an NERC postgraduate studentship. PT thanks K. Meneer and M. Whitehouse-Yeo for assistance with palaeomagnetic measurements and chemical analyses.

References

ALLEN, J. R. L. 1963. The classification of cross-stratified units with notes on their origin. *Sedimentology*, **2**, 93–114.

ARCHER, A. A. 1965. Red beds in the Upper Coal Measures of the western part of the South Wales coalfield. *Bull. geol. Surv. Gt Br.* **23**, 57–64.

ASHLEY, D. A. & PEARSON, M. J. 1978. Mineral distributions in sediments associated with the Alton marine band near Penistone, South Yorkshire. *In:* MARTLAND, M. M. & FARMER, V. C. (eds) *International Clay Conference. Developments in Sedimentology*, **27**, 311–21. Elsevier, Amsterdam.

BLOOMFIELD, C. 1964. Mobilisation and immobilisation phenomena in soils. In: NAIRN, A. E. M. (ed.) *Problems in Palaeoclimatology*, 661–5. Wiley, New York.

BOARDMAN, E. L. 1978. The Blackband Ironstones of the North Staffordshire coalfield. *N. Staff. J. Field Stud.* **18**, 1–13.

BREWER, R. 1964. *Fabric and Mineral Analysis of Soils*. Wiley, New York. 470 pp.

BROADHURST, F. M., SIMPSON, I. M. & HARDY, P. G. 1980. Seasonal sedimentation in the Upper Carboniferous of England. *J. Geol.* **83**, 639–51.

BUTTERWORTH, M. A. & SMITH, A. H. 1976. The age of the British Upper Coal Measures with reference to their miospore content. *Rev. Palaeobot. Palynol.* **22**, 281–306.

CALVER, M. A. 1969. Westphalian of Britain. *C.r. 6me Cong. int. Strat. Geol. Carb.* (Sheffield 1967) **1**, 233–54.

COLEMAN, J. M. 1966. Ecological changes in a massive fresh-water clay sequence. *Trans. Gulf-Cst Ass. Geol. Soc.* **16**, 159–74.

CZYSCINSKI, K. S., BYRNES, J. B. & PEDLOW, G. W. III 1978. In situ red bed development by the oxidation of authigenic pyrite in a coastal depositional environment. *Palaeogeogr. Palaeoclim. Palaeoecol.* **24**, 239–46.

DOWNING, R. A. & SQUIRREL, H. C. 1965. On the red and green beds in the Upper Coal Measures of the eastern part of the South Wales Coalfield. *Bull. geol. Surv. Gt Br.* **23**, 45–56.

DUNLOP, D. J. 1971. Magnetic properties of fine particle haematite. *Annls Géophys.* **27**, 269–93.

—— 1979. On the use of Zijderveld diagrams in multicomponent palaeomagnetic studies. *Phys. Earth planet. Int.* **20**, 12–24.

FREYTET, P. 1971. Paléosols résiduels et paléosols alluviaux hydromorphes associés aux depôts fluviatiles dans le Crétacé Supérieur et l'Éocène

Basal du Languedoc. *Revue Géogr. phys. Géol. dyn.* **13**, 245–68.

FRIEND, P. F. 1966. Clay fractions and colours of some Devonian red beds in the Catskill Mountains, U.S.A. *Q. Jl geol. Soc. Lond.* **122**, 273–88.

—— 1978. Distinctive features of some ancient river systems. *In:* MIALL, A. D. (ed.) *Fluvial Sedimentology. Mem. Can. Soc. Petrol. Geol.* **5**, 531–42.

GIBSON, W. 1905. Geology of the North Staffordshire coalfields. *Mem. geol. Surv. Gt Br.* HMSO, London.

HO, C. & COLEMAN, J. M. 1968. Consolidation and cementation of recent sediments in the Atchafalaya Basin. *Bull. geol. Soc. Am.* **80**, 183–92.

HOLDRIDGE, D. A. 1959. Compositional variation in Etruria Marls. *Trans. Br. Ceram. Soc.* **58**, 301–28.

HUDDLE, J. W. & PATTERSON, S. H. 1961. Origin of underclays and related rocks. *Bull. geol. Soc. Am.* **72**, 1643–60.

IRVING, E. 1977. Drift of the major continental blocks since the Devonian. *Nature*, **270**, 304–9.

KRYNINE, P. D. 1949. The origin of red beds. *Trans. N.Y. Acad. Sci.* **11**, 60–8.

—— 1950. Petrology, stratigraphy, and origin of the Triassic sedimentary rocks of Connecticut. *Bull. Conn. St. geol. nat. Hist. Surv.* **73**, 247 pp.

MOHR, E. J. C., VAN BAREN, F. A., VAN SCHUYLENBORGH, J. 1972. *Tropical Soils.* Mouton, The Hague. 481 pp.

MORISON, C. G. T., HOYLE, A. C. & HOPE-SIMPSON, J. F. 1948. Tropical soil-vegetation catenas and mosaics. A study in the South Western part of the Anglo Egyptian Sudan. *J. Ecol.* **36**, 1–84.

MYKURA, W. 1960. The replacement of coal by limestone and the reddening of Coal Measures in the Ayrshire coalfield. *Bull. geol. Surv. Gt Br.* **16**, 69–109.

NAGETGAAL, P. J. C. 1966. Scour and fill structures from a fluvial piedmont environment. *Geol. Mijnb.* **45**, 342–54.

NOLTIMIER, H. C. & ELLWOOD, B. B. 1977. The coal pole: palaeomagnetic results from Westphalian B, C and D coals, Wales. *Eos. Trans. Am. geophys. Un.* **58**, 375.

PAIJMANS, K., BLAKE, D. H., BLEEKER, P. & McALPINE, J. R. 1971. Land Resources of the Morehead—Kiunga Area, Territory of Papua and New Guinea. *CSIRO Australia, Land Res. Ser.* **29**, 124 pp.

PEARSON, M. J. 1979. Geochemistry of the Hepworth Carboniferous sediment sequence, and origin of the diagenetic iron minerals and concretions. *Geochim. cosmochim. Acta*, **43**, 927–41.

POOLE, E. G. 1966. Trial boreholes on the site of a reservoir at Eymore Farm, near Bewdley, Worcestershire. *Bull. geol. surv. Gt Br.* **24**, 151–6.

—— 1970. Trial boreholes in Coal Measures at Dud-ley, Worcestershire. *Bull. geol. Surv. Gt Br.* **33**, 1–41.

PULLAN, R. A. 1967. A morphological classification of lateritic ironstones and ferruginised rocks in Northern Nigeria. *Nigerian J. Sci.* **1**, 161–73.

RAMSBOTTOM, W. H. C. *et al.* 1978. A correlation of Silesian rocks in the British Isles. *Spec. Rep. geol. Soc. Lond.* **10**, 82 pp.

RETALLACK, G. J. 1977. Triassic palaeosols in the Upper Narrabeen Group of New South Wales II: classification and reconstruction. *J. geol. Soc. Aust.* **24**, 19–36.

ROESCHMANN, G. 1971. Problems concerning investigations of palaeosols in older sedimentary rocks, demonstrated by the example of Wurzelboden of the Carboniferous System. *In:* YAALON, D. H. (*ed.*) *Paleopedology*, 311–20. International Society for Soil Science and Israel University Press, Jerusalem.

ROY, J. L. 1977. La position stratigraphique déterminée palaeomagnetiquement de sediments Carbonifères de Minudie Point, Nouvelle Écosse, a propos de l'horizon repère magnetique du Carbonifère. *Can. J. Earth Sci.* **14**, 1116–27.

RZÓSKA, J. 1974. The Upper Nile swamps, a tropical wetland study. *Freshwat. Biol.* **4**, 1–30.

SCOTT, A. C. 1979. The ecology of Coal Measure floras from northern Britain. *Proc. geol. Ass.* **90**, 97–116.

SLAGER, S. & VAN SCHUYLENBORGH, J. 1970. Morphology and geochemistry of three clay soils of a tropical coastal plain (Surinam). *Agric. Res. Rep.* **734**, Wageningen.

TAUXE, L., KENT, D. V. & OPDYKE, N. D. 1980. Magnetic components contributing to the NRM of Middle Siwalik red beds. *Earth planet. Sci. Lett.* **47**, 279–84.

VAN HOUTEN, F. B. 1964. Origin of red beds—some unsolved problems. *In:* NAIRN, A. E. M. (ed.) *Problems in Palaeoclimatology*, 647–61. Wiley, New York.

—— 1968. Iron oxides in red beds. *Bull. geol. Soc. Am.* **79**, 399–416.

—— 1972. Iron and clay in tropical savanna alluvium: a contribution to the origin of red beds. *Bull. geol. Soc. Am.* **83**, 2761–72.

WALKER, T. R. 1967. Formation of red beds in modern and ancient deserts. *Bull. geol. Soc. Am.* **78**, 353–68.

—— 1974. Formation of red beds in moist tropical climates: a hypothesis. *Bull. geol. Soc. Am.* **85**, 633–8.

——, WAUGH, B. & CRONE, A. J. 1978. Diagenesis in first-cycle desert alluvium of Cenozoic age, southwestern United States and northwestern Mexico. *Bull. geol. Soc. Am.* **89**, 19–32.

ZIJDERVELD, J. D. A. 1967. AC demagnetization of rocks: analysis of results. *In:* COLLINSON, D. W., CREER, K. M. & RUNCORN, S. K. (eds) *Methods in Palaeomagnetism*, 254–86. Elsevier, Amsterdam.

B. M. BESLY, Department of Geology, University of Keele, Staffordshire ST5 5BG; present address: Shell UK Exploration and Production, Shell-Mex House, London WC2R ODY.

P. TURNER, Department of Geological Sciences, The University of Aston, Birmingham.

DURICRUSTS: CALCRETES, SILCRETES AND GYPCRETES

Environment of silcrete formation: a comparison of examples from Australia and the Cologne Embayment, West Germany

H. Wopfner

SUMMARY: Silcretes are subdivided into three main groups, depending on whether their matrix has been formed by: I crystalline quartz growth; II cryptocrystalline quartz aggregation or III formation of cristobalitic-tridymitic silica (opal CT). Groups I and II silcretes are usually associated with intense kaolinisation and formed in low-pH environments, where Al_2O_3 solubility exceeds that of SiO_2. Palludal regimes in moist/warm climates provide such conditions. Group I silcretes are purely diagenetic, whereas pedogenic processes play an essential role in the formation of Group II silcretes. Group III silcretes, composed primarily of cristobalite lepispheres, developed primarily under oxidising conditions in arid, alkaline environments. These genetic models, originally developed from observations on Australian silcretes, are applied with Tertiary silcretes of the Cologne Embayment in western Germany. SEM investigations of grain surfaces and fabrics of these so-called 'lignite quartzites' show that they are comparable with the Australian examples, with the exception that cristobalite-lepispheres are not exclusive to Group III silcretes, but occur also in porcellanitic interlayers of Group I silcretes, where the parent material consists of very fine grained siliciclastics.

The origin of silcretes is still a somewhat controversial topic, although some convergence of viewpoints has occurred in recent years. Based on extensive field observations and laboratory investigations of Australian silcretes, I suggested a preliminary classification in which mineralogical matrix composition, macroscopic habit and profile development are considered the main criteria (Wopfner 1978). In this classification (see Table 1), three basic groups of silcretes are distinguished and assigned to specific genetic models. These are briefly reiterated and checked against results of new investigations of silcretes occurring in the Tertiary of the Cologne Embayment.

Silcretes of Groups I and II

Although quite different in appearance, both silcrete types have one feature in common: they are invariably associated with an horizon of extensive bleaching and kaolinisation. As I have stressed repeatedly (Wopfner 1964; Wopfner & Twidale 1967) this zone of kaolinisation plays a crucial role in the formation of these silcretes. Chemical analyses and petrological investigations of continuous profiles comprising unaltered rock, kaolinised zone to silcrete (Wopfner 1978) have shown that:

(1) The silcretes consist of 97–99% of SiO_2, the main constituent being quartz (see Table 1). TiO_2 was also enriched, making up the remaining percentages.

(2) They are almost entirely depleted of Fe, Al, Ca, K, Mg and P.

These features suggest the following formational conditions:

(a) silicification formed at or near the top of a constant ground water level;

(b) to achieve the observed accumulation of silica, the solubility of Al_2O_3 must have exceeded that of SiO_2;

(c) SiO_2 was provided by the release of silica during kaolinisation of feldspar and other silicates.

These constraints require a very low pH-environment that is only likely in paludal conditions.

The differences between Group I and Group II silcretes resulted form the first being formed below the zone of pedogenic influences, whereas the development of Group II silcretes took place within a palaeosoil profile. Grain size and permeability were a further influencing factor, as is discussed below.

Group I silcretes from the Cologne Embayment

Occasionally, small silcrete bodies become exposed during mining operations in the large lignite open cut at Fortuna-Garsdorf, situated about 24 km west of Cologne. The silcretes occur within the upper parts of the Neurath Sand, a Miocene interseam deposit which separates the main lignite seam from the Garzweiler

TABLE 1. *Preliminary classification scheme of silcretes. Accordingly, a silcrete would be identified by the appropriate combination of numerals and letters. For example a silcrete with cryptocrystalline, pseudo-pebbly matrix and columnar habit would be identified by the code II-2a (after Wopfner 1978)*

Group I

Matrix:		crystalline quartz
	(1)	irregular crystalline
	(2)	optical continuity with framework quartz

Habit:		
	(a)	blocky
	(b)	bulbous—pillowy

Retention of host rock texture: in 1a and 2a
Profile: Overall kaolinisation of underlying rocks

Group II

Matrix:		cryptocrystalline quartz
	(1)	massive
	(2)	pisolitic—pseudopebbly
	(3)	laminated—'schlierig'

Habit:		
	(a)	columnar, polygonal—prismatic
	(b)	platy
	(c)	botryoidal
	(d)	pillowy ('Knollenstein')

Retention of host rock texture: None
Profile: Intense kaolinisation of underlying rocks, usually with zone of brecciation between kaolinised portion and silcrete

Group III

Matrix:		amorphous (opaline) cristobalitic—tridimitic (opal C-T)

Habit:		
	(a)	breccious
	(b)	conglomeratic
	(c)	replacements and infillings

Retention of host rock texture: invariably perfect
Profile: No specific profile but association with gypsum and alunite

Seam in the northern portion of the Cologne Embayment (Teichmüller & Teichmüller 1968). The sands are medium grained (Md 0.3–0.4 mm) and they are very well sorted with a Trask sorting coefficient in the vicinity of 1.3. They are very similar to bleached Miocene sands below lignite horizons in the Netherlands (van den Broek & van der Waals 1967). They are possibly shallow lacustrine fan deposits, but they have also been interpreted as estuarine or marginal marine sediments. The lower three quarters are dark brown, due to the presence of humic material, and they also contain appreciable amounts of clay and very fine mica.

The upper portion of the sand is invariably white, almost without any interstitial clay. This zone of bleached sand immediately underlies the Garzweiler Lignite Seam (Fig. 1). The demarcation between the white and the dark brown sands is fairly sharp, although there is no obvious change of physical properties at the boundary between the two colours; grain size, porosity and permeability are the same, demonstrating that its position was governed by hydrological factors. A considerable number of hypotheses have been advanced to explain the colour change (e.g. Karrenberg 1958), not all of them very plausible. In my opinion, it is related to the overlying lignite seam. I therefore consider the bleaching to be due to leaching of the sands by low pH waters that characterize swamps.

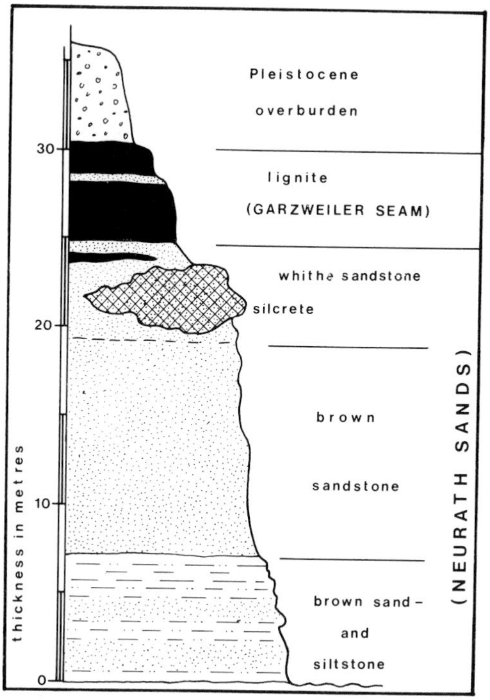
Pleistocene
overburden
lignite
(GARZWEILER SEAM)
whithe sandstone
silcrete
brown
sandstone
brown sand-
and
siltstone
(NEURATH SANDS)
thickness in metres
30
20
10
0

FIG. 1. Generalized profile of silcrete-bearing Miocene sequence in open cut lignite mine Fortuna.

Scattered throughout the leached zone of the Neurath Sand are irregular bodies of immature, Group I silcretes (Fig. 1). They may be up to 7 m thick and attain a width in excess of 15 m. Their shape is quite variable but there is a definite tendency towards bulbous forms as shown in Fig. 2. Nevertheless, the actual contact between the silcrete and the surrounding unconsolidated sand is quite abrupt. The original bedding, however, continues unchanged from the friable sand into the resistant silcrete, a feature typical of Group I silcretes (Wopfner 1978; Wopfner *et al.* 1974). None of the silcretes of the Neurath Sand, observed so far by the author, are completely indurated. They are still semi-friable on the margins and somewhat harder and resistant in the core, but very much to the dismay of the mine management they are too hard to be handled by even the largest of the bucket wheel excavators.

The proximity of friable parent material and indurated silicified bodies allows mineralogical features to be compared between the two. The low coalification state of the soft lignite immediately above the zone of silcrete formation (Fig. 1) demonstrates beyond doubt that the processes leading to the observed changes must have taken place at low, near surface temperatures and at less than 100 m overburden pressure.

Investigations were carried out with a Cambridge Instruments scanning electron microscope, with an EDAX microanalyser attached

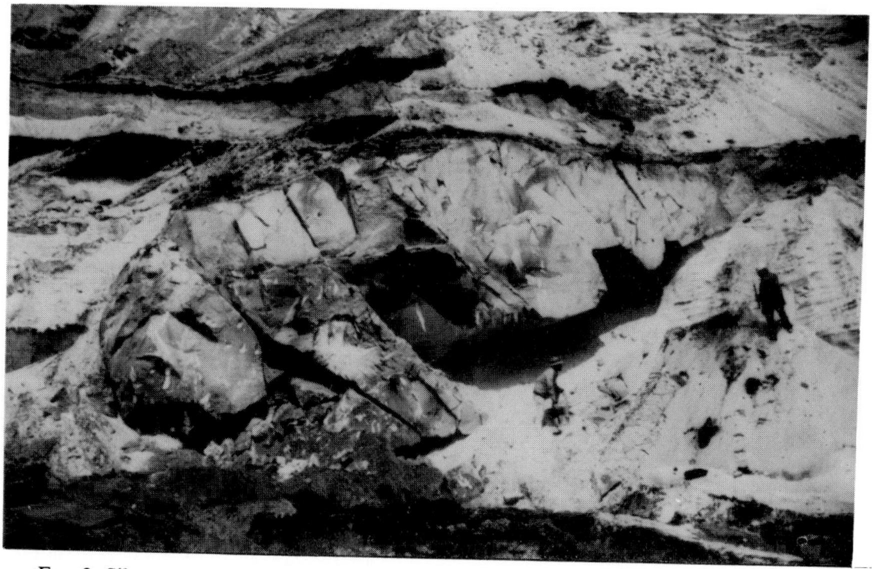

FIG. 2. Silcrete exposed in upper Neurath Sand in Fortuna lignite mine in June 1975.

for rapid element identification. Freshly broken chips or loose grains from unconsolidated material were fixed to sample carriers without any additional treatment except conventional coating procedures. All investigations were carried out on samples collected from the exposure depicted in Fig. 2. Samples were taken from the unconsolidated sand at 4 and 2 m lateral distance from the edge of the silcrete, directly at the lower margin of the silcrete and from the central portion of the silcrete.

The sample taken at 4 m distance from the silcrete consists of subrounded to well rounded clastic quartz grains, with generally smooth surfaces. At greater magnification, however, lunate impact patterns and V-shaped depressions are clearly visible, as depicted in Fig. 3. According to Krinsley & Doornkamp (1973) such markings are characteristic for subaqueous transport. On some surfaces minute triangular etch figures are recognizable (Fig. 4). The sample taken 2 m from the silcrete shows essentially the same features as those described above, although some quartz overgrowth is discernible in rare instances.

Different features are observed in the sample from the silcrete contact. Quartz overgrowth, or

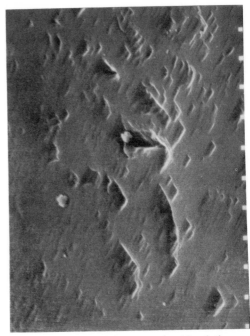

FIG. 4. Surface of quartz grain with solution marks. Distance between scale-bars 3 μm.

FIG. 3. SEM photograph of surface of well rounded quartz grain showing crescent and V-shaped impact marks. Distance between scale-bars 10 μm.

FIG. 5. Oriented quartz growth on grain at base of silcrete. Scale-bars 10 μm.

rather quartz growth in crystallographic continuity with the original clastic grains, has transformed them to crudely shaped quartz crystals, and their surfaces are commonly covered by prolific crystal faces, as demonstrated in Fig. 5.

The samples from the central portion of the silcrete, exemplified by Fig. 6 show an interlocking fabric of well-developed quartz crystals, which is typical of Group I silcretes. Thin-section investigations clearly indicate that crystal growth occurred in optical continuity with the original clastic grain. Retention of open pore space indicates that the silicification process has not been completed, so the silcrete is immature.

Identical silicification phenomena, although not identified as silcretes, have been described from Miocene sands of South Limburg in Holland (van den Broek & van der Waals 1967; Riezebos 1974). As in the lignite open cut in Germany, the silicified layers are again restricted to white, leached sands.

One long standing controversy concerns the origin and nature of the silica. The type of crystal structures depicted in Figs 5 and 6 leaves little doubt that silica was available in true solution. As to the origin of the silica, any transport direction (up, down or lateral) has

been invoked. Surface corrosion, as evidenced by etch-figures like those in Fig. 4 could be, and indeed has been, quoted as evidence for the derivation of the silica from quartz solution within the leached or kaolinised zone. That this model is untenable, in this example at least, is demonstrated by Figs 7 and 8.

Fig. 7 shows a grain from the friable sand 4 m distant from the silcrete. The lower part of the grain shows fractures, impact and solution marks on the surface, whereas a conchoidal fracture, due to mechanical impact is evident in the top half. The upper edge of the conchoidal fracture is slightly rounded, demonstrating clearly that the fracture was inflicted during transport and not later during compaction. Fig. 8 shows an enlargement of the fracture. On the very bottom impact markings are still visible on the shoulder of the fracture. The surface of the fracture itself, however, is completely unmarked, which proves that no solution of quartz surfaces took place during compaction and diagenesis.

So where did the silica originate? Data from Australian silcrete profiles demonstrate that all the silica required for the formation of a solid sheet of 2–3 m thick silcrete could easily be derived by the liberation of silica from the

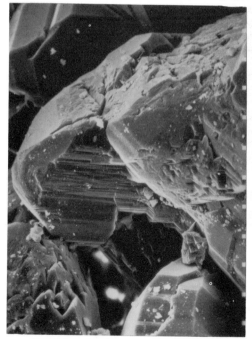

FIG. 6. Interlocking fabric of quartz crystals from central portion of silcrete. Scale-bars 30 μm.

FIG. 7. Clastic quartz grain with conchoidal fracture. Scale-bars 100 μm.

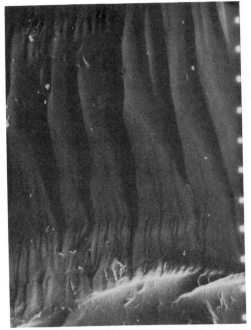

FIG. 8. Detail of conchoidal fracture shown in Fig. 7, lacking any trace of solution marks. Scale-bars 3 μm. Compare with Fig. 4.

FIG. 9. Lepispheres of bladed cristobalite from fine grained silcrete (porcellanite). Scale-bars 3 μm.

kaolinisation of feldspar and other silicate minerals within the kaolinised zone (Wopfner 1978). There is no reason why the same model should not be applied here. Admittedly, the amount of feldspar and other silicate minerals contained within the unaltered brown sands is not very high, but it is quite sufficient to account for the silcrete accumulation. The dearth of suitable source material for the liberation of silica may well be responsible for the patchiness of the silcrete occurrences.

Group I silcretes with cristobalite

Fully matured Group I silcretes are exposed on the south-eastern margin of the Cologne Embayment, a few kilometres east of Bonn. The parent material consists of fluviatile coarse sands and quartz conglomerates with interbedded pelitic material. Silicified wood is also abundant. The sediments are thought to be of late Oligocene age. The rocks are now completely indurated to a dense, hard silcrete without any remnant porosity. Original sedimentary textures are retained, indicating that the silicification process evolved without volume change,

primarily by quartz overgrowth. However, porcellanites derived from the pelitic interbeds consist of lepispheres of bladed cristobalite (Fig. 9). The lepispheres are recognizable only when they occur in cavities, otherwise they form an interlocking fabric.

So far I have observed cristobalite only in Group III silcretes (see Table 1) but obviously, under certain circumstances, cristobalite may also form under conditions of Groups I and II silicification. It is suggested that grain size and composition of the parent material may influence the type of silicia precipitated. Possibly the dominance of clay minerals in the pelitic parent material fostered the formation of opal-CT (cristobalitic-tridymitic silica). This observation also shows, that no reliance can be placed on cristobalite lepispheres as indicators for a marine environment, as has been suggested, for example, by Blair (1978).

Group III silcretes

The main differences between Group III silcretes and the preceding types are a lack of a distinctive profile and the retention of appreciable amounts of Fe_2O_3, on the average about

3%, but attaining more than 9% in some specimens (Wopfner 1978). So far only Australian occurrences have been studied by the author. They are predominantly breccias, usually composed of white, sharply angular components floating in a red, dense, siliceous matrix. The name 'pudding stone' aptly describes these rocks. They are primarily silicified red-earth soils or regoliths, but conglomerates and other surficial deposits may serve as parent material also, provided sufficient interstitial clay is present. Silica transport must have been lateral, the invading silica gently displacing original mineral matter. However, in contrast with Groups I and II silcretes, Fe_2O_3 and some Al_2O_3 (around 1%) were retained. TiO_2 was slightly enriched whereas Ca, K, Mg and P were substantially depleted (Wopfner 1978).

X-ray diffraction studies show that the dominant constituent of these silcretes is opal-CT. Investigations under the SEM show that the dense material is formed by complete fusion of pseudo-hexagonal blades of tridymite (Fig. 10). The iron appears to be retained within spherical relict structures, one of which is depicted in Fig. 11. They were apparently formed during soil processes preceding silicification.

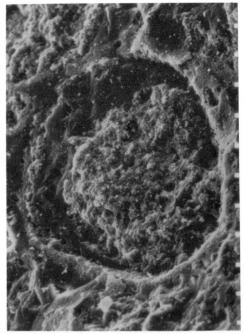

FIG. 11. Iron-rich, spherical relict particle from parent material. Same sample as Fig. 10. Scale-bars 30 μm.

FIG. 10. Tridymite and cristobalite in Group III silcrete from Frome Embayment, South Australia. Scale-bars 10 μm.

Geomorphological association of Group III silcretes with the margins of the Plio-Pleistocene precursors of Lake Eyre and the common occurrence of gypsum under these silcretes suggests an evaporitic and alkaline environment for the formation of Group III silcretes.

This process of silicification, however, should not be confused with the formation of chert after magadiite as described by Eugster (1980) from some alkaline lakes of the East African rift system.

Conclusions

Studies of Miocene silcretes of the Cologne Embayment fully support the formational model of Group I and II silcretes previously derived from investigations of Australian silcretes. A paludal, low pH and reducing environment for these silicification processes is conclusively established. The basic mineralogical composition of Group III silcretes has been verified by SEM investigations. The proposed arid-alkaline environment is regarded as valid.

References

BLAIR, W. N. 1978. Gulf of California in Lake Mead area of Arizona and Nevada during late Miocene time. *Bull. Am. Ass. Petrol. Geol.* **62** (7), 1159–70.

EUGSTER, H. P. 1980. Lake Magadi, Kenya, and its precursors. *In*: NISSENBAUM, A. (ed.) *Hypersaline Brines and Evaporitic Environments*, 195–232. Elsevier, Amsterdam.

KARRENBERG, H. 1958. Die 'Neurather Sande' als Sonderausbildung des Zwischenmittels zwischen den Flözen Garzweiler und Frimmersdorf am Nordende der Ville. *Fortschr. Geol. Rheinhld Westf.* **1**, 151–8.

KRINSLEY, D. H. & DOORNKAMP, J. C. 1973. *Atlas of Quartz Sand Surface Textures*. Cambridge University Press. 91 pp.

RIEZEBOS, P. A. 1974. Scanning electron microscopical observations on weakly cemented Miocene sands. *Geologie Mijnb.* **53**, 109–22.

TEICHMÜLLER, M. & TEICHMÜLLER, R. 1968. Cainozoic and Mesozoic coal deposits in Germany. *In:* MURCHISON, D. & WESTOLL, T. S. (eds) *Coal and coal-bearing strata,* Oliver & Boyd, London.

VAN DEN BROEK, J. M. M. & VAN DER WAALS, L. 1967. The late Tertiary peneplain of South Limburg (The Netherlands)—Silicifications and fossil soils; a geological and pedological investigation. *Geologie Mijnb*, **46**, 318–32.

WOPFNER, H. 1964. Tertiary duricrust profile on Upper Proterozoic sediments, Granite Downs area. *Geol. Surv. S. Aust. Quart. Geol. Notes. No. 12*, 1–3.

——, 1978. Silcretes of northern South Australia and adjacent regions. *In*: LANGFORD-SMITH, T. (ed.) *Silcrete in Australia*, 93–141. University of New England Press, Armidale.

—— & TWIDALE, C. R. 1967. Geomorphological history of the Lake Eyre Basin. *In: Landform Studies from Australia and New Guinea*, 118–43. Australian National University Press, Canberra.

——, CALLEN, R. & HARRIS, W. K. 1974. The Lower Tertiary Eyre Formation of the southwestern Great Artesian Basin. *J. geol. Soc. Aust.* **21**, 17–52.

H. WOPFNER, Geologisches Institut, Universität zu Köln, Zülpicher Strasse 49, D 5000 Köln 1, West Germany.

Silcrete in Western Australia: geomorphological settings, textures, structures, and their genetic implications

W. J. E. van de Graaff

SUMMARY: Tertiary silcrete and laterite are extensively preserved in Western Australia in low-relief landscapes. Silcrete is best developed on topographic highs developed on fluviatile sands which overly radiolarites or deeply kaolinised claystones. Laterite occurs topographically lower and is generally developed on fine-grained bed rock. Lateral transition of silcrete to laterite indicates that these duricrusts form a soil catena. Cone structures in the silcretes indicate strongly evaporitic, dry conditions, whereas solution pipes in duricrust profiles indicate humid conditions. The silcrete/laterite catena thus indicates a strongly seasonal, hot, and humid climate, with rainfall probably in excess of 1000 mm yr^{-1}.

Silcrete occurrences were studied in the Gibson and Great Victoria Deserts and in the Carnarvon Basin area of Western Australia (Fig. 1), as part of regional mapping projects of the Geological Survey of Western Australia (NB 1: 250,000 series maps referred to in the figure captions are available from the GSWA). Tertiary duricrusts are well preserved in the desert areas, but are extensively dissected in the Carnarvon Basin. Thus silcrete petrography and profiles are best known from the coastal region, whereas gross morphological settings are best established for the two desert areas. This paper describes some salient features of silcrete occurrences and their probable genetic implications.

Petrographic features of silcretes

Three types of highly siliceous surficial deposits are distinguished: (1) *floating-fabric silcrete*, (2) *preserved-texture silcrete*, and (3) *laminar silcrete*. Although locally occurring in close association with each other (e.g. Fig. 4b,g), no transitional types have been recognized.

(1) *Floating-fabric silcrete*, which is the most common silcrete duricrust, has a cryptocrystalline quartz matrix supporting 'floating' detrital quartz grains. Grains and matrix occur in sub-equal amounts. Quartz grains range from very fine, angular sand, to well rounded granules to cobbles. Mineralogically these silcretes consist of up to 99% quartz, 0.3–2% TiO_2, and 0.2–1.5% $Fe_2O_3 + FeO$. At outcrop they may be sheet, hummocky, cylindrical, or cone shaped. Primary textures, structures, and mineralogy of the host rock are obliterated. Figs 2(b,d),

3 (i), 4 (b–j, l–m) show various aspects of this type of silcrete.

(2) *Preserved-texture silcrete* has a cryptocrystalline quartz and/or opaline matrix, and is rare except in radiolarian-rich sediments. Primary textures, structures, and locally, mineralogy of the host rock are preserved. This type of silcrete invariably underlies type 1 silcrete, and evidently formed lower in the weathering profile. Figs 1 (f), 2 (b,d), 4 (k) show aspects of this type of silcrete.

(3) *Laminar/pisolitic silcrete* has distinct laminar and/or pisolitic fabrics, and is composed of quartz and disordered kaolinite. It either overlies floating-fabric silcrete (e.g. Fig. 4g), or occurs as solution-pipe fills in silcrete profiles. Laminar/pisolitic silcrete is only known from the Carnarvon Basin, where it is volumetrically unimportant. Figs 2 (d), 3 (e–h), 4 (b,g,n) show features of this type of silcrete.

The cone-shaped habit (Figs 2d, 3i, 4b–g) of floating-fabric silcrete suggests an evaporitic origin. Pebbles and cobbles commonly form the nuclei of the geopetally oriented cones (Fig. 4c–h), which at first glance look like erosional earth pillars. These silcrete cones, however, formed *on top* of large clasts and are obviously concretionary features. This is also shown by the presence of small cones *within* larger cones (Fig. 4c,d). In carbonates such preferential cementation on top of clasts has been experimentally produced, on a smaller scale, under conditions of intense evaporation (Badiozamani, Mackenzie & Thorstenson 1977).

By contrast very humid conditions are indicated by the presence of solution pipes within the siliciclastic bedrock of the silcrete profiles.

159

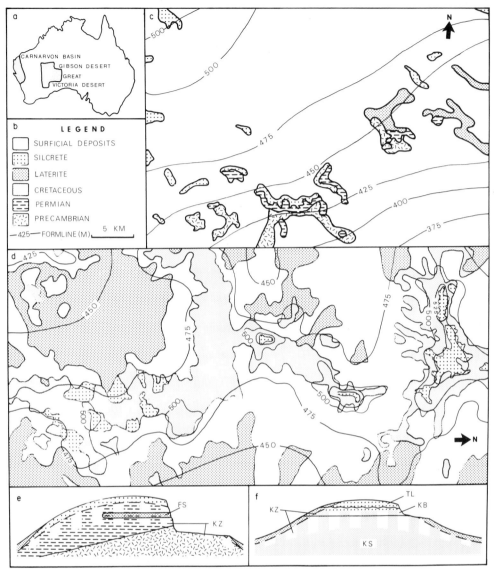

FIG. 1. (a) Location map. (b) Legend for (c,d,e & f). (c) Distribution of silcrete and laterite in south-western Great Victoria Desert around 28° 40′ S, 123° 45′ E (RASON 1: 250,000 sheet area; SH 51–3). (d) Distribution of silcrete and laterite in central Gibson Desert around 25° 10′ S, 125° 05′ E (BROWNE 1: 250,000 sheet area; SG 51–8). (e) Schematic cross-section of (c) showing topographic control on duricrust distribution, KZ—kaolinised zone, FS—ferruginised sandstone within Permian silt/claystone. (f) Schematic cross-section of (d) showing topographic and bedrock control on duricrust distribution, TL—fluviatile sandstone of Tertiary Lampe Beds, KB—radiolarian silt/claystone of Cretaceous Bejah Claystone, KS—sand/silt/claystone of Cretaceous Samuel Formation, KZ—kaolinised zone.

FIG. 2. (a) Distribution of silcrete and laterite on Dairy Creek Station around 25° 21′ S, 115° 50′ E (Carnarvon Basin, GLENBURGH 1:250,000 sheet area, SG 50–6). (b) Fluviatile conglomeratic sandstone of Tertiary Pindilya Formation erosively overlying soil profile developed in Permian claystone, both Pindilya Fm and soil profile silcreted, floating fabrics in Pindilya Fm and preserved textures in pre-Pindilya soil (Carnarvon Basin at 26° 22′ 39″ S, 115° 47′ 24″ E, BYRO 1: 250,000 sheet area, SG 50–10). (c) Sketch map and cross-section showing topographic control on distribution of silcrete and laterite. Note lateritisation at base of silcrete. (GLENBURGH sheet area, yard grid reference 349 777). (d) Silcrete types observed in Carnarvon Basin: (1) kaolinitic silcrete with laminar and/or pisolitic fabrics; (2) small cones of 'floating fabric' silcrete in unsilicified host (cones may have draped caps of laminar silcrete, cf. Fig. 4g) and siliceous rhizoconcretions; (3) silicified mar/calcilutite underlying (2); (4) botryoidal structures at base of silcrete; (5) nodular 'floating

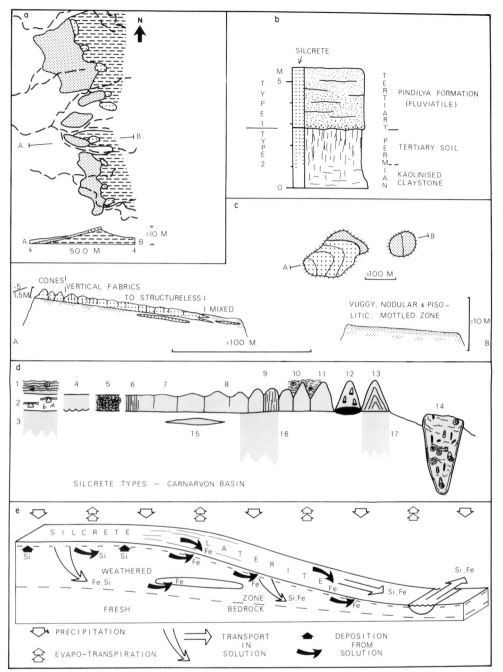

fabric' silcrete with pisolitic coatings; (6) silcrete with pronounced, sub-parallel, planar vertical joints; (7) 'floating fabric' silcrete with sheet-like geometry; (8) silcrete with pronounced hummocky upper surface, cf. Fig. 4 (j); (9) silcrete with distinct cylindrical structure, cylinders may have internal smaller-scale vertical cylindrical fabrics, cf. Fig. 4 (h,i). (Note that types 7, 8 and 9 are the most common forms of silcrete in the Carnarvon Basin area); (10) type (3) silcrete over well-developed 'floating fabric' silcrete cones, cf. Fig. 4 (b); (11) cone-shaped 'floating fabric' silcrete, cf. Fig. 4 (b); (12) silcrete cones formed on top of pebbles/cobbles in conglomeratic host sediment, small cones may be present within larger cones, cf. Fig. 4 (c,d,e & f); (13) silcrete cones with concentric layering, cf. Fig. 4 (g); (14) solution pipe fill with edgewise conglomerate, 'floating fabric' silcrete breccia, laminar silcrete, capped pisoliths, cf. Figs 3 (b–h) & 4 (n); (15) lens/sheet shaped silcrete in kaolin-ised zone below main silcrete layer; (16) pervasively silicified Permian bedrock, e.g. sandstone, tillite, primary textures and mineralogy preserved/recognizable; (17) porcellanised radiolarian chert/siltstone of Cretaceous Windalia Radiolarite, cf. Fig. 4 (k). (e) Conceptual model of silcrete/laterite soil catena with inferred transport/deposition pattern of Si and Fe.

These karst pipes are locally up to 5 m deep, and are commonly filled with laminar/pisolitic silcrete (Figs 2d, 3b–f). The pisoliths are geopetally oriented, and in some pisoliths the geopetal cap has progressively changed orientation (Fig. 3g–h). This indicates slow rotation of the pisolith during the formation of the laminar coating. The rotation of pisoliths is thought to be due to subsidence of the solution pipe fill, and thus indicates continued deepening of the pipe through solution concomitant with (re)precipitation of silica as laminar silcrete in the pipe fill.

The laminar/pisolitic silcrete looks similar to some types of calcrete which are considered to have formed under the influence of soil organisms (e.g. Klappa 1979). The shape and structure of some capped pisoliths (Fig. 3h) resemble stromatolites. However, no organic filaments have been recognized in thin section or by SEM (Fig. 4n). Any soil organisms would have been unusual in that they precipitated SiO_2 + disordered kaolinite rather than $CaCO_3$. As yet neither an organic nor an inorganic origin for this type of silcrete can be considered as established. It is nevertheless clear that laminar/pisolitic silcrete formed within the soil profile (e.g. Fig. 4g—draped laminar silcrete deposited within the host sediment; Fig. 3g—laminar silcrete formed at the base of 2 m thick vermiform/pisolitic laterite).

Geomorphological and stratigraphic setting of silcrete

Floating fabric silcrete occurs in the higher parts of relict duricrusted landscapes. Typically it formed in the fluviatile gravels and sands of the Tertiary Pindilya Formation and Lampe Beds, which are mainly preserved as terrace remnants on the crests of interfluves of palaeodrainage systems (Van de Graaff et al. 1977). Silcretisation has also affected Permian sandy deposits and granitic weathering residues in similar topographic positions. In other words, silcrete occurs on the oldest parts of the landscape. In addition to topography and the presence of a sandy host sediment, silcrete development is also controlled by bedrock type. Floating fabric silcrete is best developed where Tertiary sandstones to conglomeratic sandstones overlie Cretaceous radiolarian-rich cherts and silt/claystones, or thick Permian claystones. Both rock types evidently formed a source of soluble silica, either from radiolarian opaline silica or from kaolinitisation of claystone. The Cretaceous radiolarian-rich deposits are commonly porcellanised for several metres below the floating-fabric duricrust. This porcellanised zone is the most common type of preserved-texture silcrete.

In contrast, laterite covers the topographically lower, somewhat younger landforms (Figs 1c–f, 2a,c, 3a). It is best developed on fine-grained siliciclastics of Cretaceous age (Figs 1f, 3b). Silcrete is commonly surficially ferruginised, and where silcrete and laterite occur in a single profile, poorly to moderately developed laterite typically overlies silcrete. Locally, however, lateritisation of the basal part of the silcrete has occurred (Figs 2c, 4i). Only rarely is silcretisation of lateritic material observed (Figs 3g, 4f). Away from the crests of the interfluves, laterite overlying silcrete is progressively better developed and eventually replaces the silcrete completely (Figs 1c–f, 2a,c). A transition zone has only been observed in one outcrop (Fig.

FIG. 3. (a) Low-relief laterite duricrust landscape in Gibson Desert, BROWNE sheet area, looking towards Mt Beadell. S indicates silcrete on hills of radiolarian-rich sediments. (b) Mottled zone of laterite profile with breccia-filled solution pipes (sp) in claystone bedrock, Great Victoria Desert, TALBOT sheet area (SG 52–9), yard grid 441 665. (c) Breccia-filled solution pipe (sp) in Permian sand/silt/claystone, note sagging of bedrock layers, GLENBURGH sheet area, yard grid 383 791, Diddit Bore. (d) Edgewise conglomerate as solution-pipe fill, laterite profile on Cretaceous sandstones of Yarraloola Fm. (e) Oblique cross-section of solution-depression fill formed in Permian sandstone, walls and base coated with laminar, kaolinitic silcrete, same material forms pisolitic coating on breccia fragments with preferential development on top side, specimen embedded in plaster of paris (p), GLENBURGH sheet area, yard grid 357 784, GSWA sample 44571. (f) Detail of (e) (arrow) showing 'capped' pisoliths and discordant relationships within laminar silcrete (d). (g) Laminar kaolinitic silcrete from basal part of laterite duricrust. Pisolith cores are mostly laterite fragments; note rotated 'capped' pisolith (r), this is indicated by progressive change in orientation of laminar pisolith cap, BYRO sheet area, yard grid 351 767, GSWA sample 44563 a. (h) Geopetal 'capped' pisoliths of laminar silcrete, note rotated pisoliths (r), pisolith cores consist of bedrock claystone (b) or 'floating fabric' silcrete (s), locality as for Fig. 3 (e), GSWA sample 44571 c, cf. Fig. 4 (n). (i) Exhumed silcrete cones on hill side, silcreted sandstone abruptly overlies kaolinised Permian claystone, Carnarvon Basin, AJANA sheet area (SG 50–13), yard grid 316 586.

2c), as it is a weak part of the duricrust cover and thus most readily destroyed by erosion. In this exposure a 'mixed' zone occurs where iregular silcretisation and lateritisation have both taken place. Topographically higher, well-developed silcrete is partly lateritised at its base.

Silcretisation generally preceded laterite formation as is indicated by the common partial lateritisation of silcrete. However, the lack of mechanical reworking of silcrete in the associated laterites, the evidence of lateral transitions between the two duricrust types (Fig. 2c), and the rare evidence of silcretisation post-dating lateritisation (Figs 3g, 4f), suggest that the age difference is minimal, and that laterite and silcrete are to be considered as a soil catena. In this catena, silcrete formed in well drained, topographically high, sandy soils, whereas laterite formed on topographically lower, less well drained, generally more clayey soils.

Floating fabric silcrete generally formed close to the original land surface as is indicated by the complete loss of primary textures typical of soils, and by the presence of cone structures that indicate strong evaporitic effects. Laminar/pisolitic silcrete, which overlies floating-fabric silcrete, is interpreted as having formed during the mature phase of profile development (cf. calcrete profiles). At that stage of silcretisation, drainage characteristics of the profile had deteriorated to the extent that solution-pipe formation and/or lateritisation could occur. This interpretation implies both lateral and vertical mobility of Si and Fe within the weathering profile, with large-scale precipitation within the near surface zone. In addition to surficial accumulation of Fe, deposition of Fe also occurred in sandstones within the deeper parts of the weathering profile (Figs 1e, 2e). These deep accumulations, which formed in palaeo-aquifers and at permeability contrasts, are the lateritic equivalent of preserved-texture silcrete.

Age and palaeoclimatic setting of duricrusts

Kemp (1978) concludes that prior to the Oligocene, humid and warm conditions prevailed in the Australian region, with rain forests extending to the presently arid centre of the continent. Oxygen isotope data indicate a

---

FIG. 4. (a) Kaolinised Permian claystone (kP) erosively overlain by fluviatile sandstone of Tertiary Pindilya Formation (Tp) which is partly silcreted in upper part; AJANA sheet area, yard grid 326 582. (b) Silcrete cones overlain by laminar/pisolitic silcrete; AJANA sheet area, yard grid 326 581. (c) Silcrete cones formed in conglomeratic host sediment; note that quartz/quartzite pebbles/cobbles (q) form foundation of both the small cones within the larger cone, and the large cone on the left; BYRO sheet area, yard grid 370 735, GSWA sample 44585. (d) Detail of (e). (e) Large silcrete cone on quartzite cobble, cone fallen over, photo R. M. Hocking. (f) Silcrete cone on quartzite pebble (q), note iron-rich (lateritic) peds within silcrete cone; BYRO sheet area, yard grid 361 759, GSWA sample 44582. (g) Cone-shaped and spherical silcrete concretions within texturally identical but non-silcreted, very poorly sorted sandstone; note concentric structure of largest cone and laminar kaolinitic silcrete draping pre-existing concretions; Carnarvon Basin, YARINGA sheet area, yard grid 218 763, GSWA sample 32026. (h) Well-developed small scale cylindrical to elongate nodular structures in cylinder-shaped silcrete; BYRO sheet area, yard grid 366 739. (i) Silcrete cylinders/indistinct cones overlying pisolitic/nodular laterite (L); silcrete forms cores of laterite pisoliths/nodules; note dark lateritic cutans on silcrete cylinders; BYRO sheet area, yard grid 353 752. (j) Small-scale hummocky topography of top surface of silcrete; Carnarvon Basin, WOORAMEL sheet area, Meedo Pool, yard grid 302 802. (k) SEM photograph of porcelanised radiolarian chert (Windalia Radiolarite—Cretaceous), note microcrystalline matrix enclosing radiolarian; bar scale 100 μm; arrow indicates enlarged part of radiolarian shown as inset with bar scale 3 μm; XRD indicates opal-C/opal-CT composition (pronounced tridymite peaks and subdued quartz peaks); WOORAMEL sheet area, yard grid 273 808, GSWA sample 32021. (l) SEM photograph of 'floating fabric' silcrete; note detrital quartz (q) and two crystal sizes of matrix; bar scale 30 μm; EDAX of outlined area indicates SiO_2 composition (1.7 keV peak = Si, 2.1 & 9.7 keV peaks = Au); BYRO sheet area, GSWA sample 44574. (m) SEM photograph of silcrete that formed on granitic bedrock; note floating fabric of detrital quartz grains (q) in micro-crystalline matrix; bar scale 100 μm; EDAX of outlined area indicates silica, anatase, and trace iron oxide (1.7 keV peak = Si, 4.5 & 4.9 keV peaks = Ti, 6.4 keV = Fe, XRD confirms high anatase content); Great Victoria Desert, MINIGWAL sheet area, GSWA sample 29097. (n) SEM photograph of laminar, kaolinitic silcrete from capped pisolith (Fig. 3h); note comparatively porous matrix which consists of 0.1–15 μm sized spherulites, which are locally rich in SnO_2; bar scale 30 μm; EDAX of whole field of view indicates silica (1.7 keV), Al (1.5 keV), Sn (3.4 & 3.7 keV), and traces Cl (2.6 keV) and Ti (4.5 keV); XRD indicates quartz and disordered kaolinite composition with trace anatase, GSWA sample 44571 c.

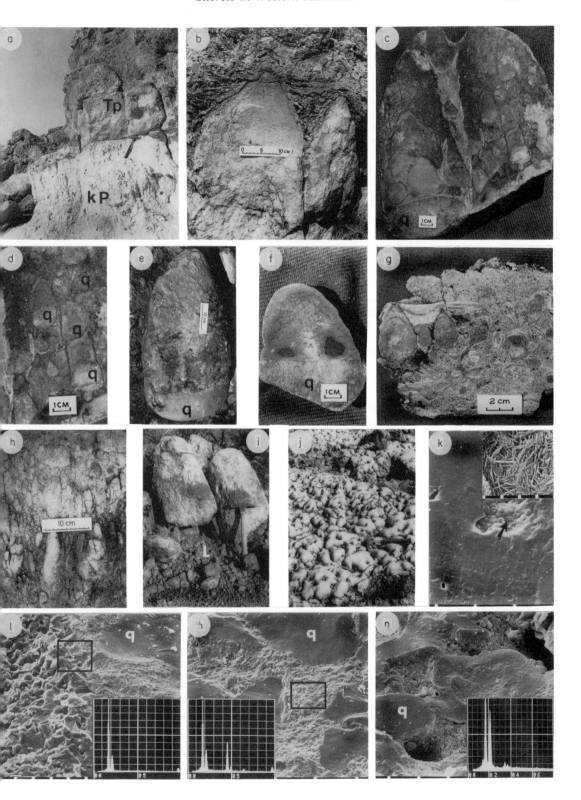

temperature drop during the Oligocene which was accompanied by a reduction in precipitation. During the Miocene, a change to greater aridity is indicated, with circulation patterns close to those of the present day becoming established. This change to aridity is also indicated by geomorphological evidence (Van de Graaff *et al.* 1977). Thus it is probable that lateritisation in the desert regions ceased by the mid-Miocene at the latest. As silcrete pre-dates laterite it is not possible to invoke late Tertiary aridity to account for silcrete formation (cf. Ollier 1978, p. 17). A late Eocene to Oligocene age is therefore favoured for the desert duricrusts, but in the more coastal Carnarvon Basin duricrust formation probably continued into the Miocene.

Present-day karstification of siliceous rocks is only known from tropical regions. White, Jefferson & Haman (1966) describe an active quartzite karst which forms in a humid, hot, very seasonal climate with annual rainfall in the range 1000–7500 mm depending on elevation. Such a wet, seasonal climate appears to be compatible with the observed evidence of strong evaporation (cone structures), and the high rainfall indicated by the karst pipes in siliciclastics. An estimate of annual rainfall of the order of a 1000 mm or more, is also consistent with the size of the palaeoriver systems that are preserved by the duricrusted landscape (Van de Graaff *et al.* 1977). Using statistical relations between climatic and tectonic setting, drainage area, and river discharge established by U. Seemann (pers. comm.) discharge figures can be estimated. For example, the Percival Palaeoriver occupied a drainage basin of some 350,000 km^2 and would have had a likely discharge of 6000 m^3 sec^{-1}. This implies an annual rainfall of 540 mm if *all* precipitation was discharged. Actual rainfall during duricrust formation would therefore have been of the order of a 1000 mm yr^{-1} as a minimum.

Discussion and conclusions

The silcrete features described are not unique to the Western Australian occurrences. Wopfner (1978) described many identical features from South Australian silcretes, and at least the silcretes of the two desert areas are thought to correlate with Wopfner's Cordillo Surface silcretes. Wopfner (1978) does not relate the distribution of silcrete to that of laterite, and concludes that silcretisation occurred in featureless, swampy, river flood-plains, a completely different environment from the well-drained upland soils inferred in this paper to be the site of silcretisation.

Langford-Smith & Watts (1978) review examples of co-existing siliceous and ferruginous weathering products, and stress that no single model can be universally applied. The proposed model of combined topographic, host sediment, and bedrock control on the distribution of silcrete and laterite in a setting of low-relief landscapes and hot, humid, very seasonal palaeoclimate, should therefore be carefully evaluated before being applied elsewhere.

ACKNOWLEDGMENTS: This paper is published with the permission of the Director of the Geological Survey of Western Australia and of Shell Research B.V.

References

BADIOZAMANI, K., MACKENZIE, F. T. & THORSTENSON, D. C. 1977. Experimental carbonate cementation: salinity, temperature and vadose-phreatic effects. *J. sedim. Petrol.* **47**, 529–42.

KEMP, E. M. 1978. Tertiary climatic evolution and vegetation history in the Southeast Indian Ocean region. *Palaeogeogr. Palaeoclim. Palaeoecol.* **24**, 169–208.

KLAPPA, C. F. 1979. Calcified filaments in Quaternary calcretes: organo-mineral interactions in the subaerial vadose environment. *J. sedim. Petrol.* **49**, 955–68.

LANGFORD-SMITH, T. & WATTS, S. H. 1978. The significance of coexisting siliceous and ferruginous weathering products at select Australian localities. *In:* LANGFORD-SMITH, T. (ed.) *Silcrete in Australia*, 143–65. University of New England, Australia.

OLLIER, C. D. 1978. Silcrete and weathering. *In:* LANGFORD-SMITH, T. (ed.) *Silcrete in Australia*, 13–7, University of New England, Australia.

VAN DE GRAAFF, W. J. E., CROWE, R. W. A., BUNTING J. A. & JACKSON, M. J. 1977. Relict Early Cainozoic drainages in arid Western Australia. *Z. Geomorph.* **21**, 379–400.

WHITE, B. W., JEFFERSON, G. L. & HAMAN, J. F. 1966. Quartzite karst in southeastern Venezuela. *Int. J. Speleol.* **2**, 309–14.

WOPFNER, H. 1978. Silcretes of northern South Australia and adjacent regions. *In:* LANGFORD-SMITH, T. (ed.) *Silcrete in Australia*, 93–141. University of New England, Australia.

W. J. E. VAN DE GRAAFF, c/o Brunei Shell Petroleum (DPG), Seria, State of Brunei.

Geochemistry of weathering profile silcretes, southern Cape Province, South Africa

M. A. Summerfield

SUMMARY: Silcrete of Cenozoic age associated with deep weathering profiles occurs on residual surfaces along the coastal belt of southern Cape Province. Petrographic and geochemical evidence indicates loss of aluminium and enrichment of silica and titanium during silcrete formation. Silica released locally within the weathering profile was apparently precipitated in a zone of restricted drainage close to the water-table where a low pH environment allowed the removal of aluminium and the migration and concentration of titanium. Silcrete formation probably occurred in a humid tropical or subtropical environment with minimal local relief.

The numerous reports of silcrete now available in the literature indicate the wide range of sedimentological and environmental settings with which it is associated (Langford-Smith 1978; Summerfield 1983). Work in southern Africa and north-west Europe has indicated that silcretes associated with kaolinitic weathering profiles possess a typical suite of petrographic and geochemical characteristics, including authigenic glaebules (Brewer 1964, pp. 259–60), colloform features and relatively high concentrations of TiO_2 ($>0.2\%$), which are not present in non-weathering profile occurrences (Summerfield 1978, 1979, 1982).

In southern Africa, non-weathering profile silcretes, including silicified sands, pan sediments, calcrete and bedrock, occur predominantly in the Kalahari Basin in Botswana, northern Cape Province and eastern Namibia (Summerfield, 1982). This paper describes the weathering profile silcretes of southern Africa, which are confined to a relatively narrow coastal belt (Cape coastal zone) extending from the Oliphants River valley in the west to the Transkei in the east (Fig. 1). A limited number

of non-weathering profile silcretes also occur within this area but these are considered elsewhere (Summerfield 1981). There have been a number of previous studies of silcrete in the Cape coastal zone, but none of these have provided detailed geochemical data on associated weathering profile materials (Bosazza 1936, 1939; Frankel 1952; Frankel & Kent 1938; Mountain 1946, 1951). Moreover, the interpretations of silica geochemistry in these studies were based on earlier erroneous notions about the nature and behaviour of silica in earth surface environments (Summerfield 1981). A more recent preliminary investigation of a number of occurrences by Smale (1973) also appears to have been influenced in its genetic interpretations by the earlier ideas of Frankel & Kent (1938), who emphasized the role of capillary rise and the presence of percolating soil waters containing NaCl in the formation of silcrete.

As with most silcrete occurrences, the age of the Cape coastal zone silcretes has only been estimated by uncertain stratigraphic correlation with fossiliferous deposits. On the basis of cor-

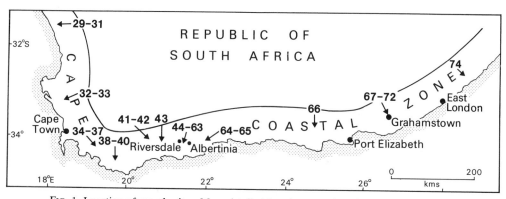

FIG. 1. Location of sample sites. More detailed locations are given in Summerfield (1981).

167

relation with the marine Alexander Beds, there is some concensus in the literature for a Mio-Pliocene age for most of the occurrences considered here (Summerfield 1981), although the Grahamstown silcretes may have formed rather earlier (Mountain 1980).

Field investigations and analytical methods

More than 40 sites were investigated, but in view of the presence of only poorly exposed natural outcrops in many localities attention was focused on a number of quarry and road cutting exposures in the Riversdale-Albertinia and Grahamstown areas. Both silcrete and associated indurated and non-indurated weathering materials were sampled to evaluate profile characteristics. Field estimates of degree of induration of profile materials were provided by the Schmidt hammer (Day & Goudie 1977).

Petrographic observations are based on examination of 92 thin-sections of silcrete and associated materials with supplementary scanning electron microscopy. Silcrete mineralogy was determined by X-ray diffraction. Major element bulk chemical composition of 74 silcretes, calcretes, ferricretes and associated weathering profile clays was determined by X-ray fluorescence from samples diluted in lithium borate glass discs (Norrish & Hutton 1964; Norrish & Chappell 1967). Normative mineralogy calculations for profiles are based on the bulk chemical analyses and X-ray diffraction identification of minerals present. Detailed procedures are documented in Summerfield (1978). Isovolumetric weathering and silica diagenesis calculations are based on chemical data and bulk density determinations from field samples. As no unweathered Bokkeveld Shale was exposed directly underlying any silcrete profiles examined, an unweathered sample collected in the same area was used as a basis for calculations. The chemical analysis for this sample compares closely with other reported analyses (Bosazza 1939; Visser 1937).

Results

Field occurrence

Silcrete occurs predominantly in the Cape coastal zone as cappings, up to 5 m or more in thickness, on a considerably dissected residual surface. Valleyside outcrops are also found but appear to be less significant. In the

Riversdale-Albertinia area the silcrete-capped residual surface is bounded by the Langeberg Mountains to the north and approximately by the Riversdale-Mossel Bay road to the south (Fig. 1). This surface slopes at an average gradient of 0.25° from an elevation of 300 m to the north of Riversdale to 120 m towards the south and east. Over virtually the whole of this area the bedrock is Bokkeveld Shale. Silcrete outcrops occurring on, or within, deeply weathered bedrock, stand at very similar altitudes on adjacent residuals suggesting the existence of an originally much more extensive silcrete layer (Fig. 2).

In the Grahamstown area silcrete has developed in association with weathering deposits overlying a variety of lithologies, including tillites, shales and quartzites (Frankel & Kent 1938; Mountain 1951, 1980). It occurs as a capping on a dissected plateau partially surrounding the town at an altitude of between 625 and 660 m. The silcrete attains a maximum thickness of 7 m and occurs at depths of up to 6 m within weathering profiles. Borehole evidence indicates considerable variation in profile thickness (Mountain 1951).

Multiple silcretes, consisting of up to four distinct silcrete units were recorded in both the Riversdale-Albertinia and Grahamstown areas (Fig. 3). In some cases they exhibit considerable lateral variation in thickness (e.g. Site 60). Composite profiles comprising silcrete in association with calcrete or ferricrete are also found (Fig. 3). In exposures north of Albertinia calcrete overlies silcrete and is locally brecciating it. The silcrete horizon was fully developed before the formation of the calcrete because the lowest few centimetres of the calcrete is a laminar unit produced by the silcrete acting as a plugged horizon. The age relations of silcrete and ferricrete are less clear.

The silcrete itself exhibits a variety of forms in outcrop. Well-developed vertical to subvertical jointing, giving a columnar appearance with subsidiary horizontal components is characteristic (Fig. 4) and has been widely reported for silcrete occurrences elsewhere (e.g. Langford-Smith 1978; Summerfield 1983; Thiry 1978). Both massive and glaebular silcretes are common throughout the area, though the latter are apparently less so in the Grahamstown district. In some profiles silcrete occurs as discrete glaebules, up to 10 cm or more across, which in some cases grade through coalescence into glaebular aggregates and ultimately into fully indurated horizons of glaebular silcrete. Although generally occurring in laterally extensive horizons, massive silcrete is locally present

FIG. 2. View westwards from Site 44, north of Riversdale, showing silcrete capped residuals comprising the remnants of a deeply dissected surface. Langeberg Mountains in background to right.

as vertically elongated bulbous, mammilated masses up to 50 cm across lying within partially silicified weathering profile clays below the main silcrete horizon. A more detailed description of field occurrences is given in Summerfield (1981).

Petrography

All weathering profile silcretes examined are of the F- (floating) or M- (matrix) fabric type (see Summerfield 1983 for details of petrographic classification). The F-fabric silcretes consist of predominantly quartz skeletal grains (>30 μm) floating in a microquartz to cryptocrystalline silica matrix (Fig. 5). Solution embayments on skeletal grains are abundant in many samples. The skeletal grain component is highly variable. M-fabric silcretes (Fig. 6) which constitute about 35% of the samples examined, differ from F-fabric types only in their skeletal grain component which is defined as less than 5%. Fabric type may be only indirectly related to bedrock characteristics since extensive dissolution of detrital quartz during formation of the weathering profile host material may occur prior to silicification. The occurrence of F- and M-fabric silcretes may consequently

reflect the degree of such pre-silicification skeletal grain dissolution as well as the relative abundance of detrital quartz in the original bedrock.

In a number of exposures north of Albertinia M-fabric silcretes are the predominant type and scanning electron microscopy shows that they are composed of a particularly even-textured microquartz matrix, a property which renders them suitable for the manufacture of refractory bricks. Locally these otherwise extremely homogeneous silcretes contain zones up to ~50 cm across of more densely cemented and indurated material (Schmidt hammer mean R-(rebound) value of 64.0 compared with 45.2 ($n = 10$) for adjacent more porous silcrete).

Glaebules were recorded in ~35% of samples examined (Fig. 7). They occur in association with both major fabric types but are more common in F-fabric silcretes. Both nodules (undifferentiated internal fabric) and concentrations (concentric fabric) (Brewer 1964, pp. 266, 268) are found. In most cases glaebules are distinguished by darker coloration due to concentration of brownish earthy aggregates of microcrystalline anatase with iron oxide staining in some samples. Most glaebules are strongly adhesive, but weakly adhesive examples are further differentiated from the

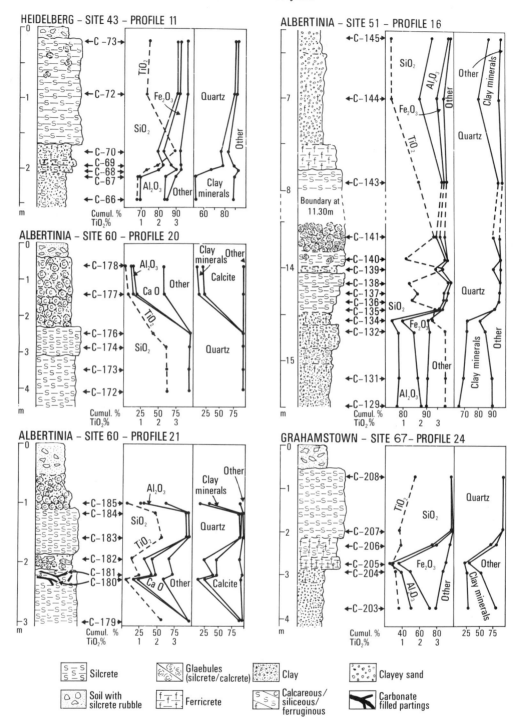

FIG. 3. Selected profiles of silcrete and associated weathering materials. Profile irregularities indicate within-profile relative degree of induration. Profile sample locations shown by sample numbers. Graphical plots show cumulative per cent chemical analyses and normative mineralogy. TiO_2 given as individual percentage.

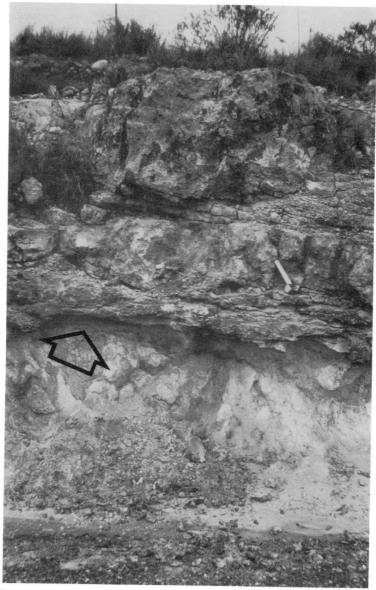

FIG. 4. Silcrete unit approximately 1.5 m thick overlying deeply weathered Bokkeveld Shale located west of Heidelberg. Profile 11, Site 43. The weathered material is partially silicified but the contact (arrowed) with the overlying silcrete is sharp.

interglaebular matrix by bounding curved planar voids. In a small number of samples anatase and iron oxide is concentrated in the inter-glaebular matrix rather than in the glaebules themselves. Concretions are generally poorly developed commonly exhibiting a disordered concentric or more rarely a quasi-lamellar fabric.

Colloform features (Fig. 8) were recorded in more than half the samples examined. They comprise cusp-like laminations formed from concentrations of anatase, and less commonly iron oxide, and usually occur in multiple vertical sequences morphologically similar to striotubules (Brewer 1964, p. 241). They are geopetal structures being invariably orientated

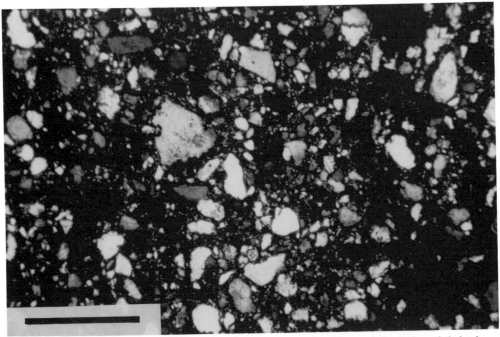

FIG. 5. F-fabric massive silcrete. High authigenic content with evidence of dissolution of skeletal quartz grains. Sample 31, Site 35. Crossed polarizers. Scale bar is 500 μm.

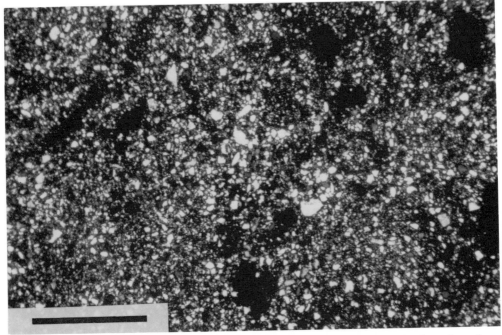

FIG. 6. M-Fabric massive silcrete. Very fine-grained with maximum skeletal grain dimensions of 100 μm but with irregular and partially smoothed metavughs up to 300 μm across. Sample C-175, Profile 20, Site 60. Crossed polarizers. Scale bar is 500 μm.

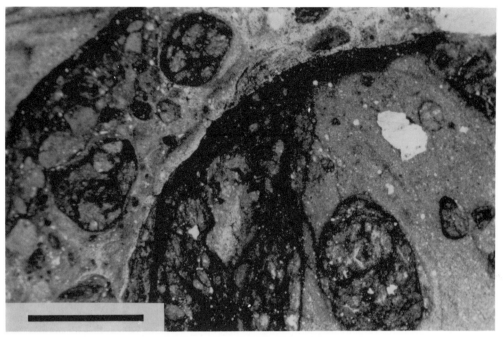

FIG. 7. F-fabric glaebular silcrete containing compound spherical to ellipsoidal titaniferous/sesquioxidic concretions (glaebular aggregates). Sample C-163, Site 52. Plain light. Scale bar is 2 mm.

FIG. 8. Vertically elongated (top to right) colloform feature (striotubule?) (arrowed) in flecked fabric anatase-rich matrix). M-fabric massive silcrete. Sample C-180, Profile 21, Site 60. Plain light. Scale bar is 2 mm.

concave-upwards, and are possibly formed from
the shrinkage and collapse of late-stage dehyd-
rating rhymically precipitated void-fills of silica
and anatase.

Geochemistry

Major element bulk chemical analyses of 51
weathering profile silcretes (Table 1) give a
mean SiO_2 concentration of 94.70% (range
86.79–98.51%). TiO_2 is the second most
abundant component (mean 1.82%; range
0.30–3.36%). Fe_2O_3 and Al_2O_3 are present in
minor but variable concentrations, the former
most commonly present in late-stage void-fills.
The high titanium content of silcretes has been
noted by several authors (e.g. Frankel & Kent
1938; Williamson 1957), although this appears
to be confined to weathering profile occurrences
(Summerfield 1983).

There is no consistent pattern of variations in
element concentrations within individual silcrete horizons (Profiles 13, 14 and 27, Table 1).
In weathering profiles the transition from silcrete to adjacent clay is chemically sharp, confirming field observations of a marked boundary in terms of induration (Fig. 3). Although
there is characteristically an increase in TiO_2 in
silcrete horizons, high titanium concentrations
are also evident in weathering profile clays adjacent to silcrete units (see separate plot in Fig.
3). Indeed in Profile 11 the maximum TiO_2
concentration is attained in clay immediately
beneath silcrete. Within silcrete horizons TiO_2
varies erratically. The lowest TiO_2 concentrations in silcrete are associated with skeletal
grain-rich F-fabric silcretes overlying quartzitic
bedrock.

Mineralogy

X-ray diffraction analyses confirm petrographic observations that silica is present as
well-crystallized quartz generating sharp diffractogram peaks. Opal-CT, which occurs in
some silcretes in the Kalahari region, was not
identified in the Cape coastal zone samples.
Titanium occurs as anatase, as has been found
for Australian silcretes (Hutton *et al.* 1972;
Senior & Senior 1972). Weathering profile clay
minerals are predominantly illite and kaolinite.

Normative mineralogy

The most significant deduction from normative mineralogy calculations (Fig. 3) is the
abundance of quartz in weathering profile clays.
This is most notable in the Riversdale-

Albertinia area where clays characteristically
contain abundant clay and fine-silt sized detrital
quartz (Murray & Heckroodt 1979) although
partial induration of some weathering profile
clays recorded in the field is consistent with the
presence of authigenic silica producing incipient
silicification.

Profiles

The chemical data show that, in comparison
with associated weathering materials, silcretes,
in relative terms, are enriched in silica, generally but not invariably enriched in titanium, and
depleted in aluminium and other major elements (Table 1). Relative iron enrichment is
evident in transitional silcretes-ferricretes, or
where late-stage surface weathering has been
marked. In order to estimate absolute budgets
for element transfers weathering studies have
frequently assumed that one element, usually
aluminium or titanium, is immobile. In this
study petrographic evidence clearly indicates
the mobility of titanium. Moreover, silicification within weathering profile clays without the
removal of aluminium is not possible as the very
considerable volume expansions implied by the
required 'dilution' of Al_2O_3 is incompatible
with field evidence (Summerfield 1978).

An alternative approach employs the isovolumetric assumption (Gardner, Kheoruenromne &
Chem 1978; Hendricks & Whittig 1968). This has
previously been applied to weathering profile
clays associated with silcrete to calculate the
amount of silica released per unit volume of
bedrock during the weathering of Dwyka tillite
(Mountain 1951) and is appropriate where
there is evidence of undistorted remnant bedrock structures. Extending this to the formation
of silcrete itself is less certain, but may be justified in view of the lack of petrographic or field
evidence for substantial volume changes during
silcrete development. Moreover, Wopfner
(1978) has presented evidence for
isovolumetric silcrete formation for some
Australian occurrences.

Employing isovolumetric calculations for the
conversion of Dwyka tillite to a predominantly
kaolinite weathering material, Mountain
(1951) determined that sufficient silica could
have been provided to form the observed thickness of silcrete at Makanna's Kop, Grahamstown. Extending such isovolumetric calculations to silcrete formation at this site gives an
absolute aluminium loss of ~95% with respect
to unweathered bedrock. Silica depletion of
40–50% in kaolinite and ferricrete is com-
plemented by an absolute accumulation of

TABLE 1. *Major element bulk chemical analyses of silcretes and associated weathering profile materials normalized to 100%. Key to associated bedrock: BS—Bokkeveld Shale; DT—Dwyka tillite; KD—Karroo dolerite; PS—phyllitic shale, Nama System; WQ—Witteberg Quartzite; WS—Witteberg Shale.*

Location and Site No.	Associated Bedrock	Sample Description	Sample No.	Profile No.	Depth[†] m	SiO_2	Al_2O_3	TiO_2	Fe_2O_3	MgO	CaO	K_2O	MnO	P_2O_5	L.O.I.
Oliphants, R. (28)	?	Silcrete	C-5	-	-	96.45	0.36	1.79	0.48	-	0.06	0.02	0.01	0.04	0.79
Hopefield (33)	PS	"	C-18	-	-	97.23	0.56	0.70	0.55	0.25	0.02	-	0.01	0.03	0.65
Caledon (35)	**BS**	"	C-26	-	-	96.16	0.57	1.95	0.61	-	0.01	-	0.01	0.05	0.66
Swellendam (42)	BS	"	C-62	-	-	96.83	0.61	1.27	0.17	-	0.02	0.01	0.01	0.02	0.99
Riversdale (44)	BS	"	C-81	-	-	95.16	0.70	1.60	1.44	-	0.04	0.02	0.01	0.11	0.93
Riversdale (45)	BS	"	C-90	-	-	96.45	0.44	1.72	0.81	-	0.01	-	0.01	0.03	0.53
Riversdale (49)	BS	"	C-123	-	-	96.84	0.22	1.75	0.42	0.18	0.01	-	0.01	0.01	0.60
Riversdale (49)	BS	"	C-124	-	-	97.14	0.21	1.72	0.30	-	0.01	0.01	0.01	0.02	0.60
Riversdale (52)	BS	"	C-157	-	-	97.35	0.23	1.39	0.31	0.28	0.02	0.01	0.01	0.02	0.68
"	BS	"	C-163	-	-	95.97	0.25	1.71	0.89	0.28	0.14	0.01	0.01	0.01	0.77
Grahamstown (70)	WQ	"	C-225	-	-	98.51	0.17	0.78	0.13	-	0.01	-	-	0.02	0.37
Grahamstown (73)	WQ	"	C-234	-	-	98.03	0.20	0.37	0.46	-	0.05	-	0.01	0.01	0.88
Kentani (74)	KD	"	C-237	-	-	90.86	0.26	3.34	4.96	-	0.01	-	0.03	0.03	0.51
Heidelberg (43)	BS?	Silcrete	C-73	11	0.20	91.74	0.49	1.53	4.48	-	0.02	0.01	-	0.03	1.68
"	"	"	C-72	"	0.95	90.86	0.52	1.31	5.56	-	0.03	0.01	0.01	0.05	1.64
"	"	Ferruginous clay	C-70	"	1.75	79.50	2.72	2.92	10.53	-	0.02	0.10	0.01	0.16	4.06
"	"	clay	C-69	"	1.90	82.95	4.81	1.95	5.44	0.51	0.04	0.30	-	0.09	3.96
"	"	"	C-68	"	2.00	80.04	10.09	1.27	2.45	-	0.02	0.36	0.01	0.07	5.70
"	"	"	C-67	"	2.10	69.32	14.75	0.85	5.82	0.40	0.03	0.27	0.01	0.11	8.46
"	"	"	C-66	"	2.40	69.26	14.23	0.80	1.38	0.15	0.03	0.36	0.01	0.04	13.76
Riversdale (45)	BS	Silcrete	C-97	13	0.35	91.92	1.18	1.80	2.82	-	0.02	0.01	0.01	0.03	2.22
"	"	"	C-98	"	1.10	91.82	0.59	1.78	3.07	0.15	0.01	-	0.01	0.05	2.54
"	"	"	C-99	"	1.55	96.17	0.34	1.71	1.12	0.22	0.01	-	0.01	0.01	0.42
"	"	"	C-100	"	1.80	96.39	0.27	1.59	1.05	-	0.01	-	-	0.02	0.67
"	"	"	C-102	"	2.45	95.52	0.41	1.60	1.34	-	0.04	0.02	0.01	0.06	1.01
"	"	"	C-103	"	2.90	96.68	0.24	1.71	0.24	0.57	0.02	-	0.01	0.05	0.50
"	"	"	C-104	"	3.25	97.41	0.34	1.31	0.09	-	0.02	0.01	0.01	0.12	0.71
"	"	"	C-106	"	4.25	96.86	0.45	1.40	0.51	-	0.03	0.02	0.01	0.07	0.65
Riversdale (49)	BS	Silcrete	C-119	14	0.40	96.30	0.30	1.57	1.04	-	0.06	0.01	-	0.01	0.72
"	"	"	C-118	"	1.45	91.51	1.03	1.32	3.72	0.31	0.01	-	0.02	0.01	2.08
"	"	"	C-117	"	2.15	88.30	1.00	1.65	6.57	-	0.02	0.02	-	0.02	2.42
"	"	"	C-116	"	2.60	86.86	0.91	1.85	6.77	0.94	0.01	-	-	0.03	2.63
"	"	"	C-115	"	2.75	93.24	0.32	1.96	3.43	-	0.03	0.01	-	0.01	1.00
"	"	"	C-114	"	2.90	95.47	0.34	2.26	0.80	0.15	0.02	-	0.01	0.03	0.93
"	"	"	C-113	"	3.20	88.98	0.40	2.19	6.04	0.73	0.04	0.03	-	0.03	1.56
"	"	"	*C-112	"	3.55	91.67	1.00	2.31	2.89	0.34	0.02	0.01	0.01	0.03	1.73
"	"	"	*C-111	"	4.10	96.54	0.31	2.08	0.37	0.14	0.02	0.01	0.01	-	0.54
Albertinia (51)	BS	Clayey sand	C-145	16	6.30	91.77	5.96	0.34	0.46	-	0.07	0.28	0.01	0.03	1.08
"	"	"	*C-144	"	7.00	86.79	6.00	0.30	2.83	0.20	0.07	0.26	0.01	0.04	3.52
"	"	Silcrete	*C-143	"	7.90	95.81	1.04	1.94	0.18	-	0.03	0.01	0.01	0.03	0.98
"	"	"	C-141	"	13.65	92.88	1.36	2.81	1.91	-	0.04	0.14	-	0.04	0.82
"	"	"	C-140	"	13.90	97.24	0.26	1.19	0.10	-	0.02	0.01	0.01	0.01	1.19
"	"	Siliceous clay	C-139	"	14.00	92.33	2.02	3.36	0.36	0.24	0.03	0.14	0.01	0.01	1.51
"	"	Silcrete	C-138	"	14.15	97.50	0.34	1.39	0.14	-	0.02	0.01	0.01	0.01	0.61
"	"	"	C-137	"	14.20	96.83	0.48	1.84	0.11	-	0.02	0.01	0.01	0.03	0.68
"	"	"	C-136	"	14.30	95.59	0.28	1.59	0.14	-	0.01	0.02	0.01	0.02	2.35
"	"	" *	C-135	"	14.40	93.48	1.42	2.87	0.29	0.29	0.01	0.23	0.01	0.18	1.25
"	"	Ferruginous clay	C-134	"	14.55	76.27	3.51	2.68	10.12	0.82	0.05	0.57	-	0.10	5.81
"	"	Silty clay	C-132	"	14.70	78.51	5.45	3.46	6.85	0.54	0.06	1.17	-	0.07	3.88
"	"	Siliceous clay	C-131	"	15.20	75.89	8.24	3.43	3.00	1.30	0.04	1.82	0.01	0.05	3.53
"	"	"	C-129	"	15.50	77.74	9.86	3.42	1.36	0.56	0.05	2.07	0.01	0.06	3.61
Albertinia (60)	BS	Calcrete	C-178	20	0.60	11.06	3.90	0.14	1.01	1.59	42.42	0.34	0.01	0.05	39.48
"	"	"	C-177	"	1.40	14.97	4.23	0.27	0.93	2.50	39.18	0.40	0.02	0.06	37.44
"	"	Silcrete	C-176	"	2.45	97.01	0.40	1.76	0.10	-	0.12	-	0.01	0.03	0.58
"	"	"	C-174	"	2.85	96.30	0.53	2.41	-	-	0.01	0.01	0.01	0.03	0.72
"	"	"	C-173	"	3.45	96.63	0.69	2.54	0.09	0.43	0.03	0.01	0.01	0.03	0.54
"	"	"	C-172	"	4.05	96.21	1.09	2.56	0.13	-	0.03	0.01	0.01	0.05	0.91
Albertinia (60)	BS	Calcareous clay	C-185	21	1.00	21.31	8.08	0.24	3.99	0.94	32.03	1.07	0.16	0.07	32.14
"	"	Silcrete	C-184	"	1.20	91.35	1.03	1.93	0.30	0.29	2.07	0.06	-	0.02	2.95
"	"	"	C-183	"	1.60	93.40	0.42	2.26	0.07	-	1.74	0.01	-	0.04	2.07
"	"	Calcareous clay	C-182	"	1.95	33.02	9.21	0.57	5.07	1.50	23.78	1.31	0.24	0.04	25.26
"	"	"	C-181	"	2.25	44.45	3.34	0.99	1.06	1.30	25.14	0.38	0.01	0.02	23.32
"	"	"	C-180	"	2.30	16.67	5.43	0.20	1.52	2.13	36.97	0.66	0.01	0.03	36.40
"	"	Silcrete	C-179	"	3.00	96.64	0.26	2.30	0.06	-	0.03	0.01	0.01	0.04	0.66
Grahamstown (67)	DT	Silcrete	C-208	24	0.70	94.60	0.69	1.73	0.70	-	0.03	0.01	0.01	0.04	2.22
"	"	"	C-207	"	1.90	96.59	1.44	0.85	0.71	0.21	0.02	-	0.01	0.02	0.16
"	"	Siliceous ferricrete	C-206	"	2.30	74.21	3.34	0.92	16.52	0.12	0.17	0.03	-	0.04	4.76
"	"	Ferricrete	C-205	"	2.75	25.23	5.59	0.40	59.52	0.19	0.04	-	0.02	0.18	8.82
"	"	Ferruginous clay	C-204	"	2.95	39.03	10.85	0.65	37.16	0.30	0.05	0.07	0.01	0.16	11.73
"	"	Clay	C-203	"	3.75	57.73	19.81	0.81	5.73	-	0.05	0.06	0.01	0.03	19.66
Grahamstown (72)	WQ/S	Silcrete*	C-232	27	0.50	96.41	0.39	2.13	0.46	-	0.02	0.01	0.01	0.06	0.52
"	"	"	C-231	"	1.30	94.01	0.57	2.91	1.09	0.31	0.02	-	0.01	0.08	1.02
"	"	"	C-230	"	2.05	94.75	0.71	2.86	0.53	0.23	0.02	-	0.01	0.05	0.85
"	"	"	C-229	"	3.30	97.46	0.12	1.68	0.22	-	0.02	0.01	-	0.03	0.47
Albertinia (58)	-	Bokkeveld Shale	C-169			61.46	19.49	1.00	6.83	1.93	0.28	3.76	0.02	0.17	5.07

*indicates weakly indurated silcrete.
†depth given for profiles is from surface.

30–40% in the overlying silcrete. Titanium shows more marked, though variable, absolute enrichment of up to 120% in the silcrete. Similar trends are evident in profiles overlying Bokkeveld Shale (Profiles 11 and 16) using sample C-169 as the base analysis. Comparison of kaolinite-illite weathering profile clays with Bokkeveld Shale indicates a maximum silica yield of 55 g per 100 g of bedrock. In both profiles there is an absolute silica accumulation in silcrete ranging up to 60–75%, whereas aluminium depletion approaches 100%.

Discussion

The TiO_2/SiO_2 ratio of silcretes analysed (mean ~0.02) exceeds that of the underlying bedrock, even in the case of the relatively TiO_2-rich shaley lithologies (e.g. sample C-169, Table 1). Segregation of titanium and its precipitation as anatase in both glaebules and colloform features in intimate association with silica precipitation indicates its mobilization and concentration contemporaneously with silcrete formation. Although highly insoluble in most surface environments, at very low pH values (~ <3.75) titanium becomes increasingly more mobile as it forms $Ti(OH)_2^{2+}$ and $Ti(OH)_3^+$ complexes (Baes & Mesmer 1976, p. 150, fig. 8.1). Above a pH of ~2 quartz surfaces carry a positive charge, whereas below a pH of 4.7–6.7 TiO_2 surfaces are negatively charged and consequently in a low pH environment titanium hydroxyl complexes can be absorbed on to silica surfaces (Watts 1977).

The large-scale removal of aluminium during silcrete formation suggested by isovolumetric calculations is also consistent with the evidence provided by the behaviour of titanium for a low pH environment, at least locally, within the weathering profile since the solubility of aluminium increases markedly below pH 4. Such an acidic environment would encourage the breakdown of silicates and the release of silica, although aluminium would be relatively much more soluble, and consequently more likely to be removed from the immediate weathering environment. In addition to initial bedrock weathering, conversion of illite to kaolinite would release large quantities of silica.

The alternating banding of titanium-rich and titanium-poor zones in colloform features suggests that pH may have been fluctuating in the range 3.5–4.0 as this would provide periodic phases favouring successively titanium mobility and precipitation. Similarly, at least some glaebules are indicative of periodic precipitation of anatase and iron oxide minerals,

although their precise mode of development is unclear (Summerfield 1978; Thirty 1978, 1981).

As quantities of silica sufficient to produce the silcrete thicknesses observed in the field can apparently be generated by weathering processes within the profile (although this does not eliminate the possibility of some lateral transfer of silica) it is necessary to evaluate mechanisms whereby this silica could be retained and precipitated. The normative mineralogy calculations (Fig. 3) show that a considerable portion of the silica in the weathering profiles examined occurs as fine silt to clay size quartz, rather than being incorporated in clay minerals. Such fine quartz fragments would provide suitable substrate for the precipitation of silica moving in solution through the profile. Gradual replacement of clay minerals by silica through the removal of aluminium could occur where the pH was below 4 or, additionally, by the action of chelating agents. The precise mechanism is uncertain but may be similar to that suggested by Bisque (1962) who envisaged initial silicification of clays by precipitation of silica polymerizing the clay fraction. Introduced silica can apparently be stabilized in a continuous three-dimensional polymer, possibly through simple condensation polymerization. Continued silicification would eventually generate F-fabric silcretes where the host material contained a relative abundance of skeletal grains (>30 μm), and M-fabric silcretes in the case of host materials with a very low skeletal grain component.

Concentration of silica implies some local impeding of percolating pore waters. Additionally, the horizontal alignment of most silcrete layers, including for example, the silcrete megaglaebules located at a depth of over 5 m within the profile at Site 51, suggests some water-table control over silicification. Silica percolating down through the profile would encounter a zone of impeded drainage at the watertable. More localized zones of low permeability could also lead to silica concentration and precipitation. This model implies that silicate horizons presently at the surface could not have formed there.

Conditions favouring the development of glaebular rather than massive silcrete are uncertain, although glaebule formation implies the existence of numerous discrete foci within the host material favouring silica precipitation. The cross-cutting structures observed in some glaebular silcretes suggest a complex history possibly involving multiple phases of precipitation and dissolution (Summerfield 1978; Thiry 1978, 1981; Watts 1978).

The model outlined here suggests a fairly narrow range of environmental conditions. The field relations of outcrops together with the likely presence of restricted drainage and a relatively shallow watertable indicate minimal local relief during silcrete development. The existence of a low pH environment, at least locally, within the weathering mantle, giving rise to aluminium and titanium mobilization implies a high rate of organic activity with abundant production of humic acids (Summerfield 1978). This would be most likely under a humid tropical or sub-tropical climate, in contrast to the arid to semi-arid environment proposed for the inland silcretes of southern Africa (Summerfield 1978, 1982). A very similar model has been proposed by Wopfner (1978) for silcretes associated with kaolinitic weathering profiles in Australia.

Conclusions

Silcretes associated with weathering profiles on residual surfaces in the Cape coastal zone have formed through silicification of kaolinitic-illitic clays. Silica released during the decomposition of silicates has apparently migrated locally through the profile and been precipitated in a zone of impeded drainage in the vicinity of the watertable. A very low pH environment is implied by the concentration of titanium in silcrete and adjacent clay horizons and this would have enabled the removal of aluminium in solution. Both F- and M-fabric silcretes have developed depending on the skeletal grain content of the host material. Glaebular silcretes formed where silica precipitation was initiated at discrete foci. The geochemical and field evidence is compatible with silcrete formation in a humid tropical or sub-tropical environment of minimal local relief.

ACKNOWLEDGMENTS: I would like to thank Drs A. S. Goudie, K. Cox and B. Butler who supervised this work, which was carried out under a Natural Environment Research Council Research Studentship. Supplementary finance was kindly provided by the Marico Mineral Co. Pty who also granted access to quarry exposures. I would also like to thank Professor E. D. Mountain for assistance in the field. The diagrams were drawn by Miss Margaret Loveless.

References

BAES, C. F. (Jr) & MESMER, R. E. 1976. *The Hydrolysis of Cations.* Wiley, New York.

BISQUE, R. E. 1962. Clay polymerization in carbonate rocks: a silicification reaction defined. *Clays Clay Miner.* **9,** 365–74.

BOSAZZA, V. L. 1936. Notes on South African materials for silica refractories. *Trans. geol. Soc. S. Afr.* **39,** 465–78.

—— 1939. The silcretes and clays of the Riversdale-Mossel Bay area. *Fulcrum, Johannesburg,* **2,** 17–29.

BREWER, R. 1964. *Fabric and Mineral Analysis of Soils.* Wiley, New York.

DAY, M. J. & GOUDIE, A. S. 1977. Field assessment of rock hardness using the Schmidt Test Hammer. *Tech. Bull. Brit. geomorph. Res. Grp.* **18,** Shorter Technical Methods (II), 19–29.

FRANKEL, J. J. 1952. Silcrete near Albertinia, Cape Province. *S. Afr. J. Sci.* **49,** 173–82.

—— & KENT, L. E. 1938. Grahamstown surface quartzites (Silcretes). *Trans. geol. Soc. S. Afr.* **40,** 1–42.

GARDNER, L. R., KHEORUENROMNE, I. & CHEM, H. S. Isovolumetric geochemical investigation of a buried saprolite near Columbia, SC, U.S.A. *Geochim. cosmochim. Acta,* **42,** 417–24.

HENDRICKS, D. M. & WHITTIG, L. D. 1968. Andesite weathering. II. Geochemical changes from andesite to saprolite. *J. Soil Sci.* **19,** 147–53.

HUTTON, J. T., TWIDALE, C. R., MILNES, A. R. &

ROSSER, H. 1972. Composition and genesis of silcretes and silcrete skins from the Beda Valley, Southern Arcoona Plateau, South Australia. *J. geol. Soc. Aust.* **19,** 31–40.

LANGFORD-SMITH, T. (ed.) 1978. *Silcrete in Australia.* University of New England, Australia.

MOUNTAIN, E. D. 1946. *The Geology of an Area East of Grahamstown.* Geological Survey, Pretoria.

—— 1951. The origin of silcrete. *S. Afr. J. Sci.* **48,** 201–4.

—— 1980. Grahamstown peneplain. *Trans. geol. Soc. S. Afr.* **83,** 47–53.

MURRAY, L. J. & HECKROODT, R. O. 1979. South African kaolins. *In:* MORTLAND, M. M. & FARMER, V. C. (eds) *International Clay Conference 1978,* 601–8. Elsevier, Amsterdam.

NORRISH, K. & CHAPPELL, B. W. 1967. X-ray fluorescence spectrography. *In:* ZUSSMAN, J. (ed.) *Physical Methods in Determinative Mineralogy,* 161–214. Academic Press, London.

—— & HUTTON, J. T. 1964. Preparation of samples for analysis by X-ray fluorescent spectrography. *Divl Rep. Div. Soils CSIRO 3/64.*

SENIOR, B. R. & SENIOR, D. A. 1972. Silcrete in southwest Queensland. *Bull. Bur. Miner. Resour. Geol. Geophys. Aust.* **125,** 23–8.

SMALE, D. 1973. Silcretes and associated silica diagenesis in southern Africa and Australia. *J. sedim. Petrol.* **43,** 1077–89.

SUMMERFIELD, M. A. 1978. *The nature and origin of*

silcrete with particular reference to Southern Africa. Unpubl. D.Phil. Thesis, University of Oxford.

—— 1979. Origin and palaeoenvironmental interpretation of sarsens. *Nature*, **281**, 137–9.

—— 1981. Nature and occurrence of silcrete, southern Cape Province, South Africa. *School Geogr. Oxford Res. Pap.* **28**.

—— 1982. Distribution, nature and probable genesis of silcrete in arid and semi-arid southern Africa. *In:* YAALON, D. H. (ed.) *Aridic Soils and Geomorphic Processes, Catena Suppl.* **1**, 37–65.

—— 1983. Silcrete. *In:* GOUDIE, A. S. & PYE, K. (eds) *Chemical Sediments and Geomorphology.* Academic Press, London (in press).

THIRY, M. 1978. Silicification des sediments sablo-argileux de l'Ypresien du Sud-est du Bassin du Paris; genèse et évolution des dalles quartzitiques et silcrètes. *Bull. Bur. Réch. Géol. Minér. Sér. 2, Sect. 1.* 19–46.

—— 1981. Sédimentation continentale et alterations associées: calcitisations, ferruginisations et silicifications les Argiles Plastiques du Sparnacien du Bassin de Paris. *Mém. Sci. Géol.* **64**.

VISSER, D. J. L. 1937. The ochre deposits of the Riversdale district, Cape Province. *Bull. geol. Surv. Pretoria,* **9**.

WATTS, S. H. 1977. Major element geochemistry of silcrete from a portion of inland Australia. *Geochim. cosmochim. Acta,* **41**, 1164–7.

WATTS, S. H. 1978. A petrographic study of silcrete from inland Australia. *J. sedim. Petrol.* **48**, 987–94.

WILLIAMSON, W. O. 1957. Silicified sedimentary rocks in Australia. *Am. J. Sci.* **255**, 23–42.

WOPFNER, H. 1978. Silcretes of northern South Australia and adjacent regions. *In:* LANGFORD-SMITH, T. (ed.) *Silcrete in Australia,* 93–141. University of New England, Australia.

M. A. SUMMERFIELD, Department of Geography, University of Edinburgh, Drummond Street, Edinburgh EH8 9XP.

Pliocene channel calcrete and suspenparallel drainage in West Texas and New Mexico

C. C. Reeves (Jr)

SUMMARY: The recent discovery of two calcium carbonate-cemented drainage channels in the Pliocene Ogallala section of the Southern High Plains, Texas and New Mexico, indicates that Ogallala drainage channels were commonly cemented with calcium carbonate, due to leaching of calcrete from adjacent Ogallala sands. Parallel drainage lines near Lovington, New Mexico, originally thought to represent calcrete incisement along ancient dune swales, are now interpreted as an example of recent drainage between a suspenparallel drainage pattern in bas-relief.

The Pliocene Ogallala section extends from South Dakota south into south-western Texas (Fig. 1), forming the physiographic unit known as the High Plains. This study deals with outcrops of Ogallala rocks confined to the southernmost physiographic unit of the High Plains in West Texas and eastern New Mexico, known as the *Southern High Plains* (Fig. 1).

Terminology

Miller (1937) proposed the shortened term *suspendritic drainage* for 'suspended dendritic

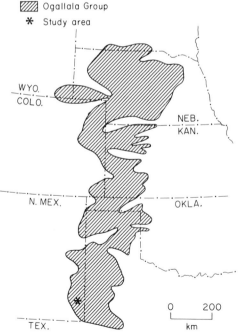

FIG. 1. Index map showing Ogallala rocks and study area.

drainage' to describe an area of relief inversion south of Darb Zubaida in eastern Saudi Arabia. Calcrete, formed in layers up to 3 m thick in an ancient dendritic drainage pattern, now caps hills and ridges over a 10 000 km^2 area. Ellis (1951) mentioned calcrete valley fills in western Australia and McLeod (1966) described an area where calcrete now forms mesa remnants, also in Western Australia. However, I was unable to find any references to any pattern of suspended drainage in the United States.

Isolated drainage channel remnants standing in bas-relief (Fig. 2) may be termed *suspended drainage*. However, use of a term like *suspendritic*, or *suspenparallel* (or any other possible combination) requires evidence of the past drainage pattern.

Southern High Plains

The Southern High Plains stands as an immense 78 000 km^2 isolated mesa-remnant of the vast Miocene-Pliocene piedmont plain formed along the east side of the Rocky Mountains. Late in Pliocene time, and later in early Pleistocene time, strong north-westerly winds scoured the Ogallala surface in south-eastern New Mexico, forming two sets of large longitudinal dunes. Later in Pleistocene time the winds shifted and strong south-westerlies stripped the dunes, distributing aeolian debris north-eastward across the Southern High Plains as the Blackwater Draw Formation. Today Ogallala rocks crop out principally along steep escarpments which define the physiographic unit known as the Southern High Plains. Most of the surface of the plains is now covered by later Pleistocene sands and lacustrine sediments.

Early (1972) satellite imagery of the Southern High Plains (Fig. 3) revealed two sets of now abandoned incised parallel drainage patterns in the Ogallala calcrete in south-eastern

179

FIG. 2. Drainage channel remnant in bas-relief in Trans-Pecos, Texas, area.

New Mexico. The oldest set, trending N45°W, is exposed in a 2500 km² area west and south of Lovington. A second but younger set, trending N65°W, covers an area of approximately 1000 km² mainly east and south of Lovington (Fig. 3).

Mapping and field study by several investigators (Price 1944; Havens 1966; Bachman 1973; Reeves 1976a) show intermittent drainages several hundred feet wide and several miles long, often with internally drained depressions (Fig. 4). The ridges are marked by hard platy calcrete which, drilling shows (Havens 1966; Reeves 1971), usually thins towards the drainages. No calcrete occurs in the drainage lines other than occasional soft vuggy masses.

Geological history

Price (1944) thought the oriented depressions around Lovington, New Mexico, resulted from etching of calcrete between ancient longitudinal dunes (Fig. 5). Havens (1966) concurred, suggesting solution of calcrete along the drainage ways (Fig. 5) during late Pliocene or early Pleistocene time. After my initial studies (Reeves 1971) I also suspected solution of the calcrete

along the drainage channels (Fig. 5). However continued investigation suggests a more complicated evolution.

A drainage channel was discovered in the lower Ogallala section near Silverton, Texas. Channel deposits of sand and gravel are tightly cemented by calcium carbonate, forming what appears to be a lenticular calcrete surrounded by Ogallala sands. A second calcium carbonate-cemented drainage channel in the Ogallala section, which is at least half a mile wide, is exposed by canyon erosion south-east of Amarillo. Again the channel is surrounded by Ogallala sands devoid of calcium carbonate.

These exposures indicate that cementation of drainageways by calcium carbonate occurred throughout Ogallala (Pliocene) time on the Southern High Plains, thus I suggest the heretofore called 'etched swales' around Lovington, New Mexico, result from relief inversion of calcium carbonate-cemented drainage channels formed along interdunal swales. Development followed the sequence shown by Fig. 6.

Dune topography with a soft young or immature calcrete (Reeves 1976a) formed during the arid to semi-arid climatic regime of late Pliocene and early Pleistocene time. Wind direction was from the north-west. Intermittent

Fig. 3. Satellite image of south-eastern New Mexico showing incised parallel drainage patterns. The star marks Lovington, New Mexico. The hachured line marks the Mescalero calcrete escarpment of the Southern High Plains. The incised parallel drainage patterns are marked by the parallel lines and, on the photo, by multiple aligned lakes particularly in the western part of the plains near the escarpment.

precipitation leached the soft powdery calcium carbonate out of the permeable dunes, the ground water then percolating into the swales where calcrete was deposited. However, by the third (Illinoian) glacial a major wind shift to the north east had occurred and the dune fields of south-eastern New Mexico were stripped by strong south-westerlies. The aeolian debris was distributed north-eastward across the Southern High Plains as the Blackwater Draw Formation (Reeves 1976c). As the sandy dunal areas were increasingly deflated the original swales, by now containing wedges of laminated calcrete from repeated wetting, drying and lateral flow of groundwater, became positive divide areas.

Palaeoclimatic and geomorphic implications

Presence of drainage channels in the Ogallala section and leaching of calcium carbonate during Ogallala time (pre-calcrete) are indicative of a moist climatic regime. However, at the end of Ogallala time multicyclic development of the massive 'caprock' calcrete implies at least semi-arid climatic conditions for an extended period of time.

Strong north-westerly winds formed the large longitudinal dunes of south-eastern New Mexico at the end of Pliocene time and with the onset of the Pleistocene glacial climate. However, by

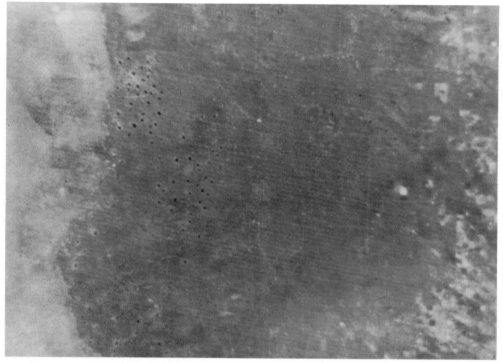

FIG. 4. Satellite image of south-eastern New Mexico showing alignment of water-filled depressions.

the third major glacial advance (Illinoian) the regional wind pattern had changed such that sandmoving winds were blowing out of the south-west desert region. This unusual wind shift, for which we (as yet) have no explanation, produced unusually high regional temperatures for the glacial period. This is reflected by a regional lack of permanent ('pluvial') lakes of Illinoian age throughouth West Texas and eastern New Mexico, formation of the aeolian sand sheet termed the Blackwater Draw Formation, and formation of a young to mature calcrete profile (Reeves 1976a) in the Blackwater Draw

section. The onset of the Wisconsin glacial period was characterized by another wind shift as strong north-westerly winds created longitudinal dunes in the recently deposited Illinoian dune debris. As the cold north-westerly winds lowered regional temperatures a profusion of so-called pluvial lakes were sustained in the area (Reeves 1976b). Termination of the glacial interval in West Texas and eastern New

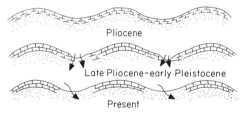

FIG. 5. Diagram illustrating possible methods of origin of aligned drainage south-eastern New Mexico.

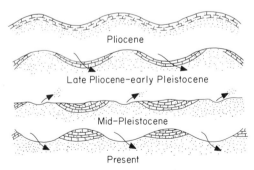

FIG. 6. Proposed sequence of development of suspenparallel drainage in south-eastern New Mexico.

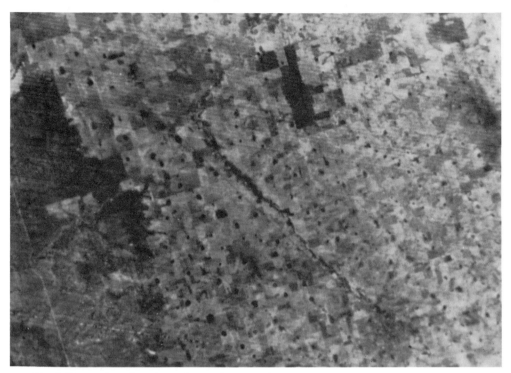

Fig. 7. The usual north-west–south-east alignment of small depressions on the Southern High Plains.
This area is 54 km east of the area discussed in this report.

Mexico was marked by a shift of sand-moving winds again out of the southwest.

Alignment of small depressions, usually with a north-west–south-east trend, occurs in several widely separated localities of the Southern High Plains (Fig. 7). Of the areas I have examined the divides are marked by hard laminated cal-crete which thins towards the depressions. Suspended drainage, and particularly *suspenparallel* drainage, is therefore more widely developed on the Southern High Plains than I originally suspected. The extent of such drainage development in other semi-arid to arid areas would also be suspected.

References

BACHMAN, G. O. 1973. Surficial features and late Cenozoic history in southeastern New Mexico. *Open-file Rep. U.S. geol. Surv.* **4339–8**, 32.

ELLIS, H. A. 1951. Report on underground water supplies in the area east of Wilana, W.A. *Am. Rep. Dep. Mines, W. Amst. 1951*, 44–7.

HAVENS, J. S. 1966. Recharge studies on the High Plains in Northern Lea County, New Mexico. *Wat.-Supply Pap. U.S. geol. Surv.* **1819-F**, F1–52.

MCLEOD, W. N. 1966. The geology and non deposits of the Hammersley Range area, Western Australia. *Bull. geol. Surv. West. Aust.* **117**.

MILLER, R. P. 1937. Drainage lines in bas-relief, *J. Geol.* **45**, 432–8.

PRICE, W. A. 1944. Greater American deserts. *Proc. Trans. Texas Acad. Sci.* **27**, 163–70.

REEVES, C. C. (Jr) 1971. Relations of caliche to small natural depressions Southern High Plains, Texas and New Mexico. *Bull. geol. Soc. Am.* **82**, 1983–8.

—— 1976a. *Geologic Atlas of Texas, Hobbs Sheet.* Bureau of Economic Geology, University of Texas, Austin.

—— 1976b. *Caliche—Origin, Classification, Morphology and Uses.* Estacado Books, Lubbock, Texas. 233 pp.

—— 1976c. Quarternary stratigraphy and geologic history of Southern High Plains, Texas and New Mexico. *In: Quarternary Stratigraphy of North America*, 312–34. Dowden, Hutchinson & Ross, Strondsburg.

C. C. REEVES (Jr), Department of Geosciences, Texas Tech. University, Lubbock, Texas 79409, USA.

Concentration of uranium and vanadium in calcretes and gypcretes

Donald Carlisle

SUMMARY: Uranium ore-bearing calcretes (in Western Australia and Namibia) are non-pedogenic. They result from lateral transport rather than vertical redistribution of components and they develop mainly in the capillary fringe along the axes of large stable drainages with low gradients under uniquely arid climates. Valley, deltaic and lake margin calcretes and dolocretes in Western Australia are part of an orderly succession from silica hardpan to playa salts. Distributions of uraniferous non-pedogenic and non-uraniferous pedogenic calcretes are mutually exclusive. Carnotite, the only ore mineral, results from: (1) evaporative concentration of U, V and/or K; (2) destabilization of uranyl carbonate complexes consequent to evaporative and common-ion precipitation of Ca/Mg carbonate; and (3) oxidation of V(IV) to V(V) in upwelling groundwater. Richest concentrations occur where groundwaters rise toward the evaporative zone. Pedogenic calcretes may inhibit ore-grade concentration. Carnotite in gypcrete is economically significant but less common.

Calcrete, dolocrete, and gypcrete host rocks for 'calcrete uranium' orebodies and prospects in Western Australia, Namibia, South Africa, and other arid regions consist of authigenic carbonate or gypsum added to alluvium, soil, or other regolith. But they are non-pedogenic. The Ca, Mg, CO_3, U and V have been added by lateral transport rather than by vertical redistribution. Nor are they evaporites in the usual sense. Uranium mineralization occurs in contiguous clay, sand, and weatherered bedrock as well as calcrete or gypcrete. Uranium and vanadium have been transported and precipitated as carnotite $[K_2(UO_2)_2(VO_4)_2 \cdot 3H_2O]$ in an entirely oxidizing environment with uranium continuously in the six-valent state.

These host rocks are nevertheless closely related to pedogenic calcrete (kunkar, croûte calcaire, caliche, nari, caprock). Their compositions and textures cover much the same range, or in some cases are indistinguishable, and their genesis, like that of all calcretes, occurs largely within the vadose zone facilitated by loss of H_2O and CO_2. But while pedogenic calcretes derive their authigenic cement only from soil, dust, loess, ash, organic debris, or the air, these uranium host rocks draw upon a large terrane of weathering rock or regolith, as well as adjacent sources. Therefore, they occur along the axes and gathering points of groundwater drainages. To become orebodies they also must have a source terrane which is large in comparison with their own size or be anomalously rich. The components must remain soluble until they reach areas of groundwater convergence where carnotite will, in fact, precipitate. As a consequence, they develop a predictable morphology and a distribution related to source terrane, near-surface groundwater flow, and the configuration of bedrock or other relatively impermeable materials. Carnotite is the sole uranium-vanadium mineral found in the deposits.

Genesis and recognition of calcrete varieties

In Fig. 1 a genetic classification of calcretes is proposed based upon relationships to subsurface water zones, depositional process and geomorphic setting. Mechanisms are indicated diagrammatically and relative favourabilities for concentration of uranium are suggested. The diagram is not meant to imply that all the varieties of calcrete form conjointly or under identical soil moisture regimes; quite the contrary as discussed below. The varieties of greatest interest here are non-pedogenic groundwater calcretes and, in particular, those labelled valley (channel), deltaic and lake margin calcrete. These form mainly in the capillary fringe directly above moving subsurface water, and in some cases slightly below the water-table. Because water-tables fluctuate and because flowing groundwater brings a continuing supply of constituents, such calcretes can be much thicker than a single pedogenic calcrete. With a shallow water-table, carbonate deposition from the capillary fringe may overlap pedogenic processes in the soil moisture zone and hybrid calcretes may result. Similarly on alluvial fans or slopes, pedogenesis is frequently combined with lateral transport. In still other places one can observe shallow non-pedogenic calcretes that

185

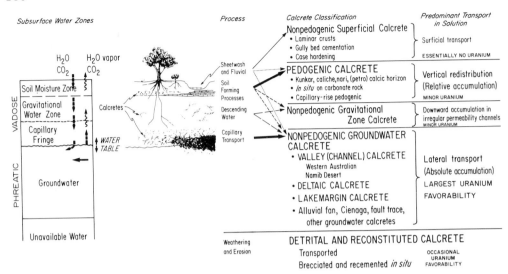

FIG. 1. A genetic classification of calcretes and their uranium favourability. A given calcrete may result from multiple processes as suggested by the arrows. Common-ion precipitation of calcite or dolomite, not shown in the diagram, is discussed in the text.

have been modified by pedogenic processes and perhaps reconstituted into pedogenic layers at the surface. But in their simplest form groundwater calcretes are diagnostically free of any profile or sequence of nodular, laminar or plugged horizons found typically in pedogenic calcretes. Their textures and structures tend rather to preserve textures and structures of the host materials.

Valley, deltaic and lake margin calcretes may differ from pedogenic calcretes in other ways: (1) they may display gradual compositional changes along their length, (2) they may have additional textures and structures related to groundwater flow and (3) they may develop optimally under conditions which do not favour pedogenic calcretes and vice versa. All of these features are well shown in Western Australia where individual uraniferous valley and deltaic calcretes are exposed over tens of kilometres (Fig. 2), where deltaic calcretes are continuing to form today (Fig. 3), and where the distributions of uraniferous non-pedogenic calcretes and non-uraniferous pedogenic calcretes are mutually exclusive (Fig. 4).

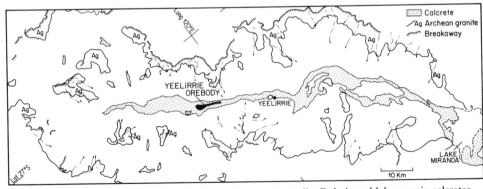

FIG. 2. Uraniferous valley calcrete, Yeelirrie, Western Australia. Deltaic and lake margin calcretes border Lake Miranda and lesser salt lakes. Geology compiled from published and unpublished sources and air photo interpretation. Orebody outline from Western Mining Corporation (1975).

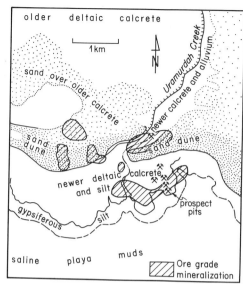

FIG. 3. Older and newer deltaic calcrete, north Lake Way, Western Australia. Geology from ground and air photo interpretation. Zones of oregrade mineralization in drill holes courtesy of Westinghouse Electric Corporation.

Uraniferous non-pedogenic calcretes in Western Australia: Yeelirrie

The valley calcrete region, morphology and age of calcretes

In Western Australia, calcrete deposits (Sofoulis 1963) are up to tens of metres thick, kilometres wide and tens of kilometres in length along the axial portions of broad extremely low-relief alluviated valleys. Comprising both calcrete and dolocrete, they cement and replace detritus which, in the uraniferous areas, is largely derived from the kaolinitic portions of laterized granitic rocks. They are earthy to aphanitic in texture but highly permeable as a result of shrinkage cracks and karstic features, especially within peculiar but abundant 'mound' structures. These non-pedogenic calcretes are limited to the 'valley calcrete region' (Fig. 4) and with rare exceptions, mainly on the edges of the region, they are not anywhere associated with primary pedogenic calcretes. The valleys, and salt lake basins in which they occur have resulted from gentle incision into the lateritic profile of the 'Old Plateau' beginning in the Tertiary, followed by alluviation during the arid trend of the last 2.5 myr

(Bowler 1976) under stable tectonic conditions. They constitute the 'New Plateau'. Today, surface water flows in the valley calcrete region only during erratic late-summer storms while temperatures and evaporation rates are very high. Ephemeral rivulets on the valley flanks infiltrate rapidly into pediments and alluvial or aeolian plains. Increasingly saline, the recharge migrates into the permeable valley calcretes which thus become prime sources of water for a unique assemblage of plants and for economic purposes. In normal years, all of the water which is not lost through evapotranspiration remains in the subsurface and eventually finds its way into clay pans or salt lakes. Gradients are from 0.5% to essentially flat. The calcretes form preferentially where converging drainages, flattening of gradients or sub-surface constrictions bring the water-table closer to the evaporative surface. On deltas and lake margins an added effect is created by less permeable lacustrine clays and hypersaline waters which help to force less saline waters upward.

Based on drilling data, *in situ* calcretes occur only in the uppermost or younger parts of the valley fill (Fig. 5a). The calcrete that hosts the orebody at Yeelirrie, for example, is a single earthy to massive but irregularly lenticular layer within the upper 4–20 m of alluvial, colluvial and aeolian sediments seldom more than 30 m thick near the orebody but at least 85 m thick in salt lakes down-drainage (Western Mining Company 1975). It grades into, but does not interfinger with, adjacent channel sediments. Nor is there any evidence that calcrete has migrated upward with sedimentation. It is a fairly recent phenomenon. Nevertheless, this and most other valley, deltaic and lake margin calcretes are obviously mature. They commonly stand 1–2 m above surrounding plains, are themselves slightly eroded and typically contain post-depositional solution cavities or collapse features. In some deltaic areas this 'older' calcrete is overlain by lake-bordering dunes which are in turn transected by active drainages leading into newly forming calcrete deltas (Fig. 3). Centres of calcrete and dolocrete development on such newer deltas ('young mounds' on Fig. 7b) show all stages of development from incipient cementation and replacement of mud to moderately well developed calcrete mounds. Therefore, given an age of 2.5 myr for beginning alluviation of valleys on the New Plateau, these relations suggest (Carlisle *et al.* 1978) that valley calcrete deposition began less than 0.5 myr ago, perhaps during the major arid phase starting at 25 000 BP (Bowler 1976). It apparently was interrupted during an extremely

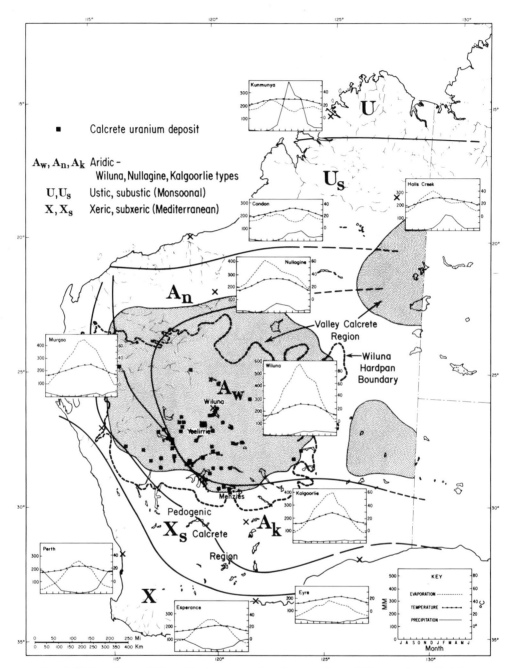

FIG. 4. Valley calcrete, Wiluna Hardpan and soil moisture regimes of Western Australia inferred from climadiagrams. Non-pedogenic valley, Deltaic and lake margin calcretes, siliceous Wiluna Hardpan and Mulga-dominant plant community all fall within the tropical cyclone belt and predominantly within a specific aridic regime (Aw) characterized by erratic late summer storms. Boundaries of valley calcrete region and Wiluna Hardpan largely after Sanders (1973) and Bettenay & Churchward (1974) respectively. Climadiagrams assembled from data supplied by Australian Bureau of Meteorology.

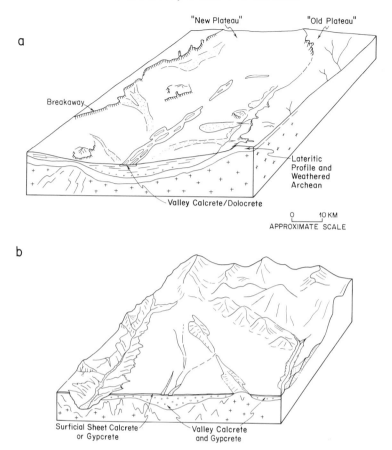

FIG. 5. Valley calcrete morphology. (a) Valley calcrete, Western Australia. (b) Valley calcrete/gyp-crete and surficial calcrete/gypcrete, Namib Desert.

arid dune-building interval and has resumed in the present day. Radiometric disequilibrium (Lively *et al.* 1979; Dickson & Fisher 1980) and carbon isotope studies (Mann & Horowitz 1979) are compatible with this inference.

Ore-bearing calcrete at Yeelirrie

At Yeelirrie (Fig. 2) calcrete crops out along the axis of the palaeochannel 30 km down-drainage from the headwater divide. Anom-alous to strongly anomalous radioactivity appears after another 15 km, the calcrete progressively widens to 5 km, and at 50 km from the divide the upper end of the ore is reached. The orebody itself, roughly 6 km long, 0.5 km wide and 8 m thick, contains 46,000 tonnes of U_3O_8 at an average tenor of 0.15%. Toward the lower end of the orebody the calcrete narrows to 2 km then con-

tinues irregularly past minor salt lakes with deltaic and lake margin calcrete and only traces of carnotite to salt Lake Miranda for a total length of 145 km. The Western Mining Corpo-ration staff (1975) has recognized two varieties of calcrete: (1) 'earthy', i.e. friable, brownish variably mature calcrete containing remnants of soil or alluvium and inherited sedimentary layering, and (2) 'porcellaneous', i.e. dense, mature, white fine-grained dolocrete without relict layering but with abundant dehydration-solution cracks and cavities and some variably replaced soil peds. The porcellaneous variety occurs as isolated domes, 'platforms' or 'mounds' surrounded by earthy calcrete (Fig. 6). These structures are extremely common in Western Australian valley calcretes (Mabbutt *et al.* 1963; Sanders 1973) and are thought by the author to represent loci of upwelling ground-water analogous to carbonate domes found around the shorelines of salt lakes in western

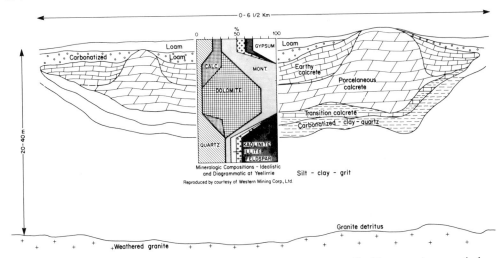

FIG. 6. Diagrammatic cross-section of valley calcrete, Western Australia. Note very large vertical exaggeration. Mineralogical composition insert and format based on drilling data from Western Mining Corporation (1975) courtesy of the authors.

United States and elsewhere. The less competent earthy calcrete appears to have compacted differentially around the mounds. A very interesting hypothesis (Mann & Horwitz 1979) suggesting substantial upward thrusting of the mounds due to displacive calcrete growth in the phreatic zone has not yet been verified to the author's knowledge by drilling data nor by the shallow excavations now available.

The conspicuous narrowing of the calcrete at the lower end of the orebody marks a constriction in the subsurface drainage. At that point the total valley width is reduced from some 45 km above the ore to about 18 km between bedrock granitic highs (Fig. 2) and the valley gradient decreases to 0.02%. The ore occurs, therefore, where a subsurface constriction impedes drainage and forces the water table closer to the evaporative surface (Fig. 7a). This relationship is causal and applies to all valley calcrete orebodies in Western Australia and Namibia. At lake margins, analogous barriers are created by low-permeability clays and hypersaline waters (Fig. 7b).

The calcrete-free drainage catchment above the ore zone is 3000 km^2 which is 20 times the area occupied by calcrete and 1000 times the area underlain by ore. Similar morphologies and relationships are characteristic of other less uraniferous valley calcretes in the region.

Carnotite occurs in both earthy and porcellaneous calcrete and also in subjacent quartzose clay. Most noticeably it rims cracks and cavities and in turn is coated with a chalcedony crust 0.1 mm thick. But this clearly epigenetic texture is predominantly, if not entirely, a result of carnotite remobilization. Supporting evidence comes from the newer calcrete deltas adjacent to salt lakes nearby (Figs 3 & 7b). Here young domes of immature muddy calcrete can be observed in the process of growth and many are anomalously radioactive. Under the microscope, clearly authigenic carnotite crystals up to 20 µm across are intimately associated with newly formed carbonate and are, with little question, contemporaneous with it.

The valley calcrete succession; groundwater chemistry; precipitation of carbonate and uranium

Throughout the valley calcrete region, calcrete, with or without carnotite, is only one component in a succession of authigenic near-surface deposits (Fig. 7) from: (1) silica hardpan (duripan) on the flanks of the basins in acidic non-saline soils devoid of calcrete to (2) calcrete in the trunk palaeodrainages with minor chalcedonic quartz, celestite and gypsum and increasingly dolomitic downdrainage, to (3) dolomitic and gypsiferous silts on lake margins, and finally (4) salines within the playa muds. Analyses of existent groundwaters by Dall'aglio, Gragnani & Locardi (1974) and Mann & Deutscher (1978) show that, except for irregularities due largely to dissolution of

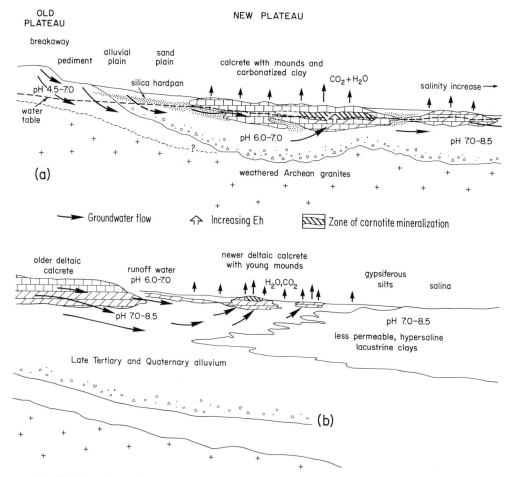

FIG. 7. Schematic models for non-pedogenic, uranium-bearing calcrete, Western Australia. (a) Valley calcrete with bedrock constriction. (b) Newer deltaic calcrete-salt lake model. Breakaways and pediments are largely within the mottled and pallid (kaolinitic) zones of the lateritic profile and in weathered Archean granitic rocks. Older valley and deltaic calcretes and included carnotite are subject to dissolution. After Carlisle *et al.* (1978); pH data after Mann (1974).

calcrete, the total salinities, pH and concentrations of Ca, Mg, Na, K, U, V, SO_4 and Cl increase downdrainage. Uranium reaches values of several hundred ppb. Total carbonate, as 'HCO_3' tends to increase before reaching the calcrete, then remains more or less steady while rising $[Ca^{2+}]$ and $[Mg^{2+}]$ promote carbonate deposition, and ultimately decreases markedly at salt lake margins. At that point the waters encounter hypersaline salt lake brines also enriched in Ca^{2+} and Mg^{2+}.

Deposition of calcrete, therefore, results not simply from: (1) loss of CO_2 and concomitant increase in pH and CO_3^{2-}/HCO_3^-, but equally or more so from (2) evaporative concentration of Ca^{2+} and Mg^{2+} along the course of the subsurface drainage and (3) from common-ion precipitation where Ca/Mg carbonate bearing waters encounter Ca, Mg sulphate or chloride brines in salt lakes or in other local highly evaporative domains. These latter two mechanisms are significant because they permit calcrete or dolomite to precipitate with ever-diminishing $a_{CO_3^{2-}}$, in essence stripping major proportions of the CO_3^{2-} from solution. This is critical to carnotite precipitation.

In carbonated groundwater uranium is transported as extremely soluble uranyl dicarbonate and tricarbonate complex ions subject to the reactions:

$$UO_2(CO_3)_2 \cdot 2H_2O^{2-} = UO_2^{2+} + 2CO_3^{2-} + 2H_2O \quad (1)$$
$$UO_2(CO_3)_3^{4-} = UO_2^{2+} + 3CO_3^{2-} \quad (2)$$

and carnotite is precipitated according to:

$$2UO_2^{2+} + 2H_2VO_4^- + 2K^+ + 3H_2O =$$
$$K_2(UO_2)_2V_2O_8 \cdot 3H_2O + 4H^+. \quad (3)$$

In a solution containing *only* the above components, removal of CO_2 drives all these equations to the right and decreases the solubility of carnotite. With $CaCO_3$ present, however, some amount of CO_3^{2-}, HCO_3^-, or H_2CO_3 is always present in solution and available to form uranyl carbonate complexes. To precipitate carnotite in this system it is necessary to first destabilize the uranyl carbonate complexes by lowering $a_{CO_3^{2-}}$. Removal of CO_2, while making $CaCO_3$ less soluble, increases pH and the ratio CO_3^{2-}/HCO_3^-. Thus depending upon the magnitude of the pH change which accompanies the CO_2 loss and probably other factors, carnotite may become either more or less, but very likely more, soluble. On the other hand, evaporative or common-ion precipitation of carbonate results in lowered $a_{CO_3^{2-}}$ with consequent destabilization of uranyl carbonate complexes and eventual precipitation of carnotite along with the carbonate.

Two other mechanisms contributing to carnotite precipitation are likely in the Western Australian environment: (1) reaction of U-V bearing groundwaters with K^+ enriched hypersaline waters in salt lakes or elsewhere and (2) as emphasized by Mann & Deutscher (1978), oxidation of four-valent vanadium in upwelling groundwater (Fig. 7a, b) to the five-valent state required for carnotite. In every case uranium remains in the oxidized or hexavalent state throughout and reduction plays no role in its deposition.

Near-contemporaneous precipitation of calcrete and carnotite by any of the above mechanisms may not in itself yield ore, but ore grades may result from the dissolution and reconcentration of carnotite indicated by epigenetic textures at Yeelirrie and by groundwater and uranium-series analyses (Mann & Deutscher 1978; Lively *et al.* 1979; Dickson & Fisher 1980). On the other hand, deposition of extensive sheets of pedogenic calcrete may well impede ore concentration by fixing carnotite in widely dispersed non-economic form.

A rationale for the distribution of uraniferous non-pedogenic calcrete in Western Australia

Non-pedogenic calcretes occur within the valley calcrete region and they contain significant carnotite, whereas pedogenic calcretes lacking significant carnotite occur to the south. Both types of calcrete are forming today.

The valley calcrete region is not coincident with the distribution of salt lakes, i.e. Salinaland, nor with the full extent of the New Plateau, nor a particular bedrock. It includes part of the Archean Yilgarn Block, Proterozoic rocks of the Gascoyne, Bangemall and Hamersley regions and some Palaeozoic and Cretaceous sedimentary rocks. It does coincide in a remarkable way, however (Fig. 4, Table 1) with: (1) a uniquely arid climate and corresponding soil moisture regime, (2) the distribution of a soil containing abundant authigenic silica, the Wiluna Hardpan, and (3) Mulga, a characteristic plant assemblage. The following explanation is suggested.

Within the valley calcrete region, rain from sporadic summer storms falls on hot, dry surfaces. Much evaporates directly and a large part infiltrates rapidly into fractured, permeable mottled or pallid zone (sublateritic) rocks on breakaways and pediments or into opentextured earthy loams. Very little remains in the soil moisture zone subject to evapotranspiration and loss of CO_2. In addition, vapour transport is toward cooler soils at depth. Consequently there is little opportunity for pedogenic accumulation of carbonate. Silica, released from the lateritic profile is continually at the point of saturation and precipitates as water evaporates in deeper soil. What little carbonate might form is dissolved away in the next storm but silica is not. Soil salinity and pH remain low on upper valley flanks rising gradually with evaporation to the point of carbonate saturation along valley axes. Similarly there is little tendency for carnotite to precipitate in the well-drained acidic soils. The highly soluble uranyl ion or, as pH rises, dicarbonate and tricarbonate complexes, are free to migrate with groundwater toward the calcreted valley axis. The entire succession of authigenic deposits from hardpan to calcrete, gypsum, and playa salts is a product of the climate, the consequent soil moisture regime, and evaporative concentration in a flowing reservoir of subsurface water. South of the valley calcrete region, on the other hand, large fractions of the rainfall occur during the winter when evaporation rates and temperatures are comparatively low but adequate, nevertheless, to cause a buildup of salts within the soil. Carbonate tends to precipitate in upper layers primarily by loss of CO_2; pH values rise toward a theoretical maximum of 9.9 and silica remains soluble even with significant evapotranspiration. Uranium, if present in the soil, may precipitate as carnotite in associ-

TABLE 1. *Contrasts between uraniferous and non-uraniferous calcrete regions, Western Australia*

Valley calcrete region	Generally south of valley calcrete
Annual rainfall 170–250 mm. Highly variable, episodic late summer thunderstorms, sporadic tropical cyclones	200–500+ mm. Predominantly winter rainfall from anti-cyclonic frontal rains or indefinite rain season
Annual potential evaporation (E_a): 3300–4200 mm	<3300 mm
Ratio E_a/P_a: 12–20	<6–16
Water balance E_a–P_a: >3000 mm	<3000 mm
Temperature; mean annual: >19° C	<19° C
Soil moisture regime: Strongly aridic and distinctive. Moderate to severe drought incidence	Aridic to Xeric. Moderate to low drought incidence
Soils: Acidic to earthy loams with siliceous Wiluna Hardpan. Calcareous on or near calcrete	Alkaline calcareous earths common. Neutral, alkaline, to acidic in humid south-west
Vegetation: Mulga (*Acacia aneura*) dominant	Mallee habit of eucalyptus dominant
Groundwater: Potable high on drainages, saline downdraingage	Extensively saline and non-potable in wells or bores

ation with pedogenic calcrete without opportunity for lateral concentration. Where observed, it is very sparsely disseminated and typically in the lower part of the carbonate.

Uraniferous non-pedogenic calcretes and gypcretes in Namibia

Carnotite-bearing valley calcrete and lesser gypcrete occur along palaeochannels on the Namib Desert platform from the Atlantic coast to the foot of the Great Escarpment 100 km inland (Fig. 5b). These host rocks differ from the Australian calcretes, notably in their late Tertiary age, the abundance of coarse detrital fragments which constitute the skeletal framework of the calcrete, the fact that carbonate is predominantly sparry calcite with neither dolomite nor opal, the presence of abundant gypsum in the upper parts of occurrences within 60 km of the Atlantic, and in the vertical gradation into Pleistocene to Holocene pedogenic sheet calcrete and gypcrete. The calcrete does not develop abundant dehydration or shrinkage cracks as does the Australian calcrete, and it is not a good aquifer, though it is appreciably permeable. In the Langer Heinrich orebody irregular solution tubules and replacement bodies about the size of a finger but sometimes much larger have developed. Commonly these have a shell of impure calcite and a sandy filling, and, under some conditions, the sandy fillings of these tubules are richly mineralized with carnotite.

Gypcrete with the same kind of aggregate as the calcrete occurs directly above valley calcrete within the belt of almost ever-present fogs from the Atlantic Ocean. Thicknesses up to 25 m have been drilled in palaeochannels on the Tumas River where a potential orebody occurs beneath Holocene surficial gypcrete. The SO_4 is from marine mist. The uranium is probably derived from uraniferous calcrete updrainage and is reconcentrated as carnotite in a sparsely gypcreted sand facies adjacent to a palaeowater-table (Carlisle 1980). An ubiquitous overlying layer of surficial gypcrete or, inland from the fog limit, surficial calcrete, covers the entire Namib Desert except only where masked by recent wind-blown sand or alluvium or where bedrock is exposed. Carnotite is found only very rarely in this superficial gypcrete or calcrete, a consequence mainly of capillary rise from subjacent mineralization.

In spite of the differences and much greater difficulty in delineating valley calcretes and gypcretes in Namibia, numerous examples support the conclusions that lateral transport of uranium in groundwater is essential to ore deposition and that bedrock barriers or constrictions which narrow the channel of subsurface flow, and thus force the water closer to the evaporative surface greatly favour the formation of uranium deposits. Carnotite precipitation mechanisms are thought to be similar to those postulated for Western Australia. Ore-grade mineralization is roughly horizontal and probably related to past or present groundwater tables (Carlisle *et al.* 1978). Source rocks include Proterozoic migmatites, pegmatites, alaskites, and granites, many appreciably anomalous in uranium.

Climate and pedogenic-non-pedogenic calcrete distribution

We must ask if any or all of the climatic parameters separating pedogenic from non-pedogenic calcrete regions in Western Australia (Fig. 4) apply elsewhere. Aridity, measured by P_a, E_a, E_a/P_a, E_a-P_a and mean temperature is substantially more severe in the valley calcrete region (Table 1). But, as suggested earlier, the most significant factor may be differences in the seasonal distribution of rainfall and consequent retention of water in the soil moisture zone. Is summer-only rainfall therefore prerequisite for valley calcrete? A large region in western South Africa and southern Namibia, largely inland from the Great Escarpment and centred upon Upington has a present climate and inferred soil moisture regime almost identical with that of the valley calcrete region of Western Australia (Carlisle 1980). It contains abundant remnants of late Tertiary to Pleistocene pedogenic calcrete including great stretches of 'Kalahari Limestone'. But with rare exceptions attributable to especially favourable circumstances, pedogenic calcretes are not forming in that

climatic region today. Instead non-pedogenic calcretes of Pleistocene(?) to Holocene age are found in shallow valleys eroded into the older calcrete. In some of these in constricted parts of the valleys, there are concentrations of carnotite, none as yet economic. Although these non-pedogenic calcretes consist very largely of eroded and recemented fragments of older pedogenic calcrete, the climatic parameters for pedogenesis vs. non-pedogenesis would seem to apply.

The same climatic parameters need not apply to valley calcrete on the extremely arid Namib Desert. Calcreted palaeochannels indicate that the carbonate-bearing groundwaters had their origin in the Great Escarpment and the Khomas Highland where relief and rejuvenation would have precluded, and still preclude, calcrete deposition. Migrating into alluviated African or post-African valleys on the gently sloping Namib platform, those groundwaters have progressively calcified the entire valley fill even while pedogenic calcrete may have been accumulating on the overlying desert surface. The evaporative succession characteristic of Western Australian valley calcretes has been distorted by the topography and marine mist, yet carnotite ores have developed.

ACKNOWLEDGMENTS: Field investigations have been supported by the U.S. Department of Energy and the University of California Los Angeles. Access to operations and unpublished data have been kindly provided by several mining corporations including Western Mining Corporation, Ltd (Yeelirrie), General Mining & Finance Corporation, Ltd (Langer Heinrich) and Anglo-American Corporation, South Africa, Ltd (Tumas River), and also by governmental entities in Australia and South Africa. The author has benefitted especially from field guidance and discussions with A. W. Mann and others of the CSIRO, Australia; E. Cameron, F. Netterberg and a great many government and mining geologists in Australia, South Africa And Namibia.

References

BETTENAY, E. & CHURCHWARD, H. M. 1974. Morphology and stratigraphic relationships of the Wiluna Hardpan in arid Western Australia. *J. geol. Soc. Aust.* **21**, 73–80.

BOWLER, J. N. 1976. Aridity in Australia. Age, origins and expression in aeolian landforms and sediments. *Earth Sci. Rev.* **12**, 279–310.

CARLISLE, D. 1980. Possible variations on the calcrete-gypcrete uranium model. *Open-file Rep. U.S. Dept Energy,* GJBX-53(80), 38 pp.

——, MERIFIELD, P. M., ORME, A. R., KOHL, M. S. & KOLKER, O. 1978. The distribution of calcretes and gypcretes in southwestern United States and

their uranium favorability based on a study of deposits in Western Australia and South West Africa (Namibia). *Open-file Rep. U.S. Dept Energy*, GJBX-29 (78), 274 pp.

DALL'AGLIO, M., GRAGNANI, R. & LOCARDI, E. 1974. Geochemical factors controlling the formation of the secondary minerals of uranium. *In: I.A.E.A., Proc. Symp. Formation of uranium ore deposits. Athens,* May 6–10, 1974.

DICKSON, B. L. & FISHER, N. I. 1980. Radiometric disequilibrium analysis of the carnotite uranium deposit at Yeelirrie, Western Australia. *Proc. Australas. Inst. Metall. No. 273*, 13–9.

LIVELY, R. S., HARMON, R. S., LEVINSON, A. A. & BLAND, C. J. 1979. Disequilibrium in the ^{238}uranium series in samples from Yeelirrie, Western Australia. *J. geochem. Expl.* **12,** 57–65.

MABBUTT, J. A., LITCHFIELD, W. H., SPECK, N. H., SOFOULIS, J., WILCOX, D. G., ARNOLD, J. M., BROOKFIELD, M. & WRIGHT, R. L. 1963. General report on lands of the Wiluna-Meekatharra area, Western Australia, 1958. *Aust. CSIRO Land Res. Ser. No. 7.*

MANN, A. W. 1974. Chemical ore genesis models for the precipitation of carnotite in calcrete, Perth, Western Australia, *CSIRO Miner. Res. Labs, Div. Mineral. Rep. FP7,* 18pp.

—— & DEUTSCHER, R. L. 1978. Genesis principles for the precipitation of carnotite in calcrete drainages in Western Australia. *Econ. Geol.* **73,** 1724–37.

—— & HORWITZ, R. C. 1979. Groundwater calcrete deposits in Australia. Some observations from Western Australia. *J. geol. Soc. Aust.* **26,** 293–303.

SANDERS, C. C. 1973. Hydrogeology of a calcrete deposit on Paroo Station, Wiluna, and surrounding areas. *Western Australia geol. Surv. Ann. Rep. 1972* 15–26.

SOFOULIS, J. 1963. The occurrence and hydrological significance of calcrete deposits in Western Australia. *Western Australia geol. Surv. Ann. Rep. 1962,* 38–42.

WESTERN MINING CORPORATION Exploration Division 1975. *A general account of the Yeelirrie uranium deposit.* Privately published and distributed to 25th int. Geol. Cong. tours, 14 pp.

DONALD CARLISLE, Department of Earth and Space Sciences, University of California, Los Angeles, Los Angeles, California 90024, USA.

Ancient duricrusts and related rocks in perspective: a contribution from the Old Red Sandstone

John Parnell

SUMMARY: Calcretes are abundant in continental Devonian (Old Red Sandstone) rocks, occurring both within the sequence and at the basal unconformity. Basal limestones in the Orcadian Basin also include algal deposits and fissure linings, all of which can be considered as a single phase of carbonate precipitation and growth. A prominent silcrete in Orkney replaces both rudites and chert-bearing lacustrine laminites. Silcrete and calcrete are examined in a wider context of chemical sedimentation in lacustrine rocks. The effects of weathering processes in these rocks may be combined or confused with those of algal processes. A silicified plant-bearing sequence at Rhynie displays the same diagenetic features as the Orkney silcrete. Limited metalliferous and bituminous enrichments are recorded in the largest ORS duricrusts. A basal sequence at Dalroy exemplifies the possible confusion between weathering concentrations and mineral deposits trapped at an unconformity surface. Barite is a common component of some duricrusts. Duricrust formation is part of a range of sedimentary replacement and cementation phenomena which may be difficult to distinguish in ancient rocks.

The interpretation of continental Devonian (Old Red Sandstone: ORS) carbonate rocks as calcretes was first proposed by Maufe (1910), who compared some Scottish nodular carbonates with the kankar of India and Central Africa. This comparison was invoked intermittently until Burgess (1961) and Allen (1965) drew more detailed analogies beween ORS carbonates and calcretes of the present day. A wide variety of ORS carbonates have been interpreted in this way, ranging from vein meshworks to massive limestones (Allen 1965, 1974a,b). Duricrusts are also represented by silcretes and ferricretes.

The calcretes of England and Wales have been described by Allen (1965, 1973, 1974a,b). Most of the discussion herein concerns calcretes and silcretes examined in the ORS of Scotland, particularly in the Orcadian Basin. The ORS calcretes in Britain are developed largely in fluvial facies of Lower ORS and Upper ORS age. The Middle ORS is represented almost solely in the Orcadian Basin (Fig. 1), a thick (up to 6 km) sequence of fluvial and lacustrine rocks on the northern Scottish mainland and in the Orkney islands. Chemical sedimentation was relatively important in the lacustrine environment, and there are many varieties of primary or secondary carbonate and silica deposit in the Orcadian Basin. Some of these deposits can be readily identified as duricrusts, some clearly do not belong to this category, and others are of uncertain affinity. The range of deposits which might be labelled as duricrusts could be varied by adopting different terminologies. This account summarizes the range of sediments in the Orcadian Basin which

are cemented or replaced by carbonate or silica, shows how they relate to each other and emphasizes that duricrust formation is not an isolated phenomenon. Massive and nodular calcretes are compared with other terrigenous carbonates, and silcretes are similarly compared with other siliceous deposits. Mineralization and organic matter within the deposits are discussed to point out the difficulty in assessing the role of duricrusts in their concentration. The problems in classification and interpretation are intensified in this Palaeozoic basin by the exposures available, which are extremely limited when compared to the widely studied duricrusts in the arid regions of the present day, such as Australia and southern Africa.

The nature of British ORS calcretes

The most primitive calcretes to be found in the ORS of Britain are clastic sediments with vague impregnations of carbonate, which grade progressively into rocks in which the carbonate becomes more and more important. Interstitial precipitation of carbonate was of prime importance and led to both displacement and replacement of the detrital grains. Simple nodules and veins of relatively pure carbonate are common forms of ORS calcrete but there are also many notable larger structures. Vertical columnar structures (Fig. 1), which rarely show evidence of plant roots, grew up to 1 m high but only a few centimetres wide and developed into massive limestones. These are concretionary or laminated, the lamination reflecting not bedding but differences in crystallinity produced during growth. The most prominent calcretes are useful as stratigraphic marker horizons, including the Lower ORS Pittendriech Limestone of the Midland Valley Basin in Scotland (Armstrong & Paterson 1970) and the Lower ORS 'Psammosteus' Limestone in the Anglo-Welsh outcrop (White 1946). The Upper ORS in the Midland Valley Basin is capped everywhere by a thick calcrete development. In Kintyre (McCallien 1927), Kinross (Geikie 1900) and Roxburghshire (Leeder 1976) this calcrete is overlain by lavas or dark shales and clearly represents an important stage of non-deposition. Elsewhere the transition into Carboniferous rocks is less clear: in Berwickshire, Stirlingshire and the Firth of Clyde, calcretes in ORS facies are found alternating with Carboniferous lithologies. But even within these sequences, the most important horizons can be locally correlated and sometimes used to define an arbitrary division between the systems.

Calcretes and other carbonates at the basin margin

In the Orcadian Basin (see Fig. 2), calcretes are found not only within the ORS sequence but also at its unconformity with the underlying crystalline basement. A good example may be seen at Buckie, where basal breccias are included in a succession of about 6 m of limestone resting on Dalradian quartzite. The limestone passes up into sandstones which are unconformably overlain by Middle ORS conglomerates. The lowermost limestone shows concretionary and dome structures (Fig. 3). Breccias and sands have a floating texture within the limestone and clasts show evidence of corrosion. The laminations in the concretionary phase mark alternating layers of micrite and fine sparite. Much of the lamination has a clotted fabric, pseudo-oolitic in parts. Pisolitic structures occur around clasts or vugs of sparry calcite. Irregular calcrete

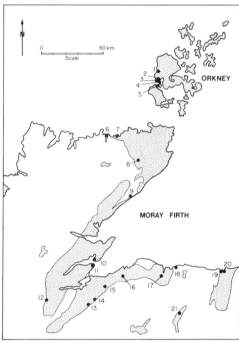

FIG. 2. Map of northern Scotland showing the localities in the Orcadian Basin referred to in text. Stippled area is Old Red Sandstone. Localities: (1) Garthna Geo, (2) Stromness Reservoir, (3) Ness, (4) Whitehouse, (5) Graemsay, (6) Baligill, (7) Red Point, (8) Dirlot, (9) Allt Briste, (10) Nigg, (11) Cromarty, (12) Aigas, (13) Dalroy, (14) Balfreish, (15) Lethen, (16) Cothall, (17) Dipple, (18) Buckie, (19) Gamrie, (20) Crovie, and (21) Rhynie.

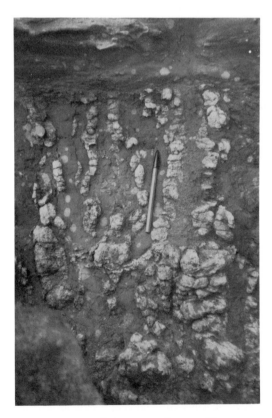

FIG. 1. Columnar calcrete, Upper ORS–Carboniferous transition zone, Millport, Buteshire (scale = 15 cm).

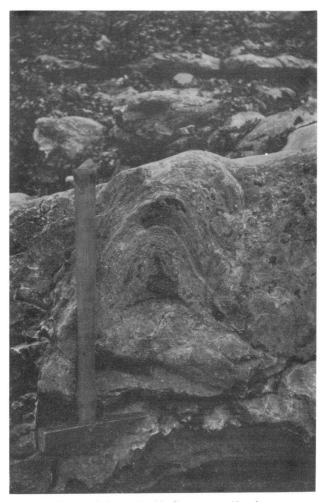

FIG. 3. Calcrete, Buckie (hammer = 40 cm).

fragments are mixed amongst the clasts. Displacive growth exploded micas and even shattered quartzite clasts, separating the fragments by up to 1 cm.

Many types of limestone other than calcrete are exposed at the basin margin. The most prominent, at Balfreish, Red Point and Baligill, contain relatively little detrital sediment. That at Balfreish is oolitic in parts, the ooliths being of mechanical origin (Parnell 1981). The detrital grains that are present have suffered displacive and replacive calcite growth. The formation of the massive marginal limestones was to some degree controlled by algae (Donovan 1975). Some show a direct transition into algal laminites.

At Red Point, isolated pebbles have oncoloid coatings and vertical laminar calcite lines fractures up to 2 cm wide in both limestone and underlying basement. Several other occurrences of oncolites in marginal sediments are known from Orkney (Parnell 1981) and Caithness (Donovan 1973, 1975). Fracture-linings also penetrate the basement at Dirlot and Garthna Geo. Across the outcrop at Garthna Geo, fracture fills in gneissic basement merge into a bed of oncoloid structures. Many of the clasts are veined rather than coated with laminar carbonate and there is no doubt that they were fractured and filled *in situ*. The carbonate extended the fractures by impregnation into the walls and the subsequent spalling off of thin slivers of basement rock. The silicate grains in the fractures have a markedly floating texture and carbonate growth may have been responsible for much of the fill. Hand specimens display what appear to be abundant lateral crack-fills, which are seen in section to result from replacement by carbonate, particularly along mica-rich layers in the gneiss. These pseudocracks, together with cross-cutting displacements, completely brecciated the basement rock in places. Sparry carbonate almost completely obliterated patches of granite-gneiss, both about the fractures and within the oncoloid rudites.

The presence of ooliths in the limestone at Balfreish suggest an environment of higher energy than is at first apparent. Evidence of extensive calcite growth supports this suggestion that clastic sedimentation was quite important though it is now obscured by the carbonate. The evidence of growth also invites comparison with calcretes in which displacement and replacement by calcite are common phenomena (N. Watts 1978). The massive carbonates lack the larger concretionary structures typical of ORS calcretes, except at Buckie, where the lower part of the limestone sequence is clearly a calcrete. The upper part appears similar to the normal massive carbonates. Vuggy structures within the Buckie calcrete are identical in petrology and pattern to those from other marginal limestones.

Several categories of carbonate precipitation seem to have been involved, although the divisions between them are not always as important as the categorization implies. All the processes may involve algae to a greater or lesser degree. The oncolites are accepted as algal by definition, but the origin of the fissure fillings with which they are associated is less obvious. Fannin (1969) and Donovan (1973) both referred them to 'tufa', a term which has no precise meaning, but is often used to describe supposedly inorganic carbonates deposited from intermittent waters. Several authors have, however, presented evidence that algae do influence the formation of such carbonates (e.g. Scholl & Taft 1964). The observed relationship between oncolites and *in situ* fissure linings can be paralleled by the cementation *in situ* of intertidal conglomerates in New Caledonia (Nelson & Rodgers 1969), where cobbles are coated with laminated micritic crusts that contain algal filaments. The clasts are only coated where they are not in contact, implying an autochthonous origin for the crusts. Nevertheless, Nelson & Rodgers thought it conceivable that limited coatings were formed before deposition. Calcretes too are increasingly being assigned an algal or bacterial component (Krumbein & Giele 1979; Wright 1981) and have been labelled as subaerial stromatolites. Even ooliths may contain significant quantities of algal filaments (Newell, Purdie & Imbrie 1960).

The intimate relationship between these categories of carbonate suggests that they can all be related to a single broad model of carbonate precipitation. In regional terms, their environment of formation was similar, and once deposited the carbonates all grew in the cemented rock in the same manner. The uncertainty in how to label these carbonates is manifest in such descriptions as 'travertine tufas occur as stromatolitic mounds' (Smoot, in Matter & Tucker 1978). Terms like 'tufa' and 'travertine', hard enough to define in modern sediments, are of less use in ancient rocks, where they may appear similar to calcretes. Such deposits are often attributable only to the broader category of terrigenous carbonates. Calcareous veins from the ORS penetrate the basement rock at several localities including the mouth of the Allt Briste, Cromarty, Aigas and other sites where the basement shows signs of contemporaneous weathering. At Dirlot, such veins grade into the laminar fracture-linings described above. These rocks could not be described as calcretes, but laminar calcretes are found lining fissures and sinkholes in Quaternary sediments (Multer & Hoffmeister 1968; Read 1974). Another account of closely related pedogenic lacustrine and lacustrine (including algal) carbonates is given by Freytet (1973) who described continental carbonate deposits from southern France and found their distinction a subtle one.

Nodular calcretes and other carbonates

In the Lower ORS calcrete complexes of the Anglo-Welsh outcrop, smaller 'crack-fill' nodular calcretes appear in plan like many ordinary sedimentary crack structures developed in Orcadian lacustrine beds. They are linear, form polygons and even show preferred orientations on a very local scale. The Orcadian structures regularly follow a preferred orientation related to subsequent joint patterns. Orientation in the Anglo-Welsh calcretes is less significant and probably reflects a more isotropic stress field during their formation. Calcretes at the present land surface do appear to fill desiccation cracks (Reeves 1970) and possibly joints (Goudie 1973).

Many Anglo-Welsh nodules are diffuse micritic bodies averaging 1 cm in diameter and formed in floodplain deposits. Similar nodules are found in Orcadian-like lithologies of the ORS in Western Norway, interpreted as lacustrine by Bryhni (1974). The Norwegian nodules are commonly asymmetric, concentrically banded and grew replacively within the sediment (Fig. 4). Calcrete formation may be directly associated with lacustrine beds (Casey & Wells 1964; Reeves 1976), including playas (Goudie 1973) upon which the Orcadian lake can be modelled (Parnell 1981). Much of the rock in the Orcadian Basin is calcareous, but in the fluvially dominated southern part of the basin there are many relatively pure carbonate nodule beds, several of them famous for their fossil fish (Fig. 5). They vary from isolated nodules within lacustrine laminites to platy limestones several centimetres thick. These nodules, like the calcretes, show much evidence of both replacive and displacive growth (Fig. 6), and often coalesced together. Expansion of the original rock by displacive growth was by up to twenty-fold.

The nodules formed around fish after temporary incursions by the Orcadian lake into the southern area. In addition several beds devoid of fossils formed in red mudstones and represent calcretes within flood-plain deposits. The sequences at Cromarty and Nigg include both types of nodule in alternating beds. The calcrete nodules are smaller and more abundant (Fig. 7) and consist of structureless micrite with diffuse boundaries. Smaller nodules coalesced to trap the host mudstone in between. All detrital grains in the nodules were replaced. Many nodules are elongate and aligned along two almost perpendicular orientations. A similar pattern is found in some nodules at Gamrie, where they formed around either fish or vertical cracks in the sediment.

The nodule bed at Dipple, although fish-bearing, includes several specimens with a fabric more akin to that of calcrete nodules. They have a secondary lamination due to iron oxide staining and contain numerous 'blotches' of sparry recrystallization, almost identical to undoubted calcretes from Crovie, which grew in red mudstones. Calcretes and fish-bearing nodules developed in a similar environment: the lacustrine regression would leave the sediments in the same situation as a fluctuating water-table would leave floodplain sediments. Remobilization and concentration of carbonate below the sediment surface occurred in both cases. Organic matter in the lacustrine rocks was not replaced by calcite and allowed most fish-bearing nodules to retain their primary texture. Where organic matter was lacking, as at Dipple, the nodule developed in the manner of calcretes.

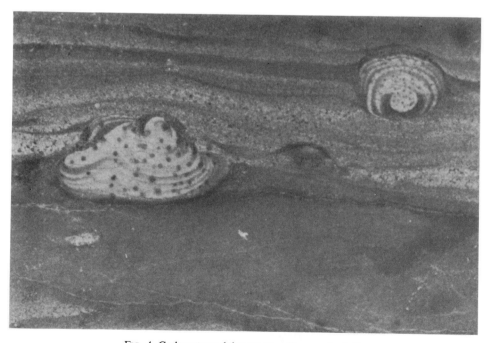

FIG. 4. Carbonate nodules, western Norway ($\times$ 0.4).

FIG. 5. Fish-bearing nodules, Lethen House quarry (hammer = 40 cm).

FIG. 6. Displacive growth of calcite in mica, nodular carbonate, Tynet Burn (thin section × 35)

FIG. 7. Nodular calcrete, Cromarty (hammer = 40 cm).

Silicrete and other silica rocks

The distinction between algal structures and duricrusts is often difficult, and several re-interpretations have been necessary (e.g. Swineford, Leonard & Frye 1958). It is conceivable, however, that an algal deposit may be superimposed by a calcrete or silcrete during subsequent exposure. Algae made an important contribution to the Orcadian lake, one particular phase of which is represented by isolated stromatolitic mounds, 60 cm high, near Stromness (Fannin 1969). A poorly exposed algal deposit over 1 km away at Whitehouse is taken to represent the same horizon (Parnell 1981). Breccias of algal mat fragments flank the mounds and can be traced away from them. A prominent laminar silcrete appears at the same horizon at Whitehouse, and chert also occurs at the top of the mounds. Silicification there could be an algal effect, but colloform structures in the silica seem to represent a diagenetic texture which is recorded in silcretes elsewhere (Smale 1973; S. Watts 1978). The banding is due to differential staining, typical of subaerial crusts, as opposed to the textural variation which is found in algal mats (Multer & Hoffmeister 1968). Mineralization occurred in both the laminar silcrete and the chert above the mounds.

Silcrete was formed on both lacustrine laminites and overlying rudites which onlapped from the basement surface. The silcrete lamination is an alternation of thin layers of microgranular quartz and thicker zones of mosaic quartz. Overgrowths on original grains grew in two preferred orientations (Fig. 8). This fabric is common in soil profiles (Brewer 1964) and has been reported from Devonian calcretes elsewhere (Allen 1973; McPherson 1979). It occurs to a limited degree in many ordinary sandstones, but where the fabric affects both authigenic clays and silica it is good supporting evidence for duricrust.

Most of the silcrete is flat-lying, but a granite boulder at Whitehouse is coated with fine laminae perfectly mimicking the surface of the boulder (Fig. 9); similar weathering skins up to 20 cm thick are found today around boulders in Australia (Hutton *et al.* 1972). Where the silcrete replaced rudites, analogous skins up to 1 mm thick are found in optical continuity around the larger clasts. The floating texture and absence of sub-sand grade sediment in these parts of the silcrete show that replace-

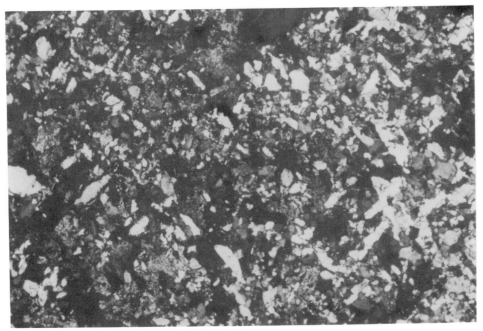

FIG. 8. Lattisepic texture in silcrete, Whitehouse (thin section × 18).

FIG. 9. Silcretised granite boulder, Whitehouse (hammer = 40 cm).

ment took place. Silica growth caused widespread buckling and doming of laminae, a feature seen in well-developed calcretes elsewhere in the basin.

Silica nodules are common in parts of the silcrete, as relics from lacustrine beds which the silcrete replaced. At Whitehouse and Stromness Reservoir, rocks still with the external appearance of laminites show a silcrete petrology in thin section. This retention of primary textures is known in modern silcretes (Watts 1977). The nodular chert superimposed by the silcrete is important in the wider context of silica diagenesis in an arid/semi-arid terrain.

The nodular cherts grew replacively and displacively in laminites rich in organic matter and phosphates. Nodules commonly show a horizontal linear orientation or polygonal pattern due to early fracturing in the sediment before nodule formation. Soft sediment intruded up these fractures. The morphology, petrology and pattern of those cherts are comparable with these of Magadiite-type cherts, produced through a sodium silicate precursor (Parnell 1981). Such cherts are known to occur as nodules produced by precipitation from a stratified lake (Hay 1968; Eugster 1969) and from groundwater processes (Rooney, Jones & Neal 1969; Maglione 1979). At Lake Magadi, Baker (1958) found calcretes directly overlying nodular chert, as the silcrete is in Orkney. Baker reported nodules with a preferred orientation, parallel to fault-lines. Eugster (1969) figured them about desiccation polygons. Orientated cherts may be analogous to linear calcrete deposits formed by the evaporation of carbonate-bearing waters along fault-planes (Cuyler 1930).

The silcrete that overprints the pre-existing chert nodules implies a temporary period of geomorphic stability at that part of the basin margin. Erosion, later or elsewhere, fed silica back into the lake waters, ready to precipitate once more as a silicate or silica gel. Silica was also derived from the basement rocks, where there is ample evidence of weathering and replacement.

Silicification is common in Scottish Upper ORS calcretes (e.g. Sedgwick & Murchison 1835; Stark 1925; Burgess 1961; Leeder 1976).

Mineralization and organic matter

Arid and sub-arid zones are typical sites for the accumulation of sedimentary metallic ores, particularly of lead, zinc and copper (Strakhov 1970). These elements are highly mobile in rocks of the arid red-bed association and are leached and transported to a site of deposition (Lur'ye 1978).

Scattered occurrences of galena and sphalerite in south-west Orkney appear to be syngenetic (Muir & Ridgway 1975). They are disseminated throughout the silcrete and adjacent basal breccias. Even the coating formed by leaching and replacement of a granite boulder is enriched. Stromatolitic mounds and breccias of algal mat debris also bear significant galena and sphalerite. Galena has been found in numerous algal carbonates elsewhere (e.g. Strakhov 1970). Thus the mineralization in the silicified algal mounds could be a concentration by an algal body or a duricrust, or even a hybrid of the two. The Orkney silcrete also contains relative concentrations of lithium, strontium and barium.

Enrichment of metals in basal sediments could have been achieved by two processes: penecontemporaneous concentration by weathering, and later concentration in which the unconformity surface acted as a barrier to mineralizing fluids. Manganese and iron ores in depressions of the basement surface at Dalroy present this problem of dating the mineralization. At this locality the basal rudites consist of two distinct units. Of two adjacent exposures, one shows impregnation of iron oxides in the lower rudites and basement whilst the other shows impregnation of only the upper part of the lower rudite. Iron oxides act as a matrix to the rudites, and immediately above the basement pure haematite ores are found. Micas have been at least partially replaced, and oxide growth was common along feldspar cleavage planes and grain boundaries. The association between the ore and the lower rudite rather than with the basement is suggestive of a weathering impregnation after deposition of the lower rudite.

Some ferricretes have a significant manganese content (Dury 1969; Goudie 1973) and these enrichments are associated with topographic depressions (Dury 1969) such as that envisaged in the schist at Dalroy. Elsewhere in the ORS, nodules of iron and manganese minerals are associated with calcretes in Dyfed (Allen & Williams 1978). The iron and manganese at Dalroy are segregated. Krauskopf (1957) suggested that repeated mixing of meteoric waters and groundwaters in arid regions would cause cyclic dissolution and reprecipitation of iron and manganese and lead to their increasing separation.

Metals also occur in trace quantities in bituminous segregations which are locally found in the Orcadian duricrusts. These include barium, chromium, iron, vanadium and zinc. Bituminous shale sequences are the principal hosts of organic accumulations in arid zones, as opposed to the coals typical of a humid climate (Strakhov 1970). The bitumens derived ultimately from organic-rich lacustrine laminites. The migration of bitumens from lacustrine laminites into duricrusts as well as other potential hosts took place after burial, but some organic matter accumulated in the crusts as they formed. Thin residual horizons similar to stylolites are found in the Orkney silcrete and may represent organic matter washed down from a soil surface, as figured by Duley (in Lapedes 1977).

A more direct relationship with organic matter might be exemplified by phoscrete (Goudie 1973). Many calcretes are found to include abundant plant remains, including *in situ* root casts, moulds and petrifactions (Johnson 1967; Klappa 1980). In the Orcadian Basin, good preservation of plants is known only from Rhynie, in a sequence of silicified sands, grits and rudites. The finer sediment has almost all been replaced, leaving a floating texture, and replacement is evident in most of the surviving grains. These sediments grade into clear silica, where complete replacement has occurred. Silica veins and vugs are parts of the sedimentary fabric, merging into the sediment and dividing it into clots. Thin laminae of microgranular quartz parallel to the bedding accentuate the floating texture. Relatively pure silica phases contain laminae which wrap around the remaining grains. The replacement silica locally developed a good lattisepic texture. Coarser silica was replaced by microgranular quartz, a process analogous to the sparmicritisation described by Kahle (1977) from Holocene calcrete. All of these features at Rhynie are typically found in ORS duricrusts, and match those of the Orkney silcrete very closely. The silica is commonly attributed a volcanic origin, but whatever its source, the style of diagenesis is that which is normally found in ORS duricrusts. The study of these rocks is still in progress.

Barite and the evaporative environment

Barite is a regular component of ORS duricrusts. The mineral is disseminated in the lime-stone at Buckie and in the rocks above and below. Barium in the calcrete occurs both as barite and in potassium-rich clays. Entrapment in clay minerals arising from weathering (Gurvich, Bogdanov & Lisitsyn 1979) is part of a wider phenomenon of barium enrichment during weathering (Ernst 1970; Gurvich *et al.* 1979). Enrichments in silcrete are reported from Australia by Walter (in Glover & Groves 1978) and are exemplified in the Orcadian Basin by the silcrete at Whitehouse where layering within the silcrete is picked out by barite. Barite commonly extended from the lamination to impregnate the whole rock and lined vugs as millimetre-scale bushes. Even the weathered basement clasts enclosed in the silcrete are impregnated.

The ferruginous basal sediments at Dalroy hold radiating barite clusters up to 5 mm across. An important duricrust within the Upper ORS, the Cothall Limestone, contains barium in a powdery green sediment, described by Sedgwick & Murchison (1835) as 'agaric mineral'. A very similar Upper ORS silicified calcrete from Milton Ness in the Midland Valley also includes local deposits of 'agaric mineral' rich in barite. Barite is recorded from other ORS calcretes in the Stirling district (Francis *et al.* 1970), Ayrshire (Burgess 1961) and Shropshire (Allen 1974a). There are also several recent examples of barite concentration in soils (Beattie & Haldane 1958; Swineford *et al.* 1958; Childs 1975; Senior 1979). Beattie (1970) described a residual caprock from New South Wales consisting almost entirely of barite.

Barite and celestite are also associated with the algal sediments of the Orcadian Basin, and with Devonian sulphate-bearing sequences on the Siberian Platform, where they were deposited from extensive shallow saline lakes remaining after a marine regression (Krylova *et al.* 1967). The regular evaporite minerals gypsum and halite appear as concentrations in the soil zone today, and can be considered to be both duricrusts and continental evaporites. The relative abundance of gypcrete as against calcrete may be dependent on aridity (Strakhov 1970), but gradations between the two are known in Africa (Goudie 1973). Silcretes are found in salt pans (Smale 1973) and in Australia frequently occur in the same profile as evaporites (Stephens 1971).

Ancient duricrusts in perspective

In the ORS of the Orcadian Basin, duricrusts form just part of a range of sedimentary rocks in which cementation or replacement by silica or carbonate occurred. Within fluvial sequences such rocks are usually readily identifiable as duricrusts, but in the lacustrine environment other processes unrelated to pedogenesis can produce deposits which appear similar. Therefore in ancient lacustrine sequences duricrusts are often difficult to identify with certainty, and in those cases all that may be deduced is that the local groundwaters were saturated. Difficulties in interpretation obviously limit the understanding of mineral deposits with which the rocks are associated. Some deposits, for instance the Magadi-type cherts, are not strictly duricrusts in the sense of the common usage of the term, but are related to them in a wider context of diagenetic processes in an arid/sub-arid environment.

ACKNOWLEDGMENTS: I thank Professors J. V. Watson and D. J. Shearman (Imperial College, London) for criticism of an early version of the manuscript; Dr I. Bryhni (Mineralogisk–Geologisk Museum, Oslo) for specimens from western Norway; and the Lethen Estate for access to the Lethen House quarry. The work was carried out whilst in receipt of a Royal Dutch Shell scholarship.

References

ALLEN, J. R. L. 1965. Sedimentation and palaeogeography of the Old Red Sandstone of Anglesey, North Wales. *Proc. Yorks. geol. Soc.* **35**, 139–85.

—— 1973. Compressional structures (patterned ground) in Devonian pedogenic limestones. *Nature,* **243**, 84–6.

—— 1974a. Sedimentology of the Old Red Sandstone in the Clee Hills area, Shropshire, England. *Sedim. Geol.* **12**, 73–167.

—— 1974b. Studies in fluviatile sedimentation: implications of pedogenic carbonate units, Lower Old Red Sandstone, Anglo-Welsh outcrop. *Geol. J.* **9**, 181–308.

—— & WILLIAMS, B. P. J. 1978. The sequence of the earlier LORS (Siluro-Devonian), north of Milford Haven, southwest Dyfed (Wales). *Geol. J.* **13**, 113–36.

ARMSTRONG, M. & PATERSON, I. B. 1970. The Lower Old Red Sandstone of the Strathmore region. *Inst. geol. Sci. Rep. No. 70/12.*

BAKER, B. H. 1958. Geology of the Magadi area. *Geol. Surv. Kenya Rep. 42.*

BEATTIE, J. A. 1970. Peculiar features of soil development in parna deposits in the Eastern Riverina, New South Wales. *Aust. J. Soil,* **8**, 145–56.

—— & HALDANE, A. D. 1958. The occurrence of palygorskite and barytes in certain parna soils of the Murrumbidgee Region, New South Wales. *Aust. J. Sci.* **20**, 274–5.

BREWER, R. 1964. *Fabric and Mineral Analysis of Soils.* Wiley, London.

BRYHNI, I. 1974. Old Red Sandstone of Hustadvika and an occurrence of dolomite at Flatskjer, Nordmøre. *Norg. geol. Unders.* **311**, 49–63.

BURGESS, I. C. 1961. Fossil soils of the Upper Old Red Sandstone of south Ayrshire. *Trans. geol. Soc. Glasg.* **24**, 138–53.

CASEY, J. N. & WELLS, A. T. 1964. The geology of the north-east Canning Basin, Western Australia. *BMR Aust. Geol. Geophys. Rep. 49.*

CHILDS, C. W. 1975. Notes on occurrence of barite in B horizons of palaeosols in loess, South Canterbury, New Zealand. *N.Z. J. Sci.* **18**, 227–30.

CUYLER, R. M. 1930. Caliche as a fault indicator. *Bull. geol. Soc. Am.* **41**, 109.

DONOVAN, R. N. 1973. Basin margin deposits of the Middle Old Red Sandstone at Dirlot, Caithness. *Scott. J. Geol.* **9**, 203–11.

—— 1975. Devonian lacustrine limestones at the margin of the Orcadian Basin, Scotland. *J. geol. Soc. London,* **131**, 489–510.

DURY, G. H. 1969. Rational descriptive classification of duricrusts. *Earth Sci. Jl* **3**, 77–86.

ERNST, W. 1970. *Geochemical Facies Analysis. Methods in Geochemistry and Geophysics 11.*

EUGSTER, H. P. 1969. Inorganic bedded cherts from the Magadi Area, Kenya. *Contr. Mineral. Petrol.* **22**, 1–31.

FANNIN, N. G. T. 1969. Stromatolites from the Middle Old Red Sandstone of Western Orkney. *Geol. Mag.* **106**, 77–88.

FRANCIS, E. H. FORSYTH, I. H. READ, W. A. & ARMSTRONG, M. 1970. The Stirling district. *Mem. geol. Surv. Gt Br.*

FREYTET, P. 1973. Petrography and paleoenvironment of continental carbonate deposits with particular reference to the Upper Cretaceous and Lower Eocene of Languedoc. *Sedim. Geol.* **10**, 25–60.

GEIKIE, A. 1900. The geology of central and western Fife and Kinross. *Mem. geol. Surv. Gt Br.*

GLOVER J. E. & GROVES D. I. (eds) 1978. Archaen cherty matasediments. *Geol. Dept Extens. Surv., Univ. West Aust. No. 2.*

GOUDIE, A. 1973. *Duricrusts in Tropical and Subtropical Landscapes.* Clarendon Press, Oxford.

GURVICH, YE. G., BOGDANOV, YU. A. & LISITSYN, A. P. 1979. Behaviour of barium in Recent sedimentation in the Pacific. *Geokhimiya,* **15,** 28–43.

HAY, R. L. 1968. Chert and its sodium silicate precursors in sodium-carbonate lakes of East Africa. *Contr. Mineral. Petrol.* **17,** 255–74.

HUTTON, J. T., TWIDALE, C. R., MILNES, A. R. & ROSSER, H. 1972. Composition and genesis of silcretes and silcrete skins from the Beda Valley, Southern Arcoona Plateau, South Australia. *J. geol. Soc. Aust.* **19,** 31–9.

JOHNSON, J. D. 1967. Caliche on the Channel Islands. *Miner. Inf. Calif. Div. Mines Geol.* **20,** 151–8.

KAHLE, C. F. 1977. Origin of subaerial Holocene calcareous crusts: role of algae, fungi and sparmicritisation. *J. sedim. Petrol.* **24,** 413–35.

KLAPPA, C. F. 1980. Rhizoliths in terrestrial carbonates: classification, recognition, genesis and significance. *Sedimentology,* **27,** 613–29.

KRAUSKOPF, K. B. 1957. Separation of manganese from iron in sedimentary processes. *Geochim. cosmochim. Acta,* **12,** 61–84.

KRUMBEIN, W. E. & GIELE, C. 1979. Calcification in a coccoid cyanobacterium associated with the formation of desert stromatolites. *Sedimentology,* **26,** 593–604.

KRYLOVA, A. K. MALITCH, N. S., MENNER, V. V., OBRUTCHEV, D. V. & FRADKIN, G. S. 1967. The Devonian of the Siberian Platform. *In:* OSWALD, D. H. (ed.) *International Symposium on the Devonian system, Calgary Vol. 1.*

LAPEDES, D. N. 1977. *Encyclopedia of the Geological Sciences.* McGraw-Hill, New York.

LEEDER, M. R. 1976. Palaeogeographic significance of pedogenic carbonates in the topmost Upper Old Red Sandstone of the Scottish border basin. *Geol. J.* **11,** 21–8.

LUR'YE, A. M. 1978. Copper migration conditions in Red Bed associations. *Geokhimiya,* **15,** 174–8.

MAGLIONE, G. 1979. Un example de sédimentation continentale. Le bassin Tchadian. *In: Depôts Évaporitiques.* Editions Technip, Paris.

MATTER, A. & TUCKER, M. E. (eds) 1978. *Modern and Ancient Lake Sediments. Spec. Publs int. Ass. Sediment.* **2,** Blackwell Scientific Publications, Oxford.

MAUFE, H. B. 1910. Summary of progress for 1909. *Geol. Surv. Gt Br.*

McCALLIEN, W. J. 1927. Preliminary account of the post-Dalradian geology of Kintyre. *Trans. geol. Soc. Glasg.* **18,** 40–126.

McPHERSON, J. G. 1979. Calcrete palaeosols in fluvial redbeds of the Aztec Siltstone (Upper Devonian), Southern Victoria Land, Antarctica. *Sedim. Geol.* **22,** 267–85.

MUIR, R. O. & RIDGWAY, J. M. 1975. Sulphide mineralisation of the Continental Devonian sediments of Orkney (Scotland). *Miner. Deposita,* **10,** 205–15.

MULTER, H. & HOFFMEISTER, I. E. 1968. Subaerial laminated crusts of the Florida Keys. *Bull. geol. Soc. Am.* **79,** 183–92.

NELSON, C. S. & RODGERS, K. A. 1969. Algal stabilisation of Holocene conglomerates by micritic high-magnesium calcite, Southern New Caledonia. *N.Z. J. Mar. Freshwat. Res.* **3,** 395–408.

NEWELL, N. D., PURDY, E. G. & IMBRIE, J. 1960. Bahamian oolitic sand. *J. Geol.* **68,** 481–97.

PARNELL, J. 1981. *Post-depositional processes in the Old Red Sandstone of the Orcadian Basin, Scotland.* Unpubl., Ph.D. thesis, University of London.

READ, J. F. 1974. Calcrete deposits and Quaternary sediments, Edel Province, Shark Bay, Western Australia. *Mem. Am. Ass. Petrol. Geol.* **22,** 250–82.

REEVES, C. C. 1970. Origin classification and geologic history of caliche. *J. Geol.* **78,** 352–62.

—— 1976. *Caliche.* Estacado Books, Lubbock, Texas.

ROONEY, T. P., JONES, B. F. & NEAL, J. T. 1969. Magadiite from Alkali Lake, Oregon. *Am. Mineral.* **54,** 1034–43.

SCHOLL, D. W. & TAFT, W. H. 1964. Algae, contributors to the formation of Calcareous tufa, Mono Lake, California. *J. sedim. Petrol.* **34,** 309–19.

SEDGWICK, A. & MURCHISON, R. I. 1835. On the structure and relationships of the deposits contained between the primary rocks and the Oolite Series in the north of Scotland. *Trans. geol. Soc. Lond., Ser. 2,* **3,** 125–60.

SENIOR, B. R. 1979. Mineralogy and chemistry of weathered and parent sedimentary rocks in south-west Queensland. *BMR Jl. Aust. Geol. Geophys.* **4,** 111–24.

SMALE, D. 1973. Silcretes and associated silica diagenesis in Southern Africa and Australia. *J. sedim. Petrol.* **43,** 1077–89.

STARK, J. 1925. The Red Sandstones of Toward. *Trans. geol. Soc. Glasg.* **17,** 169–72.

STEPHENS, C. G. 1971. Laterite and silcrete in Australia. *Geoderma,* **5,** 1–52.

STRAKHOV, N. M. 1970. *Principles of Lithogenesis Volume 3.* Oliver & Boyd, Edinburgh.

SWINEFORD, A., LEONARD, A. B. & FRYE, J. C. 1958. Petrology of the Pliocene pisolitic limestone in the Great Plains. *Bull. Stat. geol. Surv. Kansas,* **130,** 97–116.

WATTS, N. L. 1978. Displacive calcite: evidence from recent and ancient calcretes. *Geology,* **6,** 699–703.

WATTS, S. H. 1977. Major element geochemistry of silcrete from a portion of inland Australia. *Geochim. cosmochim. Acta,* **41,** 1164–7.

—— 1978. A petrographic study of silcrete from inland Australia. *J. sedim. Petrol.* **48,** 789–94.

WHITE, E. I. 1946. The Genus Phialaspis and the "Psammosteus Limestones". *Q. Jl geol. Soc. Lond.* **101**, 207–42.

WRIGHT, V. P. 1981. A subaerial stromatolite from the Lower Carboniferous of South Wales. *Geol. Mag.* **118**, 97–100.

JOHN PARNELL, Department of Geology, Imperial College, London SW7 2BP.

A process-response model for the formation of pedogenic calcretes

Colin F. Klappa

SUMMARY: The compositional and fabric evolution of pedogenic calcretes is viewed as a process-response model involving inorganic and organic soil-forming processes on the one hand, and diagenetic, rock-forming processes on the other; the resulting battle being expressed as six evolutionary stages: (1) preparation of host material; (2) soil formation and horizon differentiation; (3) accumulation of calcium carbonate; (4) profile development; (5) induration; and (6) reworking. In nature, some stages may be repeated or reversed; others may be omitted. Emphasis is given to the role of organic activity which determines largely the morphology of pedogenic calcrete profiles and which controls the direction and extent of profile development.

The origin of calcrete has been, and still is, a subject for much conjectural and intuitive thinking. Methodically unsound attempts to solve fundamental problems of calcrete genesis result largely from inadequate definition of the term 'calcrete'; progress towards a better understanding of calcretes has been hindered by the lack of an accepted framework for observation and description. The geographically widespread distribution of calcretes, the diverging interests of their investigators, the prolific and confusing terminology, and the many theories that have been proposed to account for their origin, all have contributed to the present level of organization—one of chaos. Nevertheless, beneath this camouflage of disorder, recurring factors of similarity can be recognized at various scales of observation.

This paper outlines characteristic features of pedogenic calcretes and provides a process-response model to explain what these features mean and how they are genetically related.

Modern calcretes have the following attributes: (1) are near-surface and surface accumulations of calcium carbonate; (2) show uniformity and continuity on a megascopic (km) scale; (3) are laterally extensive for tens or even hundreds of kilometres; (4) are common in areas with semi-arid to temperate climates; (5) occur on stable geomorphic surfaces with gentle (less than 25°) slopes; (6) are conformable with present-day topography; (7) are independent of substrate; (8) occur within, not merely on, any host material; (9) show a downward decrease in calcium carbonate content; and (10) rarely exceed 6 m vertical thickness (but see below).

In recent years, workers investigating calcrete have shown increasing favour toward models involving soil-forming processes. The reasons are not difficult to see, for on a megas-copic (km) scale, soils are laterally extensive surface materials which occur on stable geomorphic surfaces. Moreover, soils duplicate the topographical expression of gently undulating landscapes. On a local (cm) scale in a vertical sequence, soils show zonations from the surface, grading downwards into unaltered host material.

Soils are formed by weathering and transformation of solid rock *in situ* or weathering and/or transformation of transported materials derived from solid rocks (Brewer 1964). Soil-forming processes cause modification in sediments or weathered rocks by removal, translocation, transformation and neoformation of mobile constituents. Such changes generally produce horizon differentiation and, thus, the formation of soil profile.

The features outlined above, characteristic of soils, are common also to calcretes. Surely, this is more than mere coincidence? Most calcretes are, in essence, calcified biological soils and known generally as pedogenic or pedogenetic calcretes. They have been subjected to both soil- and rock-forming processes, that is, to both pedogenesis and diagenesis.

Objections to a pedogenic calcrete model are minimal, the outstanding one being the tremendous thickness (up to 60 m) of some calcrete profiles. However, profiles such as these can be accommodated easily by a pedogenic model if one allows for a multiple profile development. The remainder of this paper is directed toward an exposition of the pedogenic calcrete model.

Definitions and concepts

Before calcretes can be adequately described, and subsequently interpreted in terms of origin and formation, they need to be clearly defined. Unfortunately, it is practically impossible to define calcrete in a way that will satisfy all workers. This is due largely to divergent interests and use of independent terminologies which have greatly hindered communication among workers. A review of the literature (Klappa 1978a) indicates that confusion abounds. To avoid further misunderstanding it is necessary to discuss the meaning of calcrete as used in this paper.

Calcrete is an accumulation of predominantly fine-grained low magnesian calcite having formed within the meteoric vadose zone by *pedodiagenetic* alteration and replacement of any precursor host material.

Pedodiagenesis is a new term proposed here to embrace both pedogenesis (soil-forming processes) and diagenesis (processes which occur in sediments or sedimentary rocks between the time of initial deposition and the time—if ever—when the threshold of metamorphism is reached; adapted from Murray & Pray (1965). The reason for coining this hybrid term is because pedogenesis does not generally include lithification of the soil, and diagenesis usually excludes soil-forming processes as well as weathering by inorganic agencies. New names do not make new processes; nonetheless the term *pedodiagenesis* is proposed to embrace demonstrable and inferred processes which have operated during the formation of calcretes.

Soil formation is the result of physical, chemical and biological processes which transform a rock material into a soil (Robinson 1949). It follows that biological processes have played a role in the formation of pedogenic calcretes if one accepts the above definition. Accumulations of calcium carbonate which occur in materials unaffected by biological processes are *not* pedogenic calcretes.

The pedodiagenetic cycle

Soils form by the breakdown of rock into unconsolidated sediments and the incorporation of organic materials (Russell 1973). Soils generally show zonal differentiation as a result of biological and physicochemical soil-forming processes. Rocks of any composition of origin may be broken down into weathered detritus and organisms may transform this sediment into a soil. If physicochemical conditions are conducive to the accumulation of calcium carbonate, powdery of unconsolidated calcretes will form. Continued soil-forming process will modify the underlying host material (parent rock) until the

resulting regolith eventually forms a protective blanket over the unaltered substrate. Induration of the calcrete will lead to the formation of a near-surface pedodiagenetic rock. The indurated calcrete may be subjected to further pedodiagenetic processes, leading to its modification or destruction. The cycle is thus repeated. This idealized cycle is illustrated schematically in Fig. 1.

The above simplified cycle assumes that gains or losses of elements and minerals by mass transport are minimal. The cycle becomes more complicated, and more difficult to model, if net erosion or net deposition becomes significant. Erosion may result in degrading profiles or, if erosion is sufficient, may remove all weathered detritus. If on the other hand, sedimentation rates are sufficiently high, there may be insufficient time for profile development. In areas where erosion predominates, soils will tend to be thin or may be absent altogether. In areas where weathering predominates, soils will tend to be thick. Soils will also tend to be thick if material is being introduced from elsewhere by aeolian or fluvial transport.

Pedodiagenetic processes

Pedogenic and diagenetic processes determine the morphology and composition of calcretes. In calcrete profiles, diagenetic processes are superimposed on features formed by pedogenic

PEDODIAGENETIC CYCLE

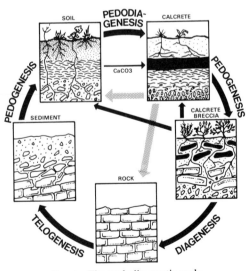

FIG. 1. The pedodiagenetic cycle.

processes, resulting in the formation of pedodiagenetic structures. In addition, diagenetic processes modify fabrics of the soil matrix, resulting in the formation of new fabrics which are diagnostic of the near-surface subaerial diagenetic environment. Diagenetic processes tend to post-date pedogenic processes, although the two processes may operate simultaneously.

Although the distinction is made here between pedogenic and diagenetic processes, in nature they cannot be separated entirely. For example, the processes of dissolution, precipitation, compaction, brecciation and neoformation are common to both pedogenesis and diagenesis. Thus, the two basic processes are treated here under the same heading.

Pedodiagenetic processes operate in an attempt to restore and maintain equilibrium. Treated simplistically, these processes lead to changes which involve additions, removals, translocations and transformations of materials. The driving mechanisms behind these changes, which are physical, chemical or biological in origin, may proceed independently or together. One of the main attributes of calcrete is its complexity of types, patterns and mutual relationships of processes. A similar viewpoint has been made by Harrison (1977) in his discussion of complex cementation patterns in caliche profiles from Barbados. Accepting the complexity of processes, the more fundamental questions—of what are the processes in detail and what causes the complexity—require answering.

Cementation, dissolution, recrystallization, brecciation and metasomatic replacements are characteristic processes of calcrete deposits (Bretz & Horberg 1949; James 1972; Esteban 1974; Goudie 1975; Klappa 1978a,b, 1979a,b, 1980a,b). However, such processes are diagnostic not only of calcretes; they have determined fabrics, textures and structures of many types of sedimentary rocks. Even so, these widespread processes have not been worked out in detail for calcrete deposits. This statement reiterates the comment of Esteban (1976) who noted (p. 2049): '...processes that produce caliche are not understood well'.

One of the main problems is that, with the exception of several experimental studies (Breazeale & Smith 1930; Thorstenson, Mackenzie & Ristvet 1972; McCauley & Roy 1974; Dumont 1975; Badiozamani, Mackenzie & Thorstenson 1977) processes are usually inferred from observations of the products. This method of study is both difficult and limited. It is difficult because of the complexity and interaction of processes; it is limited because we do not consider processes which leave no evidence of their former activity. In addition, even processes which leave clues may be overlooked if we have little understanding of them. The biological processes involved in calcretization are relevant examples. Until recently, biological processes have received scant attention by workers interested in calcrete genesis. This is unfortunate because biological processes directly influence calcrete morphology (Klappa 1978a, b, 1979a, b, 1980a, b). Recognition of structures resulting from biological activity is fundamental for making reasonable interpretations regarding the processes of pedogenic calcrete formation. Biological processes include: the dissolution of minerals as a result of secretion of organic acids and chelating compounds; plant photosynthesis and respiration which affect CO_2 and O_2 levels in the ambient microenvironment; the formation of organomineral complexes; selective uptake of ions by plants; burrowing and boring by plants and animals; brecciation and disruption of indurated hardpans by roots; calification of organic matter to form biogenic carbonate structures and; biosynthesis of new minerals. Discussions of biological processes have been given in Klappa (1978a, b, 1979a, b, 1980a, b) and will not be repeated here. The processes are listed to indicate the complexity of biological processes which influence and are themselves influenced by physicochemical soil-forming and diagenetic processes.

Physicochemical soil-forming processes include: physical disintegration and chemical decomposition of rocks and minerals; dissolution; precipitation; recrystallization and; metasomatic replacements. Grains may be translocated by illuviation, leached, dissolved, recrystallized or reorganized to form new minerals from the residues of leached minerals (neoformation). The near-surface subaerial pedodiagenetic environment is subjected to wetting and drying cycles which lead to the formation of shrinkage cracks and channels. This Pandora's box may be loosely termed subaerial weathering. The line between the above listed pedogenic processes and diagenetic processes is arbitrary. In the study of calcretes, this line is unnecessary. Pedogenic processes and diagenetic processes are so intimately related that they cannot be justifiably separated.

Sources of calcium carbonate

Possible sources of calcium carbonate have been discussed by many workers (Brown 1956;

Motts; Coque 1962; Gile, Peterson & Grossman 1966; Gardner 1973; James 1972; Yaalon & Ganor 1973). Goudie (1973) and Reeves (1976) provide useful reviews on this problem. In addition, Goudie has collated information not concerned directly with studies of calcrete formation, but nonetheless, of relevance to this topic. Here the possible sources of calcium carbonate will be evaluated and summarized in a general discussion, relying largely on information given in the above mentioned references. Examples used for illustrating points raised in this discussion are taken from materials examined during a study of calcretes from Spain (Klappa 1978a).

Breakdown of components and redistribution of calcium within soils may initiate the formation of calcretes in essentially non-calcareous host materials. Such components may include calcium-rich primary minerals derived from the host material, certain plants which contain calcium within their tissues, detrital carbonate grains incorporated into the soil by colluvial or fluvial processes and calcareous skeletal grains of soil organisms such as terrestial gastropods. For example, the land snails *Hellicella* sp. and *Rumina* sp. were found on exposed surfaces throughout coastal regions of the western Mediterranean, with densities of up to 150 m^{-2} in some places (Klappa 1978a).

Calcium carbonate may be carried in solutions of rain water and ground water. Using published data, Goudie (1973) calculated that the average calcium content of rainfall came to 6–7 ppm. However, he pointed out that this relatively low figure may be increased two-fold after rainfall has passed through the canopy of the trees or bushes. Plants also may increase the amount of calcium in the soil by removing calcium ions at depth from ground water and depositing calcium carbonate or calcium oxalate as plant residues in the litter layer. Taproots with lengths of up to 5 m were observed in Quaternary calcretes from Spain (Klappa 1980a). Goudie (1973) stated that some common tree species from southern Africa put taproots down to as much as 50 m to ground water.

Some plants accumulate carbonate within their tissues. Lichens are capable of synthesising calcium-bearing minerals (Pomar *et al.* 1975; Klappa 1979a). Other biogenic carbonate structures, including *Microcodium* (Klappa 1978b), calcified filaments (Klappa 1979b) and rhizoliths (Klappa 1980a), contribute to the buildup of calcium carbonate in pedogenic calcretes.

Atmospheric dust provides an external source of calcium carbonate in some areas. It is conceivable that fallout of wind-blown dust from the Sahara may have supplied Mediterranean soils with additional calcium carbonate.

In coastal areas, an important source of calcium carbonate may be derived from marine aerosols. James (1972, p. 829) noted that the effect of wind-blown salt spray is two-fold: first, high salinity inhibits growth of vegetation which leaves soils susceptible to erosion; and secondly, calcium carbonate is added to the surface and increases the CaCO$_3$ content of nearsurface waters.

Throughout the western Mediterranean coastal regions there are extensive outcrops of Mesozoic and Tertiary limestones and dolomites. Whether transported as alluvium, colluvium, atmospheric dust or in solution, the supply of calcium carbonate to calcrete-forming areas is not a limiting factor in these regions. Calcretes occur within *any* host material given an adequate source of CaCO$_3$.

Nature and movement of pore fluids

In a study of chemical weathering, soil development and geochemical fractionation in bedrock of the White Mountains, California, Marchard (1974, p. 379) stated: 'It is impossible . . . to clearly comprehend the transformation of bedrock into soil by studying changes in the mineral phases alone, for weathering involves important alteration of liquid and gas phases in contact with the minerals as well.' Thus, to better understand how bedrock can be transformed into calcrete, a few comments on the nature and movement of pore fluids is in order.

The moisture condition of calcretes close to or at the land surface is subject to large variations. Similarly, the composition and temperature of pore fluids fluctuate rapidly as a result of alternating periods of short, intense rainfall followed by longer periods of evaporation. Transpiration by plants, percolation into ground water and capillary rises from ground water cause changes in the amount, composition and temperature of the remaining held water within the calcrete profile. In the vadose zone (zone of aeration), the pore water pressure is less than atmospheric pressure. Above the water table, water is held at less than one atmosphere because of surface tension and adsorptive forces (Croney 1952). The intensity with which water is held by a soil can be measured in units based on the concept of soil suction. When dealing with negative pressures (soil suction), the pF scale is commonly used, where pF = log$_{10}$ (height above the water-table, h in cm).

Transpiration by plants and evaporation of held water are important processes which bring about suction gradients. Water movements take place by bulk liquid flow in plants, as rising capillary water and as descending water under the influence of gravity. After gravitational water has drained completely into ground water, stationary water is held in the vadose zone. The geometry of this immobile, held water determines fabrics of calcite crystals which precipitate within pore spaces of host materials. Occasionally, if the pores are small enough, isopachous fringes precipitate on grain surfaces rather than gravitational and meniscus cements. Thus, care is needed when interpreting isopachous, phreatic-type cements which have precipitated in the vadose zone.

In addition to calcite cements, which are by definition firmly attached to a substrate, some hyphantic (Gk. *hyphantos* = woven) fibrous and rhombic calcite crystals precipitate in the vadose zone without any substrate attachment. Hyphantic and coarse crystic fabrics which lack cementation are indicative of vadose diagenesis; their origin is attributed to rapid precipitation from evaporating pore solutions.

Apart from volumetrically insignificant but genetically important held water, a notable property of soil vadose water is its mobility. Water movements (mostly bulk liquid flow and movement as water vapour) together with the composition of these moving waters, determine the direction and intensity of pedodiagenetic processes.

Mechanisms of calcite precipitation

Possible mechanisms

To say that calcrete formation results from processes in the vadose zone which culminate in the accumulation, mainly by precipitation, of calcium carbonate tells us nothing of the mechanisms involved. To better understand the possible mechanisms of calcite precipitation, and to be in a better position to suggest the most likely mechanisms, the chemistry of the $CO_2 - H_2O - CaCO_3$ system will be discussed briefly. This system is discussed in detail by Berner (1971), Sweeting (1972), Lippmann (1973), Broecker (1974) and Bathurst (1975).

The reactions involved in the $CO_2 - H_2 - CaCO_3$ system can be summarized by the general equation:

$$CO_2 + H_2O + CaCO_3 \rightleftharpoons Ca^{2+} + 2HCO_3^-.$$

The solubility of calcium carbonate is affected by five independent variables. The solubility product $[Ca^{2+}].[CO_3^{2-}]$ decreases with increasing temperature, increases by adding salts lacking common ions, decreases by adding salts with common ions, decreases with increasing pH and increases with rising partial pressures of CO_2.

The equations which relate to the formation of calcium bicarbonate are as follows:

(1) equation of hydration $CO_2 + H_2O \rightleftharpoons H_2CO_3$,

(2) dissociation of carbonic acid $H_2CO_3 \rightleftharpoons H^+ + HCO_3^-$,

(3) combination of hydrogen and carbonic ions $H^+ + CO_3^{2-} \rightleftharpoons HCO_3^-$,

(4) dissolution of calcium carbonate $H^+ + CaCO_3 \rightleftharpoons Ca^{2+} + HCO_3^-$,

(5) ionic dissociation of water $H_2O \rightleftharpoons H^+ + OH^-$.

All these reactions are reversible. If reactions (2) and (4) move to the right, both CO_2 and $CaCO_3$ are dissolved. If CO_2 is removed, as a result of agitation, aeration of photosynthesis, $CaCO_3$ is precipitated. Loss of CO_2 may also result from solutions coming into contact with air at lower partial pressures of CO_2. Likewise, removal of H_2O, as a result of evaporation or evapotranspiration, leads to the precipitation of $CaCO_3$.

Netterberg (1971) reported that the mechanisms postulated by most authors to account for calcium carbonate crystallization from the soil solution during calcrete formation is simply evaporation. However, Netterberg pointed out that evaporation is important only in the upper metre or so of the soil profile. He suggested that the effect of changes in soil suction on the solubility of carbonate is possibly the most important mechanism in calcrete formation at all depths. Transpiration and, to a lesser extent, evaporation are considered to be responsible for bringing about increases in suction pressures.

Precipitation of calcite may occur at some point in a calcrete profile which has a downward increasing pH gradient. Multer & Hoffmeister (1968) measured a minimum pH of 5.6 in soils from Florida, increasing downwards to a maximum of 8.2 in underlying host materials.

Precipitation of calcite can be influenced or controlled by biological activities. Culture experiments by Krumbein (1968) show that microfloras, consisting of autotrophic and heterotrophic bacteria, fungi, actinomycetes, green algae and blue-green algae, were able to cause precipitation of calcite by changing pH conditions and possibly by transmission of Ca^{2+} by chelating substances generated by these micro-organisms.

With the removal of H_2O, as a result of evaporation or water uptake via plant roots, the concentration of solutions may be increased sufficiently to cause precipitation of calcite. Similarly, increasing temperatures favour calcite precipitation since the amount of CO_2 in solution decreases with increasing temperature. Removal of CO_2 also takes place during plant photosynthesis. By assimilation of CO_2 and HCO_3^-, plants may bring about the precipitation of calcite. As emphasized by Schneider (1977), this process is an inorganic precipitation of $CaCO_3$, which is promoted by the physiological activities of organisms.

In contrast to the removal of CO_2, which favours the precipitation of calcite, increasing CO_2 partial pressure leads to the dissolution of $CaCO_3$. Partial pressures of CO_2 in the soil are commonly many times greater than in the atmosphere or in the underlying bedrock (Bathurst 1975). High CO_2 partial pressures in soils result mainly from respiration of higher plants and micro-organisms. Dissolution of calcite is promoted also by secretion of organic acids from growing plants or the release of substances during the decay of organic matter. However, Berner (1968) pointed out that proteinaceous organisms, upon decay, release not only carbonic and other acids but also give off basic substances such as ammonia and amines. Thus, decomposition of organic matter may result in a rise in pH instead of a decrease. This may lead to the precipitation of $CaCO_3$.

Berner (1968) showed experimentally that bacterial decomposition of a fish led to an increase in pH and the precipitation of Ca^{2+} from solution as a mixture of calcium fatty acid salts or soaps. Berner suggested that some ancient $CaCO_3$ concretions, especially those enclosing fossils of soft-bodied organisms, may have formed initially as calcium soap (adipocere) which was later converted to $CaCO_3$.

Another source of CO_2 which may be important, both on the surface and underground, is that produced by the mixing of chemically differing natural waters (Runnells 1969). Extending the work of Bögli (1964) who introduced the term 'Mischungkorrosion' (mixture corrosion), Runnells showed experimentally that mixing of solutions which differ only in their content of dissolved electrolytes may cause either precipitation or dissolution of rock-forming minerals.

Likely mechanisms

Studies by the author concerning Quaternary calcretes from coastal regions of the western Mediterranean in particular, and pedogenic calcretes in general may provide clues as to the most likely mechanism or mechanisms for calcrete formation.

Irregular cracks and channels, suggesting that periods of wetting and drying have occurred, are common porosity types in calcretes. Hyphantic and banded needle (fibrous) calcite precipitates are characteristic of the upper parts of calcrete profiles. Crystal morphology is highly variable. Crystal sizes tend to be micron-sized or submicron-sized. Development of calcite cements tends to be irregular in distribution, type and amount. Cementation is preferential around roots of higher plants. The chalky horizon is characterized by precipitates of uncemented or weakly cemented micron- and silt-sized calcite crystals.

Evaporation is the simplest explanation for the formation of shrinkage cracks. Rapid and intense periods of evaporation following rainfall also may have been responsible for the precipitation of hyphantic and banded needle calcite crystals. James (1972) suggested that the small size of calcite needles and rhombic micrite in calcareous crust profiles from Barbados were formed from a rapidly evaporating solution. He noted further that intense evaporation of fluids in narrow pores spaces at or near the ground surface may bring solutions to very high degrees of supersaturation quite rapidly. Crystals precipitated from highly supersaturated solutions are known to exhibit special growth forms such as dendrites, spherulites and whisker crystals *s.s.* (Buckley 1951; Walton 1967).

Preferential cementation of calcite around roots has been discussed in detail elsewhere (Klappa 1980a). Removal of water during life of the plants by transpiration is one possible and likely mechanism to account for the precipitation of calcite around roots. Alternatively or in addition, organic decay of dead roots may induce calcite precipitation. The location of the precipitate will be determined largely by the distribution and amount of moisture within pore spaces of the rhizosphere. Precipitation will tend to occur on surfaces which retain moisture as held water films on and between skeletal grains, root surfaces, fungal hyphae and already precipitated calcite crystals.

The Mediterranean climate today is characterized by high temperatures which favour evaporation. Admittedly, the climatic regime may have differed considerably during earlier periods of the Quaternary, but combining the observations and interpretations made above, it is considered most likely that both evaporation and transpiration have been and still are the most important mechanisms which culminate in the precipitation of calcrete-forming pedo-diagenetic calcite.

The pedogenic calcrete profile

Although calcrete profiles commonly show rapid vertical and lateral variations within a few metres, it is possible to recognize certain patterns with sufficient repetition to allow at least some generalization. Thus, an idealized vertical calcrete profile can be drawn up (Fig. 2). The sequence illustrated in Fig. 2 is only one of many possible variations, albeit the commonest. Boundaries between designated horizons tend to be irregular and diffuse, and the horizons themselves may be repeated several times or be absent altogether.

Equivalent terms used in previous classifications are given in Fig. 2 to facilitate correlation of already published descriptions of calcrete profiles.

Compositional and fabric evolution of pedogenic calcrete profiles: a process response model

The complex interrelationships among the factors of climate, host material, topography, hydrology, organisms and time render the construction of a generalized model exceedingly difficult. Bearing this in mind, the model presented here is, of necessity, a blanket assessment of the intricacies of reality.

Stages of the model

Stage 1: preparation of host material

The formation of regolith or weathered detritus from an initially consolidated bedrock involves mechanical and biophysical disintegration, physicochemical dissolution and biochemical weathering. Whether this material accumulates *in situ* or is eroded and deposited elsewhere depends on the rate of sediment production *versus* its rate of removal. For calcrete formation to take place within a weathered host material, sediment production must equal or exceed the rate of removal.

Stage 2: soil formation

Unconsolidated sediment or weathered detritus is transformed into a soil by changes produced by the action of organisms and by changes due to the movement of water through the sediment. Vertical movements of moisture tend to reinforce the changes produced by organisms, the general result being an anisotropy shown by layers or horizons parallel to the land surface.

Stage 3: accumulation of calcium carbonate

The soil becomes progressively enriched with calcium carbonate. Pore spaces become lined or filled with calcite whereas relic quartz grains and other minerals inherited form the host material become progressively replaced by calcite. Characteristic structures composed of concentrations of pedodiagenetic calcite, such as glaebules (soil concretions) and calcitans (free and embedded grain cutans), are formed. Biological consituents of the soil may become

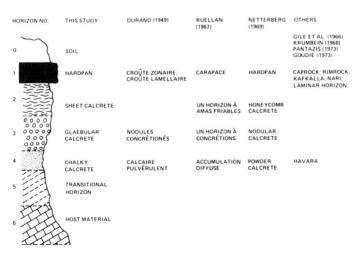

FIG. 2. Idealized calcrete profile and horizon nomenclature.

calcified, thus forming biogenic carbonate structures such as rhizoliths, calcified filaments, calcified faecal pellets, calcified cocoons and *Microcodium* aggregates. The thickness and depth of carbonate accumulation will depend on: the amount of available calcium and carbonate ions; the depth to which calcium-bearing waters can penetrate; the relative and absolute amounts of precipitation and dissolution of calcium carbonate within the soil profile; and the porosity and permeability of the host material.

Stage 4: profile development

In the early stages of calcrete development, the profile is composed of weathered materials with high porosities and permeabilities. Vertical movements of meteoric vadose water can take place relatively easily and the amount of water retained within the soil profile is insufficient to supply the requirements of all plant species. Some plants extend taproots vertically downwards to the capillary fringe above the water table or to a perched water-table overlying an impermeable substrate. Accumulations of calcium carbonate form vertically elongate glaebules and vertically oriented rhizoliths as a result of vertical water movements and the presence of taproots respectively. Roots extend downwards into fractures and joints within the host material, thereby modifying original fabrics, textures and structures. This biological modification is assisted by physicocochemical alteration of the host material, culminating in the formatin of the transitional horizon.

Precipitation of calcium carbonate, without significant cementation because of mechanical and chemical instability, forms the chalky horizon of the pedogenic calcrete profile. The chalky horizon is subjected to frequent periods of wetting and drying. Precipitation and dissolution of calcium carbonate take place concurrently. Pedoturbation (physical, chemical and biological disturbance of soil materials) precludes the formation of indurated layers.

As the accumulation of calcium carbonate increases, porosity and permeability of the profile decrease. Original constituents of the host material are progressively replaced with increasing amounts of calcite. At some point in profile development, it becomes easier for soil water to move laterally rather than vertically. By this stage, most plants form lateral root systems. The change from vertical to horizontal root systems may reflect a change of plant species as the calcrete profile evolves. Interroot distances in woody species are large compared with herbaceous species (Bowen 1973). Large (5–20 cm), isolated, vertically oriented

rhizoliths are common in glaebular horizons, whereas smaller (1–2 mm), branching, horizontally oriented rhizoliths form the bulk of sheet calcrete horizons. Thus, the development of the sheet calcrete horizon from the glaebular horizon may be a reflection of plant succession in a developing calcrete profile.

Stage 5: induration

As accumulation of calcium carbonate increases, a point will be reached when the soil organisms can no longer maintain viability. The intensity of soil-forming processes diminish and eventually cease to be important. Diagenetic processes, mainly cementation and replacement, lead to the fossilization and induration of the soil profile.

Stage 6: reworking

The indurated profile, if it remains at the land surface, is subjected to further processes which will alter or destroy the profile. Soil-forming processes, governed initially by lower plant activities (lichens, algae, fungi, bacteria) will form a protosoil. The prepared substrate allows colonization of other plants such as mosses and grasses. Eventually, the soil profile is able to support higher plants. The root systems of these plants penetrate, dissolve and fracture the indurated hardpan. Disturbance of the calcrete profile by vegetation may form tepee structures and rhizo-breccias. Further pedoturbation, carbonate dissolution and reprecipitation, lead to the formation of a reworked, recemented, breccia-conglomeratic calcrete hardpan.

Concluding remarks

As already indicated, the model presented here is simplistic. The processes involved in the rock-sediment-soil-rock cycle are exceedingly complex. The model is an outline only of sequence of possible events. Some stages may be repeated or reversed; others may be omitted. Emphasis has been given to the role of organic activity which determines largely the morphology of pedogenic calcrete profiles and which controls the direction of profile development. Little is known in detail of the exact processes of bioerosion and biolithogenesis in calcrete formation. At the present time, even less is known of the importance and ubiquity of these processes in determining fabrics, textures and structures of calcrete deposits. Physicochemical considerations alone do not explain satisfactorily the natural phenomena associated with the growth and decay of calcrete deposits. Biological considerations provide many answers to, hitherto, perplexing and unresolved problems of calcrete genesis.

ACKNOWLEDGMENTS: This paper is based on research carried out at the University of Liverpool under the direction of R. G. C. Bathurst. The author expresses gratitude to R. G. C. Bathurst, M. Esteban and M. E. Tucker for encouragement, stimulating discussion and comments on an earlier version of the manuscript, and to N. P. James and B. D. Rickets for reviewing the final draft.

Fieldwork in Spain was supported by the Natural Environment Research Council (Grant No. GT4/75/GS/131).

References

BADIOZAMANI, K., MACKENZIE, F. T. & THORSTENSON, D. C. 1977. Experimental carbonate cementation: salinity temperature and vadosephreatic effects. *J. sedim. Petrol.* **47**, 529–42.

BATHURST, R. G. C. 1975. *Carbonate Sediments and their Diagenesis.* 2nd Ed. Elsevier, Amsterdam. 658 pp.

BERNER, R. A. 1968. Calcium carbonate concretions formed by the decomposition of organic matter. *Science*, **159**, 195–7.

—— 1971. *Principles of Chemical Sedimentology.* McGraw-Hill, New York. 240 pp.

BÖGLI, A. 1964. Mischungskorrosion—ein Beitrag zum Verkarstungsproblem. *Erdkunde*, **18**, 83–92.

BOWEN, G. D. 1973. Mineral nutrition of ectomycorrhizae. *In:* MARKS, G. C. & KOZLOWSKI, T. T. (eds) *Ectomycorrhizae, their Ecology and Physiology*, 151–205. Academic Press, New York.

BREAZEALE, J. F. & SMITH, H. V. 1930. Caliche in Arizona. *Bull. Arizona Agric. Expt. Station*, **131**, 419–41.

BRETZ, J. H. & HORBERG, L. 1949. Caliche in southeastern New Mexico. *J. Geol.* **57**, 491–511.

BREWER, R. 1964. *Fabric and Mineral Analysis of Soils.* Wiley, New York. 470 pp.

BROECKER, W. S. 1974. *Chemical Oceanography.* Harcourt, Brace, Jananovich, New York. 214 pp.

BROWN, C. N. 1956. The origin of caliche in the northeastern Llano Estacado, Texas. *J. Geol.* **64**, 1–15.

BUCKLEY, H. E. 1951. *Crystal Growth.* Wiley, New York. 571 pp.

COQUE, R. 1962. *La Tunisie Présaharienne, Etude Géomorphologique.* Colin, Paris. 476 pp.

CRONEY, D. 1952. The movement and distribution of water in soils. *Geotechnique*, **3**, 1–16.

DUMONT, J. L. 1975. Les croûtes calcaires. Presentation de modèles experimentaux. *C. r. hebd. Séanc. Acad. Sci., Paris, sér D.* **280**, 2073–6.

DURAND, J. H. 1949. Essai de nomenclature des croûtes. *Bull. Soc. Sci. nat. Tunis.* **3–4**, 141–2.

ESTEBAN, M. 1974. Caliche textures and *Microcodium. Bull. Soc. geol. Ital.* **92** (*Suppl.* 1973), 105–25.

—— 1976. Vadose pisolite and caliche. *Bull. Am. Ass. Petrol. Geol.* **60**, 2048–57.

GARDNER, L. R. 1973. Origin of the Mormon Mesa caliche, Clark County, Nevada. *Bull. geol. Soc. Am.* **83**, 143–56.

GILE, L. H., PETERSON, F. F. & GROSSMAN, R. B. 1966. Morphological and genetic sequences of carbonate accumulation in desert soils. *Soil Sci.* **101**, 347–60.

GOUDIE, A. 1973. *Duricrusts in Tropical and Subtropical Landscapes.* Clarendon Press, Oxford. 174 pp.

—— 1975. Petrographic characteristics of calcretes (caliches): modern analogues of ancient cornstones. *In: Coll. 'Types de croûtes calcaires et leur répartition régionale'*, Strasbourg, 1975, 3–6.

HARRISON, R. S. 1977. Caliche profiles: indicators of near-surface subaerial diagenesis, Barbados, West Indies. *Bull. Can. Petrol. Geol.* **25**, 123–73.

JAMES, N. P. 1972. Holocene and Pleistocene calcareous crust (caliche) profiles: criteria for subaerial exposure. *J. sedim. Petrol.* **42**, 817–36.

KLAPPA, C. F. 1978a. *Morphology, composition and genesis of Quaternary calcretes from the western Mediterranean: a petrographic approach.* Unpubl. Ph.D. Thesis, University of Liverpool. 446 pp.

—— 1978b. Biolithogenesis of *Microcodium*: elucidation. *Sedimentology*, **25**, 489–522.

—— 1979a. Lichen stromatolites: criterion for subaerial exposure and a mechanism for the formation of laminar calcretes (caliche). *J. sedim. Petrol.* **49**, 387–400.

—— 1979b. Calcified filaments in Quaternary calcretes: organo-mineral interactions in the subaerial vadose environment. *J. sedim. Petrol.* **49**, 955–68.

—— 1980a. Rhizoliths in terrestrial carbonates: classification, recognition, genesis and significance. *Sedimentology*, **27**, 613–29.

—— 1980b. Brecciation textures and tepee structures in Quaternary calcrete (caliche) profiles from eastern Spain: the plant factor in their formation. *Geol. J.* **15**, 81–9.

KRUMBEIN, W. E. 1968. Geomicrobiology and geochemistry of the 'nari lime-crust' (Israel). *In:* MULLER, G. & FRIEDMAN, G. M. (eds) *Recent Developments in Carbonate Sedimentology in Central Europe*, 138–147. Springer-Verlag, Berlin.

LIPPMANN, F. 1973. *Sedimentary Carbonate Minerals.* Springer-Verlag, Berlin. 228 pp.

MARCHARD, D. E. 1974. Chemical weathering, soil development, and geochemical fractionation in a part of the white Mountains, Mono and Inyo Counties, California. *Prof. Pap. U.S. geol. Surv.* **352-J**, 379–424.

McCAULEY, J. W. & ROY, R. 1974. Controlled nucleation and crystal growth of various CaCO$_3$ phases by the silica gel technique. *Am. Mineral.* **59**, 937–62.

MOTTS, W. S. 1958. Caliche genesis and rainfall in the Pecos Valley area of southeastern New Mexico. *Bull. geol. Soc. Am.* **69**, 1737 (abstract).

MULTER, H. G. & HOFFMEISTER, J. E. 1968. Subaerial laminated crusts of the Florida Keys. *Bull. geol. Soc. Am.* **79**, 183–92.

MURRAY, R. C. & PRAY, L. C. 1965. Dolomitization and limestone diagenesis—and introduction. *In:* PRAY, L. C. & MURRAY, R. C. (eds) *Dolomitization and Limestone Diagenesis—A Symposium*, 1–2. *Spec. Publs Soc. econ. Paleont. Mineral., Tulsa,* **13**.

NETTERBERG, F. 1969. The interpretation of some basic calcrete types. *Bull. S. Afr. Archaeol.* **24**, 117–22.

—— 1971. Calcrete in road construction. *Natl. Inst. Road Res. Bull.* **10**, 1–73.

PANTAZIS, T. M. 1973. A study of the secondary limestones (havara and Kafkalla) of Cyprus. *Bull. Cyprus Geog. Ass.* 'Geographical Chronicles' **2**.

POMAR, L., ESTEBAN, M. LLIMONA, A. & FORTARNAU, R. 1975. Acción de líquenes, algas y hongos en la telodiagénesis de las rocas carbonatodos de la zona litoral prelitoral Catalana. *Inst. Invest. Geol. Univ. Barcelona* **30**, 83–117.

REEVES, C. C. 1976. *Caliche: origin, classification, morphology and uses.* Estacado Books, Lubbock, Texas. 233 pp.

ROBINSON, G. W. 1949. *Soils. Their Origin, Constitution and Classification.* 3rd Ed. Murby, London. 573 pp.

RUELLAN, A. 1967. Individualisation et accumulation du calcaire dans les sols et dépôts quaternaires du Maroc. *Cah. ORSTOM, Sér. Pédologie* **5**, 421–62.

RUNNELS, D. D. 1969. Diagenesis, chemical sediments and the mixing of natural waters. *J. sedim. Petrol.* **39**, 1188–201.

RUSSELL, E. W. 1973. *Soil Conditions and Plant Growth.* 10th Ed. Longman, London. 849 pp.

SCHNEIDER, J. 1977. Carbonate construction and decomposition by epilithic and enolithic microorganisms in salt- and freshwater. *In:* FLÜGEL, E. (ed.) *Fossil Algae*, 248–60. Springer-Verlag, Berlin.

SWEETING, M. M. 1972. *Karst landforms.* Macmillan, London. 362 pp.

THORSTENSON, D. C., MACKENZIE, F. T. & RISTVET, B. K. 1972. Experimental vadose and phreatic cementation of skeletal carbonate sand. *J. sedim. Petrol.* **42**, 162–7.

WALTON, A. G. 1967. *The Formation and Properties of Precipitates.* Wiley, New York. 232 pp.

YAALON, D. H. & GANOR, E. 1973. The influence of dust on soils during the Quaternary. *Soil Sci.* **116**, 146–55.

COLIN F. KLAPPA, Gulf Oil Australia Pty Ltd, GPO Box B88, Perth, Western Australia 6001, Australia.

Stable isotope abundances in calcretes

A. S. Talma & F. Netterberg

SUMMARY: About 300 published and unpublished measurements of carbon and oxygen stable isotope ratios of calcretes are compared in an attempt to define general trends. The $\delta^{13}C$ values of the carbonates range from -12 to $+4‰$ PDB. On a global scale the main factor responsible for the average ^{13}C content of calcrete in an area appears to be the ^{13}C content of the dominant plants in the region. No correlation could be found between ^{13}C contents and radiocarbon age. The carbonate $\delta^{18}O$ values range from -9 to $+3‰$ PDB. On a global scale, the factors discussed as possibly responsible for the ^{18}O content of a calcrete are the ^{18}O content of the local rainfall, temperature, and the extent to which the water underwent free surface evaporation before or during calcrete formation. Little or no correlation was found between ^{13}C or ^{18}O contents and rainfall, temperature, or evaporation rate on a world basis. Some relation between the ^{18}O content of calcrete and rain on a regional basis was confirmed. The range of ^{13}C and ^{18}O values suggest a diversity of origins and environments of formation.

The past decade has witnessed a remarkable revival of interest in calcretes and related materials (e.g. Netterberg 1969; Goudie 1973; Reeves 1976), and measurements of both radioactive and stable isotopes are contributing towards a better understanding of these materials. Radiocarbon contents have been studied to establish the ages, mode of formation (e.g. Brinkmann, Münnich & Vogel 1960; Gile, Peterson & Grossman 1966; Valastro, Davis & Rightmire 1968; Netterberg 1969; Bowler & Polach 1971; Williams & Polach 1971; Netterberg & Vogel 1983) and growth rates (Netterberg 1983) of calcretes. Other radioactive isotopes are also showing promise in these respects (Netterberg 1978). Measurements of stable carbon and, in some cases, oxygen isotopes have been used to assess the initial ^{14}C content and reliability of ^{14}C dates (Brinkmann et al. 1960; Valastro et al. 1968; Netterberg 1969; Williams & Polach 1971; Hendy, Rafter & Macintosh 1972) and mode of formation of calcretes (Netterberg 1969; Hendy et al. 1972; Salomons, Goudie & Mook 1978), while the use of ^{13}C to correct for isotopic fractionation during laboratory determinations of ^{14}C is now common. No measurements of silicon, calcium, magnesium, hydrogen, strontium or sulphur isotopes in calcretes appear to have been published. The variation of the first three in nature is, however, very small (Hoefs 1980) and only hydrogen, strontium and sulphur may prove useful.

It is likely that the carbon and oxygen isotopic composition of calcretes reflects that of the source materials, as well as the temperature, nature and extent of the mechanism(s) of formation and post-formational changes. Of particular interest, therefore, is the possibility of determining whether a calcrete is pedogenic or non-pedogenic in origin, the sources(s) of the carbonate(s), the mechanisms of formation (for example by CO_2- loss or by evaporation), the temperature and water salinity at the time of precipitation, and the extent of post-precipitation 'contamination' which might affect the reliability of radiometric dates and of distinguishing between calcretes and other carbonates.

In this paper a summary is presented of all the published and unpublished stable isotope measurements of calcretes known to the authors from all the continents. The data are examined for trends which might help to unravel the history of these materials.

Mechanisms of carbonate accumulation

The mechanical and chemical mechanisms of carbonate accumulation can be listed as follows (partly after Twenhofel 1939):

Organic
(1) Accumulation of skeletal and protective parts of organisms.
(2) Vital activities of organisms:
    (a) transpiration by plants;
    (b) photosynthesis by plants;
    (c) bacterial processes.

Inorganic
(1) Deposition from suspension in water or air.
(2) Mixing of solutions (common ion effect, increase in pH).
(3) Escape of carbon dioxide (pressure reduction, temperature increase).
(4) Evaporation.

Carbonate deposited or accumulated mechanically can be expected to preserve its isotopic composition. Such undissolved carbonate, which may include fragments of older calcrete, may be incorporated in a calcrete and in a few cases makes up a significant portion of it. Of the chemical mechanisms, only transpiration, evaporation and CO_2-loss are probably of general importance, although the others may be important locally (Netterberg 1969). Most authors (e.g. Salomons & Mook 1976; Salomons *et al.* 1978) only consider evaporation and CO_2-loss important. Recent years have seen renewed support for algal and bacterial origins for some calcretes, notably laminated rinds (e.g. Vaudour & Clauzon 1976).

Stable isotope geochemistry of carbonates

The stable isotope ratios $^{13}C/^{12}C$ and $^{18}O/^{16}O$ show small variations of only a few per cent in different materials in nature. Their contents are therefore reported as the relative deviation from a standard and expressed in per mille (‰):

$$\delta = \frac{S - R}{R} \times 1000$$

where R = isotope ratio of reference standard, S = isotope ratio of sample.

PDB is used as standard for *carbonate* in this paper (Craig 1957) and SMOW for *water* (Craig 1961). While SMOW is generally used to report ^{18}O values of rocks, the PDB standard remains useful in the case of carbonates: at $16.9°C$ $\delta^{18}O$ of calcite (in PDB) and water (in SMOW) in equilibrium are numerically equal. Conversion of δ values between these standards can be done with

$$\delta^{18} \text{ SMOW} = 1.03 \, \delta^{18} \text{ PDB} + 30.4‰$$

Carbon-13

The carbon-13 contents of the main carbon sources in nature are:

  marine carbonates $0 \pm 3‰$ (Degens 1967);
  atmospheric CO_2 $-7 \pm 2‰$ (Degens 1967);
  C_3 (Calvin cycle) plants (general vegetation) $-26 \pm 4‰$ (Craig 1953);
  C_4 (Hatch-Slack cycle) plants (mostly arid zone grass) $-13 \pm 3‰$ (Vogel, Fuls & Ellis 1978);
  CAM plants (mostly succulents) -32 to $-15‰$ (Hoefs 1980).

The respiration of plants and the decay of organic matter produces CO_2 concentrations of up to a few per cent in the atmosphere of unsaturated soil, compared to the 0.03%

found in the free atmosphere. The ^{13}C content of soil CO_2 usually varies between -25 and $-7‰$, depending on the depth, local vegetation and the diffusion rate to the surface (Galimov 1966; Lerman 1972; Rightmire 1978; Dörr & Münnich 1980; Talma, unpublished measurements).

Soil CO_2 dissolves in water and reacts with carbonate to form bicarbonate solution:

$$CaCO_3 + CO_2 + H_2O \rightleftharpoons Ca^{2+} + 2HCO_3^- \quad (1)$$

The ^{13}C content of the bicarbonate will be approximately midway between that of both carbon sources and can be expected to have values between about -14 and $-4‰$. After initial formation the ^{13}C content of the bicarbonate can change due to equilibration with soil CO_2, oxidation of sedimentary carbon, ion exchange, etc.

A major process for carbonate precipitation involves the loss of CO_2 from a carbonate saturated solution and reversal of reaction (1). The CO_2 loss can be caused by some of the mechanisms mentioned above. The precipitated carbonate will have $\delta^{13}C$ 1 to 2‰ higher than the bicarbonate, slightly dependent on temperature (Emrich, Ehhalt & Vogel 1970). The CO_2 removed from solution (equation 1) will be more *depleted* in ^{13}C than the carbonate will be *enriched*. The net result is that the remaining bicarbonate will show increasing ^{13}C values and so will the precipitating carbonate. For example, a 4 to 6‰ enrichment would be expected if 80% of the calcium in solution were precipitated (Salomons *et al.* 1978, fig. 1). The same degree of enrichment would be expected regardless of the mechanism of CO_2 loss.

Precipitation by common ion effects, transpiration by plants or evaporation would be expected to cause similar enrichment to that caused by CO_2 loss. Removal of CO_2 by algae and very rapid carbonate precipitation would be expected to cause further carbonate ^{13}C enrichment by up to several ‰ (Hendy 1971). Fractionation due to precipitation by bacteria would depend on whether the carbon involved originates from soil CO_2 (organic carbon) or from soil CO_2- carbonate interactions. In the former case low ^{13}C values of the precipitated carbonate ($\delta < -10‰$) would be expected, while the latter would lead to higher values ($\delta > -10‰$).

Oxygen-18

The ^{18}O content of calcrete will be directly related to that of the water from which it formed and a small temperature effect

(0.24‰/°C). The isotope content of the formation water, in turn, is that of local rain with some alteration caused by selective infiltration and evaporation. The global distribution of ^{18}O in rain water is essentially due to a depletion effect away from the evaporation zones. This results in a gradual decrease in ^{18}O from the equator towards the poles and from the coastal areas inland (Dansgaard 1964). Substantial variations exist in the annual ^{18}O content of rain at a specific locality from year to year, particularly in drier areas, but over a longer term of decades or so, these variations cancel out and regions have characteristic mean rainfall ^{18}O contents. In most areas the groundwater reflects this average value, although in arid zones some selectivity seems to exist, causing the ^{18}O content of groundwater to be lower than that of the mean annual rainfall (Vogel & Van Urk 1975). Water in the upper layers of the soil can have any ^{18}O value between that of rain and groundwater.

Enrichment of ^{18}O relative to ^{16}O occurs during evaporation from open water bodies. It implies a condition where free exchange between the vapour phase and the remaining liquid phase can take place (Ehhalt & Knott 1965). This occurs in lakes, rivers and open pans. The magnitude of enrichment depends on the fraction of water remaining, the relative humidity in the atmosphere and its ^{18}O content. During transpiration from plants, where the water transport is essentially unidirectional from soil to leaves, no isotope enrichment is found in soil water because all the water entering the plants eventually evaporates and no return flow of enriched water exists.

Salomons *et al.* (1978) have shown that where calcrete formation is produced by open-surface evaporation, *both the ^{13}C and ^{18}O contents* will increase fairly regularly: ^{18}O due to enrichment of the water and ^{13}C due to escape of CO_2 and enrichment of the remaining bicarbonate. If the release of CO_2 is the operative mechanism, *only ^{13}C will become enriched during the course of the reaction. The degree of enrichment of both isotopes can be considerable. For example, if 80% of the solution were evaporated enrichment of about 6‰ for both isotopes would be expected.

Carbon-13 and oxygen-18 measurements

Data collection

The isotopic measurements reported in this survey were mostly obtained from published

measurements, in particular from radiocarbon data lists where ^{13}C is often reported as well. The major sources are listed in the caption to Fig. 6. All of the samples are expected to be of late Tertiary or Pleistocene age, though this is not always stated and so the radiocarbon ages may be ambiguous. The materials measured were generally described as calcretes, caliches, soil carbonates, *ca.* horizons or soil concretions. Beachrocks, dune-rocks, cave deposits and finely divided loess carbonates have been excluded. A number of unpublished measurements made in the first author's laboratory of samples from South Africa and Namibia have also been used in this compilation; these include ^{18}O measurements.

Range of isotope values

The large number (322) of ^{13}C measurements from all over the world used for this study show an average of about $-4‰$ and range from -12 to $+4‰$ (Fig. 1). This may be compared with the mean of $-7‰$ and range of -18

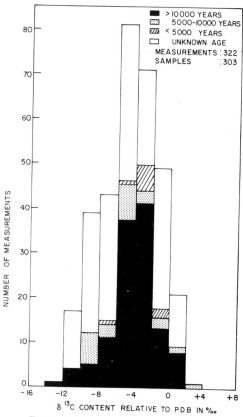

FIG. 1. Histogram of world calcrete carbonate ^{13}C contents.

to 0‰ for fresh water carbonates in general (Degens 1967). The range for calcretes is approximately that expected from the concepts for calcrete formation outlined above. The lowest value can be obtained if CO_2 from the low ^{13}C plants ($-30‰$) dissolves high ^{13}C marine limestone ($+2‰$) to yield dissolved bicarbonate of $-14‰$, from which the first carbonate to precipitate will have a ^{13}C content of $-12‰$. Cases where the calcrete formed from another calcrete or fresh water limestone and soil CO_2 both low in ^{13}C would have led to lower values (Salomons & Mook 1976), but these do not appear to be common. At the other end of the range: if carbonate precipitates from water which is in isotopic equilibrium with atmospheric CO_2 ($\delta\,^{13}C = -7‰$) at a temperature of 10°C, its ^{13}C value will be $-7 + 11 = +4‰$ (Salomons et al. 1978). Any ^{13}C values of calcretes between these two extremes can be

explained as variants of these two processes. The normal distribution of Fig. 1 suggests no particular preference for either process. The range for calcretes found in this study includes the -12 to $-5‰$ given for caliches by Degens (1967) and contains part of the range of -2 to $+5‰$ given by him for evaporite facies carbonates.

The ^{18}O measurements available for this compilation are mainly those of Salomons & Mook (1976), Salomons et al. (1978) and the present authors. The ^{18}O content of all the samples ranges from -9 to $+3‰$, average $-5‰$ (Fig. 2). This range reflects the ^{18}O content of rain water in the drier parts of the world where most calcretes are found. The low ^{18}O values are in agreement with ground and rain water found in some of the sampling areas (Vogel & Van Urk 1975; Salomons et al. 1978). The high values can reflect a degree of ^{18}O enrichment due to evaporation.

Radiocarbon remains a useful means of dating calcretes even though reservations regarding initial ^{14}C content and post-depositional exchange may persist (Netterberg & Vogel 1983). Using the measured radiocarbon content *per se*, we could find no correlation between 156 ^{13}C and eight ^{18}O measurements and radiocarbon age on a world-wide basis. One might have expected changes of ^{13}C due to vegetational changes at the end of the last ice age (Hendy et al. 1972). On a global scale this change is not evident, though obviously more detailed investigations of this nature, including ^{18}O, over smaller areas could well be fruitful (e.g. Magaritz, Kaufman & Yaalon 1981).

Climatic correlations

Because both ^{13}C and ^{18}O become concentrated relative to their lighter isotopes by evaporation (Salomons et al. 1978), it might be expected that the relative amounts of these isotopes will increase with increasing aridity. Plots of carbonate isotope measurements, separated where known according to those with radiocarbon ages of less than 5000 years, 5000–10 000 years and those over 10 000 years, were therefore made against mean annual rainfall and evaporation rates. Figs 3 and 4 show the climatic effect to be slight on a world-wide basis in both cases, though possibly somewhat more significant in the case of ^{18}O. In the arid zone (annual rainfall <250 mm) $\delta\,^{13}C$ values of less than $-6‰$ were rare (two out of 90) and $\delta\,^{18}O$ less than $-5‰$ did not occur at all. Furthermore, areas receiving less than 350 mm rainfall per annum yielded all the $\delta\,^{18}O$ values

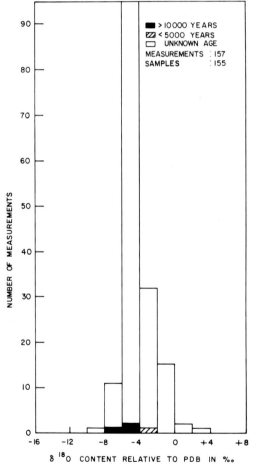

FIG. 2. Histogram of world calcrete carbonate ^{18}O contents.

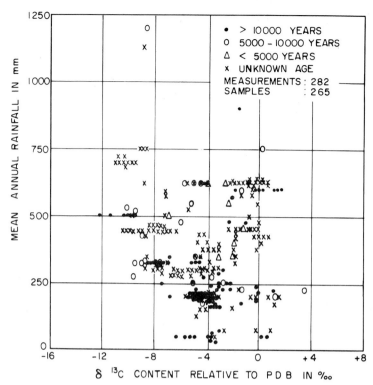

FIG. 3. Plot of world calcrete carbonate ^{13}C contents versus rainfall.

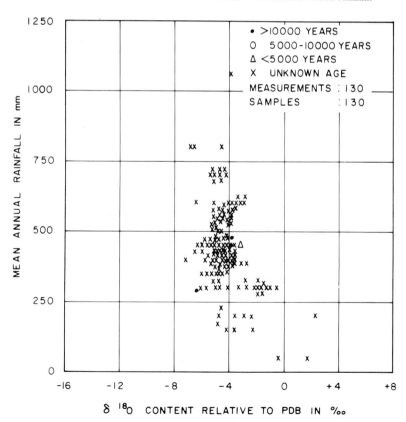

FIG. 4. Plot of world calcrete carbonate ^{18}O contents versus rainfall.

greater than −2‰. The latter values are from areas with gross evaporation rates in excess of 3.3 m *per annum*; the Southern African samples did not show a correlation between δ ^{18}O and evaporation rate (44 measurements, Fig. 5).

Other correlations unsuccessfully sought for but not presented here, were ^{13}C content against mean annual gross Class A pan evaporation rate (176 measurements world-wide), against net evaporation rate (gross minus rainfall) in Southern Africa (76 measurements) and against mean annual temperature (176 measurements world-wide). The effect of seasonal rainfall will be discussed in a later section.

Finally, although the range of 6‰ for the 102 ^{18}O measurements is nearly what might be expected from the 20°C range in temperature and the known temperature coefficient of the fractionation factor, no correlation was found between ^{18}O content and mean annual temperature on a world-wide basis (102 measurements, not shown). Temperature effects on this large scale are masked by the global ^{18}O pattern in rainwater and by evaporation effects (Salomons *et al.* 1978). Attempts to use ^{18}O

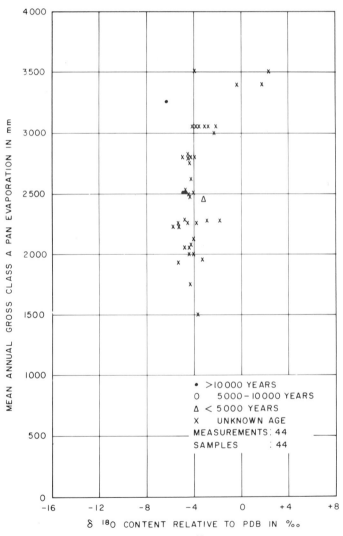

FIG. 5. Plot of southern African calcrete carbonate ^{18}O contents versus gross free water surface evaporation rate.

measurements to estimate palaeotemperatures in calcretes might be made on a more regional basis, but would require a much more detailed knowledge of the carbonate formation mechanism.

In view of the general lack of correlation between isotope content and climatic parameters on a *global* basis, a comparison with soil ^{13}C and rain ^{18}O on a *regional* basis was made.

Global isotopic variations

Carbon-13 in soil CO_2

Soil CO_2 will influence the formation of calcrete and its ^{13}C content. An attempt was therefore made to estimate the ^{13}C content of soil CO_2 in various areas from which data on calcretes are available. As only sporadic measurements of ^{13}C in soil CO_2 have been published, the indirect method of considering the local vegetation was therefore used. In the temperate areas most trees and shrubs, as well as grasses, use the C_3 photosynthetic pathway which results in these plants having an average ^{13}C content of $-26‰$ (Craig 1953). A number of grasses, mainly occurring in tropical and/or arid zones, use the C_4 pathway, resulting in average ^{13}C contents of $-13‰$ (Smith & Epstein 1971). Therefore the relative importance of these two plant types in an environment determines the ^{13}C content of the soil CO_2.

Detailed vegetation surveys of C_4 and C_3 grass species have been made in north America (Teeri & Stowe 1976), in central Africa (Tieszen *et al.* 1979; Livingstone & Clayton 1980) and in southern Africa (Vogel *et al.* 1978; Ellis, Fuls & Vogel 1980). Less detailed sampling has been reported for the northern Sahara desert (Winter, Troughton & Card 1976), the Indian desert (Sankhla *et al.* 1975) and Israel (Winter & Troughton 1978). From these data the relative occurrence of C_4 grass species relative to all grass species was estimated. For the other areas considered here, use was made of world maps depicting the distribution of the most important grass subfamilies containing C_4 (*Andropogoneae* and *Eragrostoideae*) and the C_3 grasses, (*Poa* and *Festucoideae*, Hartley 1958, 1961, 1973; Hartley & Slater 1960). This ignores the contribution of the *Panicoideae* which contain both C_3 and C_4 species. Comparison of these estimated grass C_4 abundances with measured ^{13}C in soil CO_2, where these have been reported (Galimov 1966; Rightmire & Hanshaw 1973; Wallick 1976; Rightmire 1978; Reardon, Alli-

TABLE 1. *Relation between the estimated fraction of C_4 grass species relative to all grass species and reported values of ^{13}C of soil CO_2*

Fraction C_4 grass	^{13}C of soil CO_2
Low (less than 30%)	-27 to $-19‰$
Medium (30 to 70%)	-23 to $-14‰$
High (more than 70%)	-18 to $-7‰$

son & Fritz 1979; Dörr & Münnich 1980; Talma, unpublished measurements) indicate that this approach yields a rough guide to soil ^{13}C values (Table 1).

The large ^{13}C variation within each group is probably due to the varying influence of trees and shrubs (all C_3) as well as CAM plants (succulents) which have ^{13}C contents within these two extreme values

Calcrete carbon-13 world-wide

The carbon-13 content of the calcretes considered here show some latitudinal variation. High values of -5 to $+2‰$ are found in the tropical regions and lower values (-12 to $-8‰$) in the temperate areas of 40–50° latitude (Fig. 6). This change also corresponds to the change from a mixture of C_4 grasses and other C_3 plants in the natural vegetation in India and in Africa between the 30° parallels, towards the sole C_3-type vegetation in the temperate areas and with a general trend from summer rainfall to winter rainfall (according to Köppen's climatic classification, Trewartha & Horn 1980).

The actual ^{13}C value of the soil CO_2 in areas with high C_4 occurrence will naturally depend on the relative importance of grass to the total biomass of a particular habitat. In the high C_4 areas of southern Africa soil CO_2 δ ^{13}C values between -18 and $-7‰$ have been measured (Talma, unpublished data), representing some 20–100% influence of grass. Carbonate deposited from water in equilibrium with soil gas having these ^{13}C values will possess ^{13}C contents from -8 to $+3‰$ (Emrich *et al.* 1970).

The existence of isotopic equilibrium between soil CO_2 and calcrete implies that the CO_2 forms an important solubility control during calcrete formation. Carbon-13 measurements on soil CO_2 in areas in which calcrete is currently forming should therefore be carried out. Guidelines for distinguishing between such contemporary and fossil calcretes have been prepared (Netterberg 1981).

The few samples identified as *grounawater*

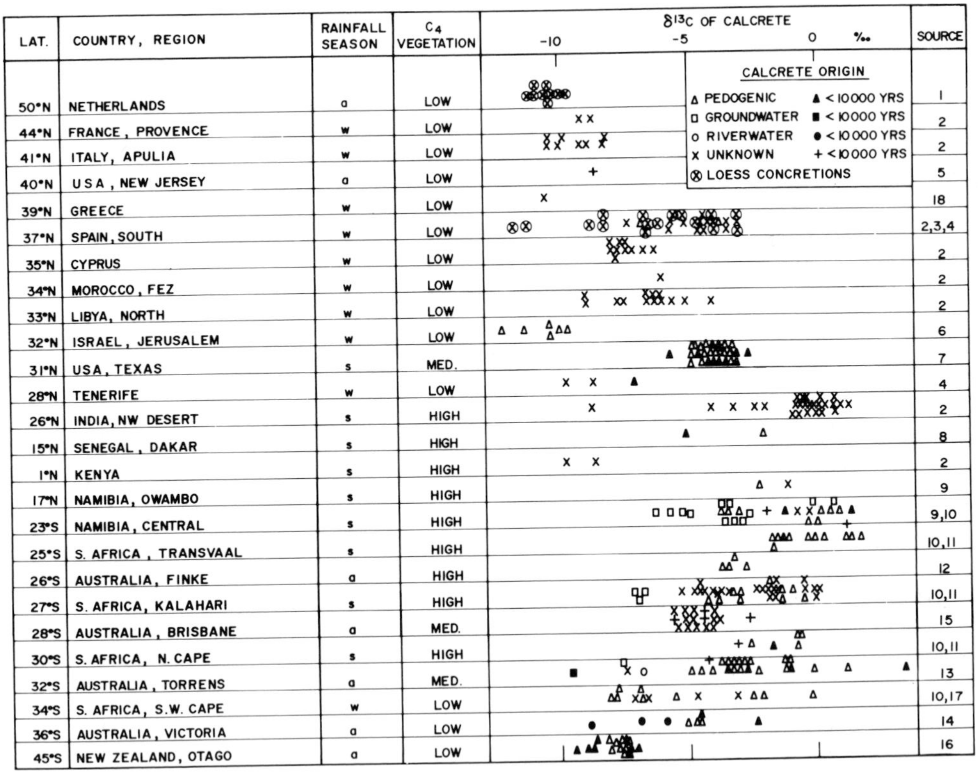

LAT.	COUNTRY, REGION	RAINFALL SEASON	C₄ VEGETATION	δ¹³C OF CALCRETE	SOURCE
50°N	NETHERLANDS	a	LOW		1
44°N	FRANCE, PROVENCE	w	LOW		2
41°N	ITALY, APULIA	w	LOW		2
40°N	USA, NEW JERSEY	a	LOW		5
39°N	GREECE	w	LOW		18
37°N	SPAIN, SOUTH	w	LOW		2,3,4
35°N	CYPRUS	w	LOW		2
34°N	MOROCCO, FEZ	w	LOW		2
33°N	LIBYA, NORTH	w	LOW		2
32°N	ISRAEL, JERUSALEM	w	LOW		6
31°N	USA, TEXAS	s	MED.		7
28°N	TENERIFE	w	LOW		4
26°N	INDIA, NW DESERT	s	HIGH		2
15°N	SENEGAL, DAKAR	s	HIGH		8
1°N	KENYA	s	HIGH		2
17°N	NAMIBIA, OWAMBO	s	HIGH		9
23°S	NAMIBIA, CENTRAL	s	HIGH		9,10
25°S	S. AFRICA, TRANSVAAL	s	HIGH		10,11
26°S	AUSTRALIA, FINKE	a	HIGH		12
27°S	S. AFRICA, KALAHARI	s	HIGH		10,11
28°S	AUSTRALIA, BRISBANE	a	MED.		15
30°S	S. AFRICA, N. CAPE	s	HIGH		10,11
32°S	AUSTRALIA, TORRENS	a	MED.		13
34°S	S. AFRICA, S.W. CAPE	w	LOW		10,17
36°S	AUSTRALIA, VICTORIA	a	LOW		14
45°S	NEW ZEALAND, OTAGO	a	LOW		16

Legend (CALCRETE ORIGIN):
- △ PEDOGENIC   ▲ < 10000 YRS
- □ GROUNDWATER   ■ < 10000 YRS
- ○ RIVERWATER   ● < 10000 YRS
- × UNKNOWN   + < 10000 YRS
- ⊗ LOESS CONCRETIONS

FIG. 6. Spread of ¹³C content of calcretes at different latitudes. Calcrete origins, age classes, Köppen rainy seasons (a = all year round, w = winter, s = summer) and C₄ abundances in grass species are shown. Sources of ¹³C measurements: (1) Salomons & Mook (1976); (2) Salomons *et al.* (1978); (3) Manze *et al.* (1974); (4) Brinkmann *et al.* (1960); (5) Buckley (1976); (6) Carmi, Noter & Schlesinger (1971); (7) Valastro *et al.* (1968); (8) Lappartient (1971); (9) Netterberg (1969); (10) Talma & Vogel (unpublished); (11) Vogel & Marais (1971); (12) Polach, Lovering & Bowler (1970); (13) Williams & Polach (1971); (14) Bowler & Polach (1971); (15) Hendy *et al.* (1972); (16) Leamy & Rafter (1972); (17) Siesser (1972); (18) Degens & Epstein (1964).

calcretes have lower ¹³C values than the more common *pedogenic* varieties in the same region (Fig. 6). This is to be expected, since groundwater at greater depth can have been influenced by the deep roots of trees (low ¹³C) compared to shallow soils where only grass (high ¹³C in C₄ areas) will be effective. As pointed out by Salomons *et al.* (1978), even higher ¹³C contents can be expected in calcretes formed close to the surface of poorly vegetated, well-aerated soils in which the ¹³C content of the soil CO₂ will approach that of the atmosphere, i.e. −7‰. Such pedogenic calcretes might be termed '*rainwater* calcretes' and should have ¹³C content similar to that of marine limestone and δ¹⁸O fairly close to that of the local rainwater.

Although Salomons *et al.* (1978) did not find a consistent upward increase in calcrete ¹³C and ¹⁸O in the four profiles investigated, both their Indian profiles have ¹³C content within the range for marine carbonates, while the only other profile which has sampled at depths shallower than 300 mm (Cyprus) does show an increase in both isotopes towards the surface. The same trend is common, though not ubiquitous, in the Texan profiles measured by Valastro *et al.* (1968). The few '*river water* calcretes', i.e. carbonate-cemented alluviums (Fig. 6), appear to fit in between the groundwater and pedogenic types. This emphasizes the importance of proper sample and profile description and identification if isotopic methods are to be utilized to their fullest extent.

Calcrete oxygen-18 related to local water

Separation of the calcrete measurements on a regional basis shows an increase, as well as a larger scatter, of ¹⁸O values southwards (Fig. 7). Most of this increase is due to the

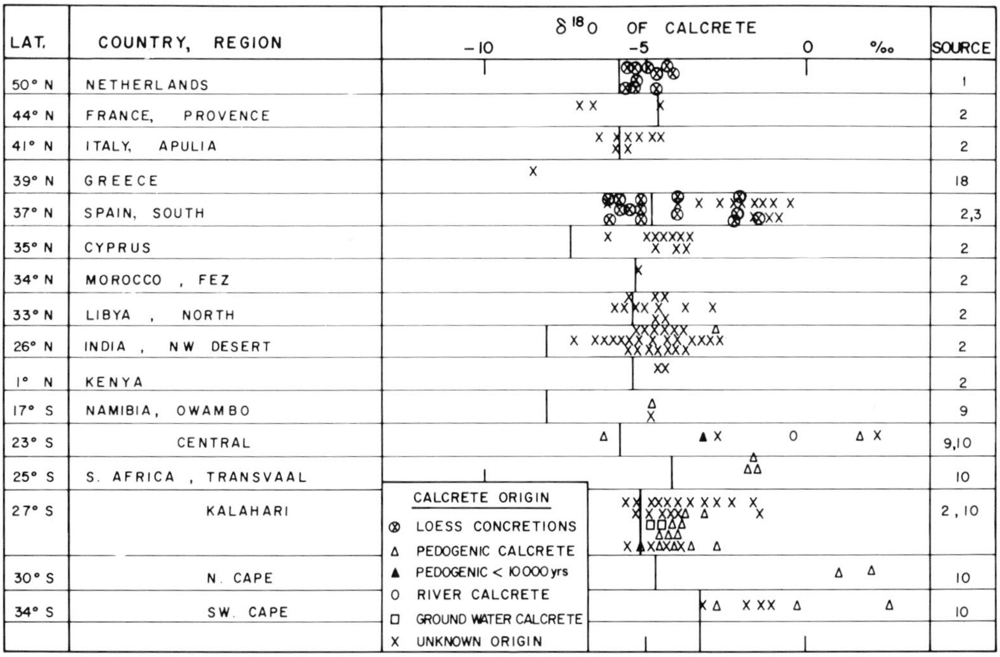

LAT.	COUNTRY, REGION	$\delta^{18}O$ OF CALCRETE	SOURCE
50° N	NETHERLANDS		1
44° N	FRANCE, PROVENCE		2
41° N	ITALY, APULIA		2
39° N	GREECE		18
37° N	SPAIN, SOUTH		2,3
35° N	CYPRUS		2
34° N	MOROCCO, FEZ		2
33° N	LIBYA, NORTH		2
26° N	INDIA, N W DESERT		2
1° N	KENYA		2
17° S	NAMIBIA, OWAMBO		9
23° S	CENTRAL		9,10
25° S	S. AFRICA, TRANSVAAL		10
27° S	KALAHARI		2,10
30° S	N. CAPE		10
34° S	SW. CAPE		10

CALCRETE ORIGIN
⊗ LOESS CONCRETIONS
△ PEDOGENIC CALCRETE
▲ PEDOGENIC < 10,000 yrs
○ RIVER CALCRETE
□ GROUND WATER CALCRETE
X UNKNOWN ORIGIN

FIG. 7. Worldwide spread of ^{18}O contents of calcretes. The vertical lines are the expected calcrete ^{18}O values calculated from mean temperature and rainfall ^{18}O content. See Fig. 6 for key to sources of ^{18}O measurements.

effect of increasing aridity on ^{18}O discussed above (Figs 4 & 5).

Salomons *et al.* (1978) have compared the measured ^{18}O content of calcrete to that expected from average rainfall ^{18}O and mean annual temperature in the countries concerned. They found either correspondence to these calculated values, or some ^{18}O enrichment in the calcrete, which was ascribed to evaporation during calcrete formation. However, in arid areas some enrichment of ^{18}O in rainfall may take place *prior* to infiltration (Vogel & Van Urk 1975) and therefore before calcrete formation. Additional southern African measurements by the present authors show a similar effect (Fig. 7) and indicate major evaporation influences in most of southern Africa, India, Spain and Cyprus. Extreme caution must therefore be exercised when this type of measurements is interpreted in terms of ^{18}O changes of water in former times. *Groundwater* calcrete should exhibit more consistent isotopic contents. In the drier regions of southern Africa groundwater is frequently only recharged by the heavier downpours which penetrate into the soil beyond the influence of evapo-transpiration. These downpours have a lower ^{18}O content than the mean annual rainfall (Vogel & Van Urk 1975) and

the carbonate forming from groundwater can thus be expected to have lower, more consistent ^{18}O (and ^{13}C) content. In addition there will be less possibility of any evaporation enrichment occurring. The small number of identified groundwater calcretes known to us, generally have stable isotope contents on the low ends of the ranges (Figs 6 & 7).

Obviously more detailed local investigations along the lines of Magaritz *et al.* (1981) are necessary to evaluate these effects more thoroughly.

Relationship between ^{13}C and ^{18}O

A scatter diagram with the two isotopic values as axes is useful to view general trends (Fig. 8). Most of the values are in the double negative quadrant indicating fresh water carbonates (Keith & Weber 1964; Hudson 1977). The ^{13}C variations are much larger than those of ^{18}O. This is due to the smaller carbon reservoir available for precipitation (soil CO_2 and HCO_3) than for oxygen and thus easier changed by various processes (Magaritz *et al.* 1981).

Another reason why even calcrete formed by the evaporation of water may show little ^{18}O enrichment is the very low evaporation rate of

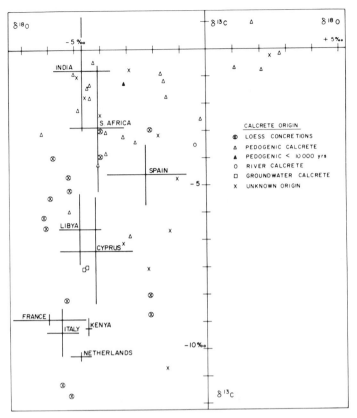

Fig. 8. Relation between ^{13}C and [18]O contents of various calcrete origins. Large crosses covering countries are from Salomons *et al.* (1978) and show 1σ spread.

soil moisture from within the soil. Already at depths of a decimetre the evaporation rate is substantially reduced (Hellwig 1973). It follows that there is very little opportunity for significant [18]O enrichment in soil water before the next rainfall will cause sufficient infiltration to swamp this effect. Calcrete formed at these depths will then be close to equilibrium with the average rainfall.

Apart from a few positive values of both δ ^{13}C and δ [18]O there is not much correlation evident (Fig. 8). There is no *a priori* reason to expect a world-wide correlation since the origin and distribution of the carbon and oxygen isotopes are quite different. A correlation can, however, be expected in a specific area where, from the same initial conditions, evaporation enrichment can cause simultaneous enrichment of both carbon and oxygen isotopes. This has been calculated by Salomons *et al.* (1978) and observed by them in a number of samples from a single site in Spain. Another possibility for correlation may be if carbonate is *very rapidly*

precipitated from water. Then kinetic fractionation effects may cause simultaneous enrichment of carbon and oxygen (Hendy 1971).

The high values of both ^{13}C and [18]O are interpreted as the result of considerable evaporation and close contact of the calcrete-forming zone with the free atmosphere: the typical situation in an arid, low-vegetation environment. Isotopically these values cannot be distinguished from those of typical marine and lacustrine carbonates (Keith & Weber 1964). Such classification schemes based on bulk rock measurement must be treated with caution because of later changes due to cementation and diagenesis (Hudson 1977).

Conclusions

The general pattern of isotope distributions emerging from this study is one of wide variations reflecting the variety of carbon and oxygen sources and the process involved in calcrete formation. The carbon isotopic content is particularly variable, both globally and regionally. There is therefore a great need for cooperative studies between pedologists and isotope geochemists in order to evaluate some of these effects in detail. These must concentrate on specific sites and establish the most likely mechanism responsible for calcrete formation. Such studies should also include soil chemistry, mineralogy and dating (^{14}C). Specific conclusions from our study are:

(1) The carbonate ^{13}C contents of calcretes on a world-wide basis range between -12 and $+4‰$, with a mean of $-4‰$.

(2) The carbonate ^{18}O contents of calcretes on a world-wide basis range between -9 and $+3‰$, with a mean of $-5‰$.

(3) There is no correlation between ^{13}C and ^{18}O contents on a world-wide basis.

(4) Calcrete ^{18}O and ^{13}C contents range from the lower values for freshwater carbonates to the maxima for marine carbonates, indicating a diversity of mechanisms ranging from that of CO_2-loss from freshwater, to evaporation under possibly saline conditions.

(5) Calcretes can often be distinguished from authigenic marine carbonates on the basis of their ^{18}O and ^{13}C contents, and these isotopes may also prove of value in distinguishing between pedogenic and other calcretes and between some calcretes and other freshwater carbonates.

(6) Little or no correlation was found between ^{13}C and ^{18}O contents and ^{14}C age, rainfall, temperature, potential evaporation or actual evapotranspiration on a world-wide basis.

(7) The ^{13}C content of calcrete increases from the temperate, C_3-plant dominant areas in high latitudes to the arid, C_4- grass dominant areas closer to the equator. This is paralleled by a change from predominantly winter to predominantly summer rainfall environments.

(8) The lowest ^{13}C values found in calcretes of a given area are consistent with equilibrium with the ^{13}C of bicarbonate dissolved in ground or soil water found there. Higher ^{13}C contents can be explained by enrichment due to CO_2-loss, equilibration with atmospheric CO_2 or other causes.

(9) The conclusion by previous authors that the ^{18}O content is determined by the ^{18}O content of the local rainfall, the local temperature and the extent of free water surface evaporation is supported.

(10) ^{18}O values in calcrete that are higher than would be expected from the ^{18}O content of local rainfall and the average temperature indicate that water evaporation caused the deposition of the carbonate unless ^{18}O enrichment occurs in the rain water prior to infiltration into the soil.

(11) Deeper calcretes formed by evaporation may still have ^{18}O values close to equilibrium with average rainwater due to the low evaporation rate of water from within the soil.

ACKNOWLEDGMENTS: This paper is published with the permission of the Directors of the National Physical Research Laboratory and the National Institute for Transport and Road Research. J. C. Vogel took part in many helpful discussions and made unpublished data available. R. P. Ellis advised on the vegetation aspects and E. L. Lursen performed some of the stable isotope analyses. D. Ventura, P. Skosana and M. Netterberg abstracted much of the raw data from the literature.

References

BOWLER, J. M. & POLACH, H. A. 1971. Radiocarbon analyses of soil carbonates: an evaluation from paleosols in S.E. Australia. *In:* YAALON, D. H. (ed.) *Paleopedology*, 97–108. Int. Soc. Soil Sci. plus Israel University Press, Jerusalem.

BRINKMANN, R., MÜNNICH, K. O. & VOGEL, J. C. 1960. Anwendung der C^{14}-Methode auf Bodenbildung und Grundwasserkreislauf. *Geol. Rdsch.* **49**, 244–53.

BUCKLEY, J. 1976. Isotopes' radiocarbon measurements XI. *Radiocarbon*, **18**, 172–89.

CARMI, J., NOTER, Y. & SCHLESINGER, R. 1971. Rehovot radiocarbon measurements I. *Radiocarbon*, **13**, 412–9.

CRAIG, H. 1953. The geochemistry of the stable carbon isotopes. *Geochim. cosmochim. Acta,* **3**, 53–92.

—— 1957. Isotopic standard for carbon and oxygen and correction factors for mass-spectrometric analysis of carbon dioxide. *Geochim. cosmochim. Acta,* **12**, 133–49.

—— 1961. Standard for reporting concentrations of deuterium and oxygen-18 in natural waters *Science*, **133**, 1833–4.

DANSGAARD, W. 1964. Stable isotopes in precipitation. *Tellus*, **16**, 436–68.

DEGENS, E. T. 1967. Stable isotope distribution in carbonates. *In:* CHILINGAR, G. V., BISSELL, H. J.

& FAIRBRIDGE, R. W. (eds) *Carbonate Rocks*, Part B, 194–208. Elsevier, Amsterdam.

DEGENS, E. T. & EPSTEIN, S. 1964. Oxygen and carbon isotope ratios in coexisting calcites and dolomites from recent and ancient sediments. *Geochim. cosmochim. Acta*, **28**, 23–44.

DÖRR, H. & MÜNNICH, K. O. 1980. Carbon-14 and carbon-13 in soil CO_2. *Radiocarbon*, **22**, 909–18.

EHHALT, D. H. & KNOTT, K. 1965. Kinetische Isotopentrennung bei der Verdampfung von Wasser. *Tellus*, **17**, 388–97.

ELLIS, R. P., FULS, A. & VOGEL, J. C. 1980. Photosynthetic pathways and the geographical distribution of grasses in South West Africa. *S. Afr. J. Sci.* **76**, 307–14.

EMRICH, K., EHHALT, D. H. & VOGEL, J. C. 1970. Carbon isotope fractionation during the precipitation of calcium carbonate. *Earth planet. Sci. Lett.* **8**, 363–71.

GALIMOV, E. M. 1966. Carbon isotopes of soil CO_2. *Geokhimiya*, **9**, 1110–8.

GILE, L. H., PETERSON, F. F. & GROSSMAN, R. B. 1966. Morphological and genetic sequences of carbonate accumulation in desert soils. *Soil Sci.* **101**, 347–60.

GOUDIE, A. 1973. *Duricrusts in Tropical and Subtropical Landscapes*. Clarendon Press, Oxford.

HARTLEY, W. 1958. Studies on the origin, evolution and distribution of the Graminae: I. The tribe Andropogoneae. *Aust. J. Bot.* **6**, 116–28.

—— 1961. Studies on the origin, evolution and development of the Graminae. IV. The genus Poa. *Aust. J. Bot.* **9**, 152–61.

—— 1973. Studies on the origin, evolution and distribution of the Graminae: V. The subfamily Festucoideae. *Aust. J. Bot.* **21**, 201–34.

——, & SLATER, C. 1960. Studies on the origin, evolution and distribution of the Graminae: III. The tribes of the sub-family Eragrostoideae. *Aust. J. Bot.* **8**, 256–76.

HELLWIG, D. R. 1973. Evaporation of water from sand, 4. *J. Hydrol.* **18**, 317–27.

HENDY, C. H. 1971. The isotopic geochemistry of speleothems: 1. The calculation of the effects of different modes of formation on the isotopic composition of speleothems and their applicability as palaeoclimatic indicators. *Geochim. cosmochim. Acta*, **35**, 801–24.

——, RAFTER, T. A. & MACINTOSH, N. W. G. 1972. The formation of carbonate nodules in the soils of the Darling Downs, Queensland, Australia and the dating of the Talgai cranium. *Proc. 8th int. Conf. Radiocarbon Dating, Wellington*, **1**, D106–26. Roy. Soc. N. Zealand, Wellington.

HOEFS, J. 1980. *Stable Isotope Geochemistry*. Springer-Verlag, Berlin.

HUDSON, J. D. 1977. Stable isotopes and limestone lithification. *J. geol. Soc. London*, **133**, 637–60.

KEITH, M. L. & WEBER, J. N. 1964. Carbon and oxygen isotopic composition of selected limestones and fossils. *Geochim. cosmochim. Acta*, **28**, 1787–816.

LAPPARTIENT, J. R. 1971. Périodes de concrétionnement calcaire dans le Quaternaire récent de Dakar (Senegal). *Bull. Soc. géol. Fr.* **XIII**, 409–15.

LEAMY, M. L. & RAFTER, T. A. 1972. Isotope ratios preserved in pedogenic carbonate and their application in palaeopedology. *Proc. 8th int. Conf. Radiocarbon Dating, Wellington*, **1**, D42–D57. Roy. Soc. N. Zealand, Wellington.

LERMAN, J. C. 1972. Soil CO_2 and groundwater: carbon isotope compositions. *Proc. 8th int. Conf. Radiocarbon Dating, Wellington*, **1**, D92–D105. Roy. Soc. N. Zealand, Wellington.

LIVINGSTONE, D. A. & CLAYTON, W. D. 1980. An altitudinal cline in tropical African grass floras and its paleo-ecological significance. *Quat. Res.* **13**, 392–402.

MAGARITZ, M., KAUFMAN, A. & YAALON, D. H. 1981. Calcium carbonate nodules in soils: $^{18}O/^{16}O$ and $^{13}C/^{12}C$ ratios and ^{14}C contents. *Geoderma*, **25**, 157–72.

MANZE, U., VOGEL, J. C., STREIT, R. & BRUNNACKER, K. 1974. Isotopenuntersuchungen zum Kalkumsatz im Löss. *Geol. Rdsch.* **63**, 885–97.

NETTERBERG, F. 1969. *The geology and engineering properties of South African calcretes*. Ph.D. Thesis, University of Witwatersrand, Johannesburg.

—— 1978. Dating and correlation of calcretes and other pedocretes. *Trans. geol. Soc. S. Afr.* **81**, 379–91.

—— 1983. Rates of calcrete formation from radiocarbon dates. *Proc. Calcrete Speciality Session, 10th int. Congr. Sed., Jerusalem, Sedimentology* (submitted).

—— 1981. Distinguishing fossil from contemporary calcretes. *Paper presented at 5th S. Afr. quat. Conf.*, Pretoria, May 1981.

—— & VOGEL, J. C. 1983. Radiocarbon dating of calcretes. *Proc. Calcrete Speciality Session, 10th int. Congr. Sed., Jerusalem, Sedimentology* (submitted).

POLACH, H. A., LOVERING, J. F. & BOWLER, J. M. 1970. ANU radiocarbon date list IV. *Radiocarbon*, **12**, 1–18.

REARDON, E. J., ALLISON, G. B. & FRITZ, P. 1979. Seasonal chemical and isotopic variations of soil CO_2 at Trout Creek, Ontario. *J. Hydrol.* **43**, 355–71.

REEVES, C. C. (JR) 1976. *Caliche*. Estacado Books, Lubbock.

RIGHTMIRE, C. T. 1978. Seasonal variation of PCO_2 and ^{13}C content of soil atmosphere. *Wat. Resour. Res.* **14**, 691–2.

—— & HANSHAW, B. B. 1973. Relationship between the carbon isotope composition of soil CO_2 and dissolved carbonate species in groundwater. *Wat. Resour. Res.* **9**, 958–67.

SALOMONS, W. & MOOK, W. G. 1976. Isotope geochemistry of carbonate dissolution and reprecipitation in soils. *Soil Sci.* **122**, 15–24.

——, GOUDIE, A. & MOOK, W. G. 1978. Isotopic composition of calcrete deposits from Europe, Africa and India. *Earth Surf. Processes*, **3**, 43–57.

SANKHLA, N., ZIEGLER, H., VYAS, O. P., STICHLER,

W. & TRIMBORN, P. 1975. Ecophysiological studies on Indian arid zone plants. V: A screening of some species for the C_4 pathway of photosynthetic CO_2-fixation. *Oecologia* (Berl.) **21**, 123–9.

SIESSER, W. G. 1972. Petrology of the Cainozoic coastal limestones of the Cape Province, South Africa. *Trans. geol. Soc. S. Afr.* **75**, 177–85.

SMITH, B. N. & EPSTEIN, S. 1971. Two categories of $^{13}C/^{12}C$ ratios for higher plants. *Plant Physiol.* **47**, 380–4.

TEERI, J. A. & STOWE, L. G. 1976. Climatic patterns and the distribution of C_4 grasses in North America. *Oecologia* (Berl.) **23**, 1–12.

TIESZEN, L. L., SENYIMBA, M. M., IMBAMBA, S. K. & TROUGHTON, J. H. 1979. The distribution of C_3 and C_4 grasses and carbon isotope discrimination along an altitudinal and moisture gradient in Kenya. *Oecologia* (Berl.) **37**, 337–50.

TREWARTHA, G. T. & HORN, L. H. 1980. *An Introduction to Climate.* McGraw-Hill, New York.

TWENHOFEL, W. A. 1939. *Principles of Sedimentation.* McGraw-Hill, New York.

VALASTRO, S., DAVIS, E. M. & RIGHTMIRE, C. T. 1968. University of Texas at Austin radiocarbon dates VI. *Radiocarbon*, **10**, 384–401.

VAUDOUR, J. & CLAUZON, G. 1976. Chronique de pédologie méditerranéenne. Les croutes ont-elles toutes une origine pedologique? *Méditeranée*, **1**, 71–81.

VOGEL, J. C. & MARAIS, M. 1971. Pretoria radiocarbon dates I. *Radiocarbon.* **13**, 378–94.

—— & VAN URK, H. 1975. Isotopic composition of groundwater in semi-arid regions of Southern Africa. *J. Hydrol.* **25**, 23–36.

——, FULS, A. & ELLIS, R. P. 1978. The geographical distribution of Kranz grasses in South Africa. *S. Afr. J. Sci.* **74**, 209–15.

WALLICK, E. I. 1976. Isotopic and chemical considerations in radiocarbon dating of groundwater within the semi-arid Tucson Basin, Arizona. *In: Interpretation of Environmental Isotope and Hydrochemical Data in Groundwater Hydrology.* IAEA, Vienna, 195–212.

WINTER, J., TROUGHTON, J. H. & CARD, K. A. 1976. $\delta^{13}C$ values of grass species collected in the northern Sahara desert. *Oecologia* (Berl.) **25**, 115–23.

—— & TROUGHTON, J. 1978. Photosynthetic pathways in plants of coastal and inland habitats of Israel and the Sinai. *Flora*, **167**, 1–34.

WILLIAMS, G. E. & POLACH, H. A. 1971. Radiocarbon dating of arid-zone calcareous paleosols. *Bull. geol. Soc. Am.* **82**, 3069–86.

A. S. TALMA, National Physical Research Laboratory, Council for Scientific and Industrial Research, PO Box 395, Pretoria 0001, South Africa.

F. NETTERBERG, National Institute for Transport and Road Research, Council for Scientific and Industrial Research, PO Box 395, Pretoria 0001, South Africa.

A geotechnical classification of calcretes and other pedocretes

F. Netterberg & J. H. Caiger

SUMMARY: Authigenic calcareous accumulations within regoliths can be simply classified for geotechnical purposes as calcareous soils, calcified soils, powder calcretes, nodular calcretes, honeycomb calcretes, hardpan calcretes, and calcrete boulders and cobbles. Each of these categories represents a particular stage in the growth or weathering of a calcrete horizon and possesses a significantly different range of geotechnical properties. A similar classification can be applied to other pedocretes.

Development of the arid and semi-arid zones has increasingly involved the use of non-traditional materials such as calcretes for construction and foundation materials. Such exploitation has often revealed inadequacies in certain geotechnical procedures developed in temperate zones as well as the necessity for studies on these materials. This paper outlines a simple, descriptive classification suitable for geotechnical use on calcretes and similar materials based on approximately 20 years of personal experience of both the authors with these materials. The classification is the latest of several earlier studies (Caiger 1964; Netterberg 1967, 1969a, 1971), and largely represents a very condensed and simplified geotechnical version of one of them (Netterberg 1980) embracing all the known morphogenetic forms of calcrete formation and weathering processes. Although based largely upon southern African experience, perusal of the literature, together with the authors' limited experience in Australia, Israel and Texas, suggests that this classification is applicable to calcretes everywhere and, with minor modifications, to other pedocretes such as ferricretes and silcretes.

Necessity for and requirements of a calcrete classification

The necessity for a calcrete classification stems from the inability of temperate zone soil classifications of the Casagrande (British Standards Institute (BSI) 1957; Bureau of Reclamation 1974; American Society for Testing and Materials (ASTM) 1980) and American Association of State Highway and Transportation Officials (AASHTO) (1978) types adequately to describe and predict the engineering performance of materials composed of cemented particles of clay, silt, sand, etc. or almost pure carbonate, and ranging in consistency from loose silt to

very strong rock and in thickness from millimetres to 100 m. Some of these materials are not rock, but they do not slake or soften greatly in water, and when excavated and broken down during compaction, they behave as soils. Only then can they be said to possess a particle size distribution and Atterberg limits. Descriptive methods intended for use on undisturbed material such as those of the ASTM (1980b), BSI (1957, 1972), Geological Society (1970, 1977a,b), Jennings, Brink & Williams (1973), and the Core Logging Committee (1978) are better in this respect, but often require lengthy descriptions to convey an adequate picture. As calcretes frequently present unusual geotechnical properties and performance, it is necessary to distinguish them from other materials (Netterberg 1969a, 1971, 1980, 1982; Horta 1980).

A calcrete classification suitable for geotechnical use should be of both geological and engineering significance, and must be applicable in the field by relatively untrained personnel, as well as satisfying certain other requirements (Netterberg 1969a, 1980). Previous calcrete classifications (reviewed by Netterberg 1980) appear to be either too simple for modern use or too complicated for geotechnical use. The most recent (Horta 1980) only considers calcrete gravels and sands.

Definitions

The extensive calcrete literature has been reviewed in recent years by Netterberg (1969a), Goudie (1973) and Reeves (1976). It is clear that the terms 'calcrete' and 'caliche' have been applied to almost any material of almost any consistency and carbonate content formed by the *in situ* cementation and/or replacement of regolith material by (dominantly) calcium carbonate precipitated from the soil water or ground water. Calcified cave soils, spring tufas, aeolianites, and beachrocks are usually

235

excluded, largely for the sake of convention, although they could be included for geotechnical purposes. The term 'calcrete' has also been used in more restricted senses for indurated materials only or for materials containing more than about 50% $CaCO_3$ equivalent, i.e. the lower limit for the term 'limestone'. This somewhat conflicting usage is accommodated here by the use of the unqualified term 'calcrete' for the widest usages only and the application of qualifying adjectives when more restricted use is intended. In the more restricted usage, calcretes generally possess more than about 50% $CaCO_3$ equivalent and, with one exception, are also indurated, more or less in accordance with the recommendation of the Speciality Session on Pedogenic Materials (1976).

The term 'soil' is used here in its wide engineering sense for practically any geological material which the engineer does not classify as rock, which requires blasting for excavation.

The classification

Basis of the classification

The classification suggested here is a simple, morphogenetic one based upon secondary (chemical) structure and sequence of development. It employs a combined geological and engineering approach, in its simplest form consisting of a genetic term such as 'calcrete', 'calcified', 'ferricrete', 'ferruginised', etc., plus a traditional engineering soil or rock term such as 'sand', 'gravel', etc., e.g. 'calcified sand', 'calcrete gravel', 'calcrete rock', as recommended by the Speciality Session on Pedogenic Materials (1976). This scheme is not dissimilar to that of Fookes & Higginbottom (1975) for the geotechnical classification of near-shore carbonate sediments. As material is often classified simply as 'rock' (requires blasting or consists of large boulders), 'hard' (requires pneumatic tools) and 'soft' (other materials) for

TABLE 1. *Stages in the development and weathering of calcretes (Netterberg 1969b, 1980)*

Stage	Host material			
0	Weathered rock	Shattered clay	Mixed texture	Clean sand or gravel
1	Calcrete soluans in cracks	Calcrete powder soluans in cracks	Scattered calcrete glaebules in host soil	Calcrete-coated grains
2	Calcified weathered rock	Powder calcrete (sandy silt or silty sand)	Glaebular calcrete (clayey, silty or sandy gravel)	Calcified sand or gravel (massive)
3			Honeycomb calcrete (partially coalesced nodules or soluans)	
4			Hardpan calcrete (rock-like horizon)	
5			Calcrete boulders, cobbles or gravel (discrete fragments formed by weathering)	

(Left margin labels: DEVELOPMENT spanning stages 1–4; WEATHERING spanning stage 5)

excavation payment purposes, the addition of such terms would represent the final descriptor in the simplest form of the classification. However, it is often necessary to use the classification together with more detailed geotechnical descriptive and particle size-plasticity classifications. The applicability and modifications required of such classifications have been considered (Netterberg 1969a, 1980, 1982; Horta 1980). Horta's (1980) suggestion of adding calcrete gravels and sands and gypcrete sands to Casagrande-type classifications should be taken even further.

Calcretes are thus classified simply into calcareous soils, calcified soils, powder calcretes, nodular calcretes, honeycomb calcretes, hardpan calcretes, and calcrete boulders and cobbles. As calcretes form more or less in this sequence (Table 1) (Netterberg 1969a,b, 1980; Goudie 1973) this classification should cover all the basic forms possible. Each of the forms listed in Table 1 represents an easily recognizable stage of growth or weathering and possesses a significantly different range of geotechnical properties. Possible correlations between this and other classifications have been discussed by Netterberg (1980). Calcrete profile log symbols have also been suggested by him, as well as a standard method for describing calcrete profiles.

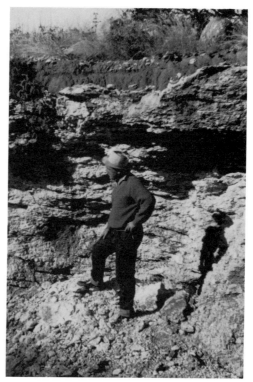

FIG. 1. Pseudobedded calcified alluvial sand (Netterberg 1980) with slight overlying hardpan development.

Calcareous soil

Calcareous soils (further described as sand, gravel, etc.) are soils with little or no cementation or development of carbonate concentrations such as nodules, but which effervesce with dilute hydrochloric acid. As, apart from ion exchange effects, the geotechnical properties of the original host soil have not been significantly altered by the carbonate (usually only 1–10% $CaCO_3$ equivalent), it is probably not necessary to distinguish this category (Stage 1, Table 1) unless the presence of even small amounts of carbonate are of significance to the works in question.

Calcified soil

A calcified soil (further described as sand, gravel, etc.) is a soil horizon (mass) cemented by carbonate usually to a firm of stiff consistency. Although often just friable, it does not usually slake in water. The carbonate is usually evenly distributed throughout the horizon as in calcified sands (Fig. 1) and gravels, but may occur as fissure-fillings as in calcified weathered rocks, although nodules are few. The amount of

carbonate (usually 10–50% $CaCO_3$ by mass) is sufficient to have significantly altered the geotechnical properties of the original soil. Calcified soils can generally be dug with a pick or a face shovel (although particularly well-cemented gravels may require more drastic methods) and compacted with rollers to yield sandly or gravelly pavement layer material. Only after excavation and processing can most calcified soils be said to possess a particle size distribution, which is very dependent on the type and amount of such processing. Most aeolianites could be classified as calcified sands with some calcrete hardpan horizons.

Powder calcrete (calcrete silt or calcrete sand)

Powder calcretes are chiefly composed of loose silt-sized and fine sand-sized carbonate with few or no visible host soil particles or calcrete nodules. Any nodules present are generally weak and friable. Powder calcrete horizons are occasionally cemented to a consistency of up to stiff but break down on working (Fig. 2). Carbonate contents often

FIG. 2. Unsuccessful use of powder calcrete as gravel road material.

exceed 70% $CaCO_3$ equivalent. Powder calcretes may develop into nodular calcretes, from which they are distinguished by having more than 75% of particles by mass finer than 2 mm (Fig. 3) or a grading modulus of less than

1.5. (The grading modulus (Kleyn 1955) is the sum of the cumulative mass percentages retained on each of the 2.00, 0.425 and 0.075 mm sieves divided by 100. A minimum value of 1.5 is often specified for rural road sub-bases in southern Africa.) Most powder calcretes also possess more than 55% finer than 0.425 mm. Many powder calcrete possess sub-base California bearing ratios (CBR). However, they are generally troublesome materials to compact and best avoided (Von Solms 1976).

Powder calcretes can also be called calcrete silt or calcrete sand (*not* silty calcrete or sandy calcrete), but the use of the term 'powder calcrete' may be more appropriate for use by unsophisticated road workers, and Fig. 3 actually represents the limiting particle-size distributions of powder and nodular calcretes visually classified in the field.

Nodular calcrete (calcrete gravel or calcrete sand)

Nodular calcretes are natural mixtures of silt-sized to gravel-sized particles of carbonate-cemented host soil particles in a matrix of usually calcareous soil (Fig. 4). More than 25% of the particles by mass are coarser than 2 mm (Fig. 3) or the grading modulus has a minimum value of 1.5. The overall consistency of the horizon is generally loose, but the nodules may vary from firm and friable to very strong. Calcrete nodules vary in shape and texture from nearly spherical and smooth, through botryoidal to irregular and rough, while platy, elongated

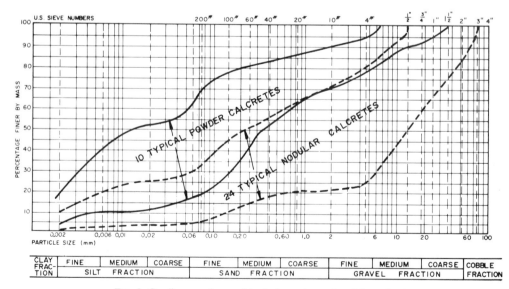

FIG. 3. Grading envelopes of typical powder and nodular calcretes.

FIG. 4. Nodular calcrete (Netterberg 1980). Calcrete cobble in lower right hand corner is a weathering relic of an older hardpan calcrete and not a nodule.

and cylindroidal forms also occasionally occur. The maximum size of individual or compound nodules very rarely exceeds 50 or 60 mm. Nodular calcretes can usually be scraper-loaded or bulldozed without ripping, and compacted to produce a good pavement layer material. Most calcretes display gap gradings by mass (Fig. 3) even after compaction. These are at least partly due to variations in particle bulk density with size and disappear or are reduced if gradings are calculated on a volumetric basis (Netterberg 1969a, 1971). The best nodular calcretes have properties comparable to those of graded crushed stone.

Geologically, the best term for nodular calcretes is really 'glaebular calcretes' (Brewer 1964). However, since calcrete glaebules other than nodules are rare (Netterberg 1969a, 1980), use of the more common term for geotechnical purposes seems sensible. Similarly, other non-glaebular, secondary structures such as pedotubules and small crotovinas can also be included under the term 'nodular calcrete' for geotechnical purposes.

Geotechnically, the best term for nodular calcrete is 'calcrete gravel'. However, many materials called nodular calcretes by field personnel classify as calcrete 'sands' according to a Casagrande type of classification (e.g. BSI 1957; ASTM 1980a) (Fig. 3). For this reason, as well as the one that, with experience, it is easy to estimate in the field when a material has a grading modulus of 1.5 or more and is thus potential road sub-base or base material, the term 'nodular calcrete' has been retained, especially at a less sophisticated level. Proper geotechnical descriptions should, however, also use the terms 'calcrete gravel' etc. as estimated by the usual criteria for the Casagrande-type classification employed.

Honeycomb calcrete

As the nodules in a nodular calcrete grow larger and more numerous, they may become partially cemented together to form a honeycomb calcrete (Fig. 5). A honeycomb calcrete is thus a stiff to very hard, open, honeycomb-textured calcrete horizon with the interstitial voids often filled with loose or soft soil. Both the voids and the individual nodules seldom exceed a diameter of about 30 mm, and are usually interconnected. Honeycomb calcretes can usually be ripped and grid-rolled to yield an excellent pavement base comparable to or even better than graded crushed stone in quality.

Another less common type of honeycomb calcrete can be formed from carbonate fissure-fillings in a weathered rock to result in a box-work structure. In both forms the soil filling the voids may be quite plastic.

FIG. 5. Honeycomb calcrete.

Although honeycomb (and boulder) calcretes can be geologically regarded as forms of hardpan, their geotechnical properties are sufficiently different to warrant classifying them separately.

Hardpan calcrete

A hardpan calcrete (Fig. 6) is formed when most of the voids in a honeycomb calcrete become cemented or the upper part of a calcified soil horizon becomes more heavily cemented than the rest of the horizon (Table 1). It is a usually stiff to very strong, relatively massive and impermeable, sheetlike horizon which normally overlies a weaker material such as nodular or powder calcrete or calcified soil. Hardpans may vary from millimetres to several metres in thickness, although individual horizons more than 500 mm in thickness are not common. They may be sandy or gravelly or nearly pure limestone, and may be nearly structureless, or pseudobedded, tufaceous, jointed, veined, brecciated or laminated, and may contain voids of various kinds. Many are capped with a thin, very hard laminated 'rind'.

Many calcrete hardpans can be ripped and grid-rolled to yield a good to excellent pave-

TABLE 2. *Summary of some properties of calcretes in comparison with calcareous and calcified soils*

Material type	Total carbonate[a] as $CaCO_3$ %	Grading modulus[b]	Classification		BSI CP 2001 (1957)	Mod. AASHO soaked CBR[b] %	<0.425 mm	
			AASHTO M 145–73 (1978)				$PI^{a,b,c}$ %	Electric. conductivity^{a,b,c,d} Sm^{-1} at 25° C
			Group	Index				
Calcareous soil	1–10?[b]	Variable	Variable	Variable	Variable	Variable	Variable	Variable
Calcified sand	10? – 50	1.0 – 1.8?	A–1–b to A–2–7	0–2	GF, GP, SU, SF	25? – 100	NP–20	0.02–0.23
Calcified gravel	10? – 50	>1.8?	A–1–a to A–1–b	0–1?	GF to GW?	>80?	<8?	<0.1?
Powder calcrete	70 – 99	0.4 – 1.5	A–2–4 to A–7–5	0–13	ML to GF	25? – 70?	SP–22	0.1–2.1
Nodular calcrete	50 – 75	1.5 – 2.3	A–1–a to A–6	0–3	GF, GP, GU	40 – >120	NP–25	0.02–0.74
Honey-comb calcrete	70 – 90	>2.0	Rock?	–	(Hard, h or Rock, r)[i]	>100	SP–16	0.01–0.1?
Hardpan calcrete	50 – 99	>1.5?	Rock?	–	(Hard, h or Rock, r)[i]	10? – >100	NP–7	0.01–0.06
Calcrete boulders and cobbles	50 – 99	>2.0	Boulders	–	Boulders and cobbles[l] (B)	>100	NP–3	0.01–0.02

[a]Without the soil between calcrete boulders and cobbles.
[b]After excavation and rolling or crushing in the case of hardpans, honeycombs, boulders, calcified gravels and some calcified sands.
[c]On the fines produced in the Los Angeles Abrasion test in the case of honeycombs, hardpans and boulders.
[d]Saturated paste method (Netterberg 1970).

ment layer material. Those which require blasting and crushing are probably best described as 'calcrete rock'. Such materials may occasionally be several metres thick.

Calcrete boulders and cobbles

Calcrete hardpans weather to boulders, cobbles and smaller fragments, usually in a matrix of non- or only slightly calcareous soil (Fig. 7). The shape and sphericity of the fragments vary from subrounded and sub-spherical to subangular and blocky, depending upon whether dissolution or disintegration was

the dominant mode of weathering. Such fragments are generally strong to very strong and are often confused with nodules, from which they can usually be distinguished by their greater strength, sphericity and size, lower grain/matrix ratios, sharper and smoother boundaries, and a frequent partial or complete skin of laminated rind. Significant amounts of gravel-sized fragments have not been observed.

Calcrete boulders and cobbles are relatively useless as pavement materials. In their natural state they are usually too coarse and gap-graded for uses other than as fill, and are generally uneconomic to crush. However, in parts of

TABLE 2 (Continued)

Natural or crushed aggregate		APT[e]			Whole mass *in situ*			
ACV %	10% FACT kN	AFV[e] %	APV[e] %	Mohs hardness[f]	Overall consistency[g]	Seismic velocity m sec⁻¹	Workability	Usual max. thickness m
Variable	Variable	Variable	Variable	Variable	Variable	300–900?	Variable	Variable
35?	18?	70?	20?	2–3	Med. dense	600?	Bulldoze,	5
–	–	–	–		—dense	–	shovel, or	
55?	70?	95?	50?		or firm–stiff	1200	rip and grid-roll	
25?	70?	90?	50?	≥3?	Med. dense	1200	Rip and	10
–	–	–	–		—very dense	–	grid-roll or	
35?	135?	100?	90?		or firm to very stiff	2450?	blast and crush	
33?	18	25	5	2–3	Loose	400	Bulldoze,	5
–	–	–	–			–	shovel, or	
55	90?	95	65		stiff	1070	scraper	
20	9	0	0	1–5	Loose	600	Bulldoze,	5
–	–	–	–			–	shovel, or	
57	178	100	90		med. dense	900	scraper	
16	80?	90?	60?	3–6	Stiff	900	Rip and	1
–	–	–	–			–	grid-roll	
35	205	100	100		very stiff	1200		
19	27	75?	30?	2–6	Stiff—very strong	900	Rip and grid-roll or	1, rarely
–	–	–	–			–	blast and crush	10
53	196	100	100			4500		
20	98	95?	70?	3.5	Very stiff—very strong	Erratic	Rip and crush	1
–	–	–	–	–				
33	205	100	100	5				

[e] APT = Aggregate Pliers Test; AFV = Aggregate Fingers Value; APV = Aggregate Pliers Value (Netterberg 1969a, 1978)
[f] Of the carbonate or silicified carbonate cement (aggregate or mass).
[g] According to methods of BSI (1957, 1972) and Geological Society (1977b).
[h] Up to 50% when many nodules present.
[i] Suggested term and symbol.

FIG. 6. Hardpan calcrete overlying nodular calcrete (Netterberg 1980).

Australia they are gathered by means of 'rock pickers' and crushed with travelling 'rock busters' for base coarse.

Geotechnical properties

The geotechnical properties of calcretes (Netterberg 1969a, 1971, 1982; Reeves 1976; Weinert 1980) depend largely upon the nature of the original host soil (e.g. whether it was

FIG. 7. Calcrete boulders and cobbles.

sand or clay) and the extent to which it has been cemented and/or replaced by carbonate. They thus vary from those of soil to those of rock (limestone), improving in a general fashion with the stage of development (Table 2).

Application to other pedocretes

Like calcretes, other pedocretes such as ferricrete and silcrete are also simply soils which have been cemented and/or replaced to a varying degree by (in this case) iron oxides and amorphous silica respectively. They therefore pass through similar stages of growth and weathering and, with minor modifications, a similar classification can be applied to them (Netterberg 1975, 1976; Weinert 1980).

Classification for other purposes

With minor modifications and amplifications the scheme suggested here should be suitable for most purposes (Netterberg 1980).

Conclusions

Traditional geotechnical classifications developed for temperate zone materials require modification and amplification in order to adequately describe the non-traditional materials of other areas. In particular, an indication of the type of geological material (e.g. calcrete, weathered dolerite, ferricrete, etc.) is essential.

Authigenic calcareous accumulations in the regolith can be simply classified for geotechnical purposes into calcareous soils, calcified soils, powder calcretes, nodular calcretes, honeycomb calcretes, hardpan calcretes, and calcrete boulders and cobbles. Each of these categories represents an easily recognizable stage in the growth or weathering of a calcrete horizon and possesses a significantly different range of geotechnical properties. A similar classification scheme can be applied to other pedocretes such as ferricretes and silcretes.

ACKNOWLEDGMENTS: This paper is published with the permission of the Director of the National Institute for Transport and Road Research and the firm of Van Wyk & Louw Inc., Consulting Engineers. Research on this project was initially financed by the Roads Branch of the South West Africa Administration and carried out under the general supervision of A. A. B. Williams and H. H. Weinert. This paper is based on part of a PhD thesis submitted to the University of the Witwatersrand, Johannesburg, under the supervision of H. H. Weinert and A. B. A. Brink.

References

AMERICAN ASSOCIATION OF STATE HIGHWAY AND TRANSPORTATION OFFICIALS 1978. AASHTO M 145-73: recommended practice for the classification of soils and soil-aggregate mixtures for highway construction purposes. *In:* Part 1 of AASHTO, *Standard Specifications for Transportation Materials and Methods of Sampling and Testing*, 184–90. AASHTO, Washington, D.C.

AMERICAN SOCIETY FOR TESTING AND MATERIALS 1980a. ASTM D 2487-69: standard method for classification of soils for engineering purposes. *In:* Part 19 of ASTM, *1980 Annual Book of ASTM standards*, 374–8. ASTM, Philadelphia.

—— 1980b. ASTM D 2488-69: standard recommended practice for description of soils (visual-manual procedure). *In:* Part 19 of ASTM, *1980 Annual Book of ASTM standards*, 379–85. ASTM, Philadelphia.

BREWER, R. 1964. *Fabric and Mineral Analysis of Soils.* Wiley, New York.

BRITISH STANDARDS INSTITUTION 1957. *Site Investigations.* British Standard Code of Practice CP 2001 (1957), BSI, London.

—— 1972. *Code of Practice for Foundations.* British Standard Code of Practice CP 2004:1972, BSI, London.

BUREAU OF RECLAMATION 1974. *Earth Manual.* 2nd Ed. U.S. Printing Office, Washington, D.C.

CAIGER, J. H. 1964. *The use of airphoto interpretation as an aid to prospecting for road building materials in South West Africa.* M.Sc. thesis (unpubl.), University of Cape Town, Cape Town.

CORE LOGGING COMMITTEE SOUTH AFRICA SECTION, ASSOCIATION OF ENGINEERING GEOLOGISTS 1978. A guide to core logging for rock engineering. *Bull. Ass. engng Geol.* **XV**, 295–328.

FOOKES, P. G. & HIGGINBOTTOM, I. E. 1975. The classification and description of near-shore carbonate sediments for engineering purposes. *Geotechnique*, **25**, 406–11.

GEOLOGICAL SOCIETY ENGINEERING GROUP WORKING PARTY 1970. The logging of rock cores for engineering purposes. *Q. J. eng. Geol.* **3**, 1–24.

—— 1977a. The logging of rock cores for engineering purposes. *Q. J. eng. Geol.* **10**, 45–52.

—— 1977b. The description of rock masses for engineering purposes. *Q. J. eng. Geol.* **10**, 355–88.

GOUDIE, A. 1973. *Duricrusts in Tropical and Subtropical Landscapes.* Clarendon Press, Oxford.

HORTA, J. C. DE O. S. 1980. Calcrete, gypcrete and soil classification in Algeria. *Engng Geol.* **15**, 15–52.

JENNINGS, J. E., BRINK, A. B. A. & WILLIAMS, A. A. B. 1973. Revised guide to soil profiling for civil engineering purposes in southern Africa. *Trans. S. Afr. Inst. civ. Engrs* **15**, 3–12.

KLEYN, S. A. 1955. Possible developments in pavement foundation design. *Trans. S. Afr. Inst. civ. Engrs* **5**, 286–92.

NETTERBERG, F. 1967. Some roadmaking properties of South African calcretes. *Proc. 4th reg. Conf. Afr. Soil Mech. Fndn Engng*, Cape Town, **1**, 77–81.

—— 1969a. *The geology and engineering properties of South African calcretes.* Ph.D. thesis, University of Witwatersrand, Johannesburg.

—— 1969b. The interpretation of some basic calcrete types. *S. Afr. archaeol. Bull.* **24**, Parts 3 and 4, 117–22.

—— 1970. Occurrence and testing for deleterious soluble salts in road construction materials with particular reference to calcretes. *Proc. Symp. Soils & Earth Structures in Arid Climates*, Adelaide, 87–92.

—— 1971. *Calcrete in Road Construction,* Counc. scient. ind. Res. res. Rep. 286. Pretoria.

—— 1975. Speciality session D: pedogenic materials. *Proc. 6th Reg. Conf. Africa Soil Mech. Fndn Engng*, Durban, **1**, 293–4, Balkema, Cape Town.

—— 1976. Convenor's introduction to Speciality Session D: pedogenic materials. *Proc. 6th Reg. Conf. Africa Soil Mech. Fndn Engng*, Durban, **2**, 195–8.

—— 1978. Calcrete wearing courses for unpaved roads. *Civ. Engr S. Afr.* **20**, 129–38.

—— 1980. Geology of southern African calcretes: 1. Terminology, description, macrofeatures, and classification. *Trans. geol. Soc. S. Afr.* **83**, (2) 255–83.

—— 1982. Geotechnical properties and behavior of calcretes as flexible pavement materials in southern Africa. *Proc. Symp. calcareous Soils geotech. Practice*, Philadelphia, American Society for Testing Mats, Philadelphia STP 777, 296–309.

REEVES, C. C. (Jr) 1976). *Caliche Origin, Classification, Morphology and Uses.* Estacado Books, Lubbock.

SPECIALITY SESSION D: PEDOGENIC MATERIALS 1976. Conclusions and recommendations. *Proc. 6th Reg. Conf. Africa Soil Mech. Fndn Engng*, Durban, **2**, 211–2.

VON SOLMS, C. L. 1976. The use of natural and lime-treated pedogenic materials in road construction in South West Africa. *Proc. 6th reg. Conf. Africa Soil Mech. Fndn Engng*, Durban, **2**, 200–10.

WIENERT, H. H. 1980. *The Natural Road Construction Materials of Southern Africa.* Academica, Cape Town.

F. NETTERBERG, National Institute for Transport and Road Research, PO Box 395, Pretoria 0001, South Africa.

J. H. CAIGER, Van Wyk & Louw Inc., Consulting Engineers, PO Box 905, Pretoria 0001, South Africa.

Karstic residual fluorite-baryte deposits at two localities in Derbyshire.

R. P. Shaw

SUMMARY: Various karst processes may rework primary mineralization producing secondary ore deposits in a variety of karstic cavities both on the surface and underground. Two surface localities, on Bonsall Moor, near Matlock, and near Castleton are filled with sediments containing locally derived fluorite and baryte clasts, in sufficient quantity to be worked as ore deposits. The associated clastic sediments are of Pleistocene fluvioglacial origin.

The lead-zinc-fluorite-baryte mineralization of the area has been described by Carruthers & Strahan (1923), Dunham (1952), Ford (1967b), Ford (1976), Worley (1978), Worley & Ford (1977). The mineral deposits consist of fissure and cavity fills, and depositional textures show evidence of episodic phases of mineralization (Ineson & Al-Kufaishi 1970; Firman & Bagshaw 1974; Bagshaw 1978).

Karstic cavities including cave systems, solution hollows, solution pipes and pre-existing 'pipe' cavities are associated with some of the mineral veins. The cavity fillings consist of locally derived minerals with allochthonous clay-silt-sand deposits; the processes of cave formation have been described by Ford (1964, 1971, 1972, 1977), Ford & Worley (1977a), Beck (1980). This type of ore deposit containing easily worked and dressed galena has been worked by Derbyshire lead miners who knew it as 'gravel ore', and it has been responsible for the 'bonanzas' found in some mines in Derbyshire.

The two localities considered are on Bonsall Moor, west of Matlock, (SK261 586) and the Portway 'Gravel' Pit near Castleton (SK127 812) (Figs 1 & 2).

At Bonsall Moor the open cast pits have or are being worked for fluorspar and the Portway Gravel Pit is being worked intermittently for nodular baryte.

Cavities and residual ores

In karst areas the formation of cavities occurs both on the surface and underground.

On the surface these cavities may take several forms including:

(1) widening of joints and other lines of weakness;
(2) differential solution of limestone at joint intersections to produce surface depressions;
(3) the development of 'swallow' holes engulfing allochthonous streams;
(4) collapse of underground cavities and the formation of surface depressions.

These depressions may become filled with sediment by a combination of processes including wind, water and ice transport, and slumping of the depression walls. Filling of the depressions and subsidence collapse may be contemporary producing slump structures in the sedimentary fill. When primary mineralization is present in the area, derived fragments may be incorporated into the sedimentary fill of the depression as insoluble residue or by mechanical erosion of an outcrop of epigenetic mineralization. In some cases the mineral vein may protrude into the cavity but usually it becomes disaggregated and locally distributed.

The 'ore' minerals may form discrete layers within the deposit or may be uniformly distributed throughout the sediment.

Description

Bonsall Moor

The host rocks are partially dolomitized Carboniferous Limestone of the Matlock Group which are Brigantian (D_2) in age. They are massive, well-jointed and contain layers rich in both crinoid and coral remains. Interbedded with the limestone are numerous 'wayboards' (altered volcanic tuffs) from 2 to 35 cm thick. The wayboards are variously composed of chlorite, kaolinite, smectite and illite (Walters & Ineson 1980).

Two rake veins have been worked in the pit, each up to 3 m wide. Both consist mainly of fluorite with minor galena and traces of sphalerite. Late calcite is present in vugs. Numerous scrins (small fissure veins) up to 3 cm wide and many mineralized joints are also present.

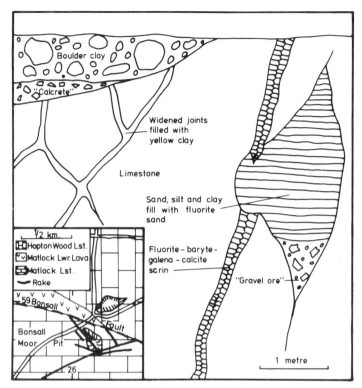

FIG. 1. Schematic section of the residual ore deposits at Bonsall Moor. Boulder clay and 'calcrete' overlying widened clay filled joints and a cavity intersecting a mineral vein containing residual ore overlain by layered sands, silts and clays.

The area has undergone deep weathering, under warmer conditions than prevailing today, prior to glaciation producing a soil layer containing a calcified horizon resembling 'calcrete'. This is overlain in places by quartz sand, derived from the Millstone Grit. In turn this is overlain by up to 6 m of boulder clay, nearly all locally derived, containing boulders up to 0.75 m across, of Millstone Grit sandstone and shale, Carboniferous Limestone and basalt with a few smaller exotic erratics (some of which are from the Lake District) (Fig. 1).

The deep weathering of the limestone has widened many of the joints to a considerable depth (over 10 m below ground level) in some parts of the pit. These joints are now filled with 5–8 cm width of stiff yellow clay of loessic origin. Any insoluble residue, such as fluorite, has accumulated in small lenses within the clay.

In places limestone solution has produced cavities, up to 2 m high and 1.5 m wide, oriented along scrins. All are filled with sandy sediments which, when finely layered, represent suspension settling from ponded waters. As well as accumulations of remnant and residual blocks of mineral vein present in the bottoms of these cavities the sand also contains derived fluorite, up to 70% by volume. The fluorite grains, up to 5 mm in length, still retain their crystal faces and show no signs of abrasion.

The non-ore-mineral fraction of the sands consists mainly of quartz (50–95%), feldspars (0–10%) and muscovite mica (2–10%). Scanning electron micrographs of quartz grain surfaces show textures associated with glacial transportation. The source rock is probably the peripheral Millstone Grit escarpments, having been derived by glacial erosion and outwash. X-ray diffraction (XRD) analysis shows the clay fraction to be composed of kaolinite and illite, with lesser chlorite and mixed layer smectite. Like the coarser fractions the source is the Millstone Grit shales with a contribution from local wayboards.

The ore-mineral fraction, which varies from 0 to 70% of the cavity fill, can be divided into two types: firstly, remnant and residual ore which consists of relatively insoluble vein material accumulated in the bottoms of the cavities. It comprises lumps of euhedral fluorite, with baryte and/or calcite. In one cavity galena is present with a ~1 mm coating of cerussite, and secondly a fluorite sand. It consists of euhedral fluorite grains (0.5–5 mm) showing no signs of abrasion indicating a local derivation. It occurs interbedded and intermixed with the quartz sand.

The deposits containing fluorite are potential fluorspar reserves, with fluorite contents of between 20 and 75%, but their clay content interferes with the froth flotation circuits of the dressing plants.

Portway 'gravel' pit

The limestones in this pit, of the Bee Low Group, are of Asbian (D_1) age. Unaltered limestone is not exposed in the pit itself but an old quarry to the north shows it to be a well-bedded, slightly crinoidal limestone. Chert nodules are well developed as layers within the limestone. Epigenetic mineralization is largely present as 'pipe' veins, linings of pre-existing cavities.

Within the pit the limestone has undergone intense silicification to produce a patchy 'quartz rock' (Ford 1967a). Two varieties of the quartz rock are present: first, partially silicified limestone containing as much as 50% acicular

quartz crystals up to 0.5 mm long; secondly, the 'maggot' rock which was originally a fossiliferous limestone composed mainly of compound corals. The matrix was selectively replaced by crystalline quartz, but the corals were resistant and remained as calcite. Later decalcification has removed the corals leaving a porous quartz rock full of coral moulds or 'maggot holes'.

In parts of the pit both types of silicified limestone have been completely decalcified and disaggregated producing a white sand composed of acicular euhedral quartz. At the margins of the silicified area unaltered limestone has silicified joints, in places decalcified to yield quartz crystal 'sand' veneer. The quartz replacement of the limestone is thought to be associated with epigenetic mineral fluids having attacked basaltic rocks lower in the stratigraphic sequence thereby extracting silica (Ford 1967a). At this locality silicification was complete before the deposition of any epigenetic minerals.

The pipe veins are solution cavities up to 3 m in diameter, lined with baryte, fluorite, galena and late calcite, often as corroded scalenohedra. Remaining voids in the pipes are filled with laminated clays and silts. The pit is worked for numerous baryte nodules present in both the epigenetic and residual fractions.

Karstification at this locality consists of enlargement of pipe vein cavities and of the

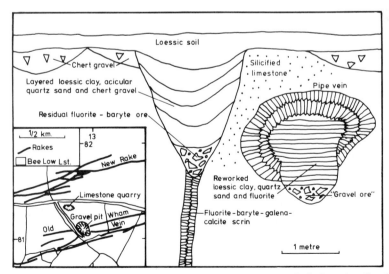

FIG. 2. Schematic section of the residual ore deposits at Portway 'Gravel' Pit. Loessic soil and pockets of chert gravel overlying partially silicified limestone containing a solution hollow developed on a mineral vein with a residual ore and layered sedimentary fill and a pipe vein with later solutional modification containing 'gravel ore' and layered sedimentary fill.

formation of solution pipes from the surface. The solution pipes are filled with surface-derived loess (40–70%) and chert gravel (0–50%) and insoluble residue from the limestone which consists of chert nodules, derived quartz rock and 'sand' (5–95%) and derived blocks of the epigenetic mineral suite (0–20%). Collapse structures and mixing caused by slumping as the limestone was dissolved are evident. Patches of laminated clay have been caught up in the collapses indicating that the process is still in operation.

In both pipe veins and solution collapse structures the 'ore' fraction consists of angular residual lumps of epigenetic pipe vein mineralization. Baryte is dominant with fluorite and some deeply etched calcite crystals present. Secondary goethite, after marcasite, is also present both in joints in the quartz rock and as small derived lumps in the fills.

Conclusions

The cavities, apart from pre-existing mineral vein and pipe cavities, were probably formed and filled with sediment during periglacial conditions when ground water circulation was restricted. The textures of the limestone walls of these cavities indicate that some solution of the limestone has occurred during the Pleistocene and recent weathering cycles. At the Bonsall Moor pit some of the joint widening occurred before glaciation.

The deposits are mainly allochthonous sand and clay fills variably containing derived limestone, galena, fluorite, baryte and calcite clasts showing no evidence of transportation. The source of the sand and clay is from the surface, probably from the Millstone Grit series via glacial action. Some of the loess may be derived from further afield.

Distinct contrasts occur between these two localities. The nature of the Pleistocene overburden is markedly different, being boulder clay at Bonsall Moor and loess at Portway. The high degree of silicification of the limestone at Portway is an important contributor of insoluble material to the cavity fills, while at Bonsall Moor silicification is insignificant. There is no evidence of much redistribution of ore minerals at the Portway pit but at Bonsall Moor redistribution has occurred though not in all the deposits and only on a local scale.

Similar deposits are known over much of the Derbyshire Orefield, such as those at Masson Hill, Matlock (Ford & Worley 1977b; Warriner, Willies & Flindall 1981), and at Hubberdale Mines, near Monyash (Worley, Worthington & Riley 1978), where they formed important lead ore deposits. Today they are of importance as numerous though small potential fluorspar reserves if the problems of mineral dressing can be overcome.

ACKNOWLEDGMENTS: I would like to thank the University of Leicester for financial support to undertake this research, which forms part of a PhD project, and Clyde Petroleum (Minerals) Limited for financial support for field work and for access to their pit on Bonsall Moor (the pit has since changed ownership). I am especially grateful to Dr T. D. Ford for his critical examination of the original manuscript and for his constructive suggestions for its improvement.

References

BAGSHAW, C. 1978. *Aspects of the mineralization of the orefield of the Derbyshire Dome*. Unpublished M.Phil. Thesis, University of Nottingham.

BECK, J. S. 1980. *Aspects of speleogenesis in the Carboniferous Limestone of North Derbyshire*. Unpublished Ph.D Thesis, University of Leicester.

CARRUTHERS, R. G. & STRAHAN, A. 1923. Lead and zinc ores of Durham, Yorkshire and Derbyshire. *Spec. Rep. Min. Res. GB Mem. geol. Surv. GB*, **26**, 114 pp.

DUNHAM, K. C. 1952. *Fluorspar* (4th Ed.). *Spec. Rep. Min. Res. GB Mem. geol. Surv. GB*, **4**, 143 pp.

FIRMAN, R. J. & BAGSHAW, C. 1974. A re-appraisal of the controls of non-metallic gangue mineral distribution in Derbyshire. *Mercian Geol.* **5**, 145–61.

FORD, T. D. 1964. Fossil karst in Derbyshire. *Proc. Brit. Spel. Ass.* **2**, 42–62.

—— 1967a. A quartz-rock filled sink hole on the Carboniferous Limestone near Castleton, Derbyshire. *Mercian Geol.* **2**, 57–62.

—— 1967b. The stratiform ore deposits of Derbyshire. *In:* JAMES, C. H. (ed.) *Sedimentary Ores, Ancient and Modern*. Proc. 15th Inter-University Geological Congress, University of Leicester, December 1967, 73–93.

—— 1971. Structures in limestone affecting the initiation of caves. *Trans. Cave Res. Group GB* **13**, 65–71.

—— 1972. Evidence of early stages in the evolution of the Derbyshire karst. *Trans. Cave Res. Group GB*, **14**, 73–7.

—— 1976. The ores of the South Pennines and Mendip Hills, England—a comparative study. *In:* WOLF, K. H. (ed.) *Handbook of Stratabound and Stratiform Ore Deposits*, **5**, *Regional Studies*, 161–95.

—— (ed.) 1977. *Limestone and Caves of the Peak District*, Geo-Books, Geo-Abstracts Ltd, Norwich.

—— & WORLEY, N. E. 1977a. Mineral veins and cave development. *Proc. 7th int. Spel. Congress, Sheffield, England, September 1977*, 192–3.

—— & WORLEY, N. E. 1977b. Phreatic caves and sediments at Matlock, Derbyshire, *Proc. 7th int. Spel. Congress, Sheffield, England, September 1977*, 194–6.

INESON, P. R. & AL-KUFAISHI, F. A. M. 1970. The mineralology and paragenetic sequence of Longrake Vein at Raper Mine, Derbyshire. *Mercian Geol.* **3**, 337–51.

WALTERS, S. G. & INESON, P. R. 1980. Mineralisation within the igneous rocks of the South Pennine Orefield, *Bull. Peak Dist. Mines Hist. Soc.* **7**, 315–25.

WARRINER, D., WILLIES, L. M. & FLINDALL, R. 1981. Ringing Rake and Masson Soughs and the mines on the east side of Masson Hill, Matlock. *Bull. Peak Dist. Mines Hist. Soc.* **8**, 65–102.

WORLEY, N. E. 1978. *Stratigraphic control of mineralization in the Peak District of Derbyshire.* Unpublished Ph.D Thesis, University of Leicester.

—— & FORD, T. D. 1977. Mississippi Valley type orefields in Britain. *Bull. Peak Dist. Mines Hist. Soc.* **6**, 201–8.

——, WORTHINGTON, T. & RILEY, L. 1978. The geology and exploration of the Hubbadale Mines, Taddington. *Bull. Peak Dist. Mines Hist. Soc.* **7**, 31–9.

R. P. SHAW, Department of Geology, University of Leicester, University Road, Leicester LE1 7RH.

Cenozoic pedogenesis and landform development in south-east England

J. A. Catt

SUMMARY: Soils are useful stratigraphically because they mark land surfaces of significant duration. They also provide valuable palaeo-environmental evidence complementary to that obtained from sedimentary successions. In south-east England both buried and unburied soils are broadly divided on field morphology and petrographic characteristics into those formed during the Flandrian Stage and others (palaeo-argillic soils) dating from the Cromerian and later interglacial stages. Even where the latter are derived from pre-Quaternary deposits, as on the Chalk dipslope, their profile characteristics and stratigraphic relationships to remnants of early Quaternary marine deposits suggest development during the Quaternary rather than in the Tertiary. Although most major landscape features were probably inherited from the late Tertiary, the entire land surface of south-east England suffered significant Quaternary erosion, and there are no definite relics of Tertiary pedogenesis.

Unfortunately there is no universally accepted definition of the word 'soil', which is used in different senses by geologists, civil engineers, botanists and farmers. In this paper, it is used to include all materials on or just beneath present and past land surfaces, which have been affected by processes of physical reorganization and/or chemical change dependent upon the presence of the atmosphere. These materials form a succession of horizons parallel to the solid earth-atmosphere interface and often unconformable with rock structure. The processes include physical disruption of rocks and mineral grains, disturbance by frost, desiccation, burrowing animals or root growth, the downward movement of solid particles by water percolating through fissures and channels (illuviation), the intimate mixing of inorganic particles with organic decomposition products, chemical weathering (hydrolysis, oxidation, carbonation), removal of soluble materials (leaching), and the accumulation of less soluble residues.

Many soil forming processes are slow and require long periods (10^3–10^6 years) of continuous or intermittent action to have any detectable effect on some rock materials. Whereas most geologists are interested in major episodes of erosion and deposition on less stable parts of the earth's surface, pedologists are concerned with processes on stable land surfaces. In terms of the complete history of landscape development in an area such as south-east England, the two approaches are complementary, but the evidence of soils is rarely considered as carefully as, for example, the composition, stratigraphy and structure of sedimentary sequences. Difficulties arise from the lateral variability of soils, from modifications following burial or environmental changes, and from removal of friable upper horizons even on the most stable land surfaces, but despite these problems soils can provide much palaeo-environmental evidence and are often useful as stratigraphic markers.

Types of evidence for past soil development

The most direct evidence for the nature of soils formed during past periods is from buried soils, the age range of which may be delimited stratigraphically by the ages of the youngest parent material of the soil and of the oldest deposit overlying it. No method of absolute dating is applicable to all soils. Radiocarbon assay of associated organic matter (if preserved) slightly over-estimates the age of burial, and can be used only for soils formed within the last 50 000 years or so. A second method of determining past soil conditions is the comparison of soils, buried or unburied, developed on deposits of increasing age and preferably similar lithology (e.g. on a sequence of river terraces or on till sheets of successive glaciations). In Britain, both these approaches are often limited by loss of diagnostically important upper soil horizons through minor periglacial or glacial erosion.

A third source of evidence for the nature of earlier soils is the composition of sediments containing the eroded soil materials. Few characteristic features of soils survive transportation by water, wind or ice, but some are preserved in solifluction or other locally derived mass movement deposits. The mineralogical

251

composition of non-marine deposits may give some idea of the dominant contemporary soil types and the general climatic conditions in which they formed, but marine sediments rarely contain useful evidence of this type, because the iron oxides and layer silicate clays that characterize many soils change readily in seawater.

Finally, it is possible to speculate about the nature of dominant soil types during past periods from palaeontological, isotopic and other geological evidence of climate, relief and vegetation, using knowledge of the ways in which soil varies in relation to these factors today. This is the least direct and satisfactory method, but is often the only way available, even for quite recent periods. Using these various lines of evidence, it is possible to reconstruct provisionally the main soil types and development processes occurring at various times within a limited region such as south-east England. As the evidence is strongest for the recent past and becomes weaker with increasing age, it is best to work backwards in time.

Flandrian pedogenesis in south-east England

In areas of Flandrian marine and estuarine alluvium, such as Romney Marsh (Green 1968) and the Fenland (Seale 1975), which include deposits ranging in age from a few hundred years (e.g. those embanked within historic times) to several thousands, it is possible to evaluate fairly precisely the effects of time, parent material composition and some environmental factors on soil development. Once the soft waterlogged sediments are exposed to the atmosphere and no longer regularly flooded, pedogenesis is initiated by various physical, chemical and biological processes collectively termed 'ripening'. Natural dehydration or artificial drainage allows oxidation of primary organic matter and diagenetic iron materials (e.g. pyrites), encourages cracking by desiccation and root penetration, promotes leaching, and stimulates aerobic microbiological activity and disturbance of the sediment by soil fauna. Leaching of water-soluble salts and brown mottling of the originally grey reduced sediments occur within a few years. In deposits up to 200 years old the sedimentary lamination is replaced by a coarse, weakly developed soil structure extending to 40–50 cm depth, and there is evidence for shallow incorporation of humus. Oxidation of pyrites (if present) causes acid conditions and rapid decalcification, but where this mineral is absent decalcification proceeds more slowly. On Romney Marsh partial decalcification of the top 25 cm has occurred on land reclaimed between 1450 and 1700 AD, but areas originating in the Roman period are fully decalcified to 50 cm. Many of the soils in the older marshlands also have uniform brown (rather than brown mottled) B horizons, with deeper incorporation of humus leading to finer, more strongly developed subsoil structures. The brown colours are usually near 10YR 5/3 on the Munsell Colour Chart.

Fenland soils formed in earlier Flandrian alluvium, deposited before 5000 BP during the main post-glacial transgression, are often decalcified to 1 m or more and have well developed structures and brighter brown subsoil colours, often approaching 7.5YR 5/6 of the Munsell Chart. However, many of these deposits were initially covered by peat, and because soil development in the mineral alluvium was delayed for an unknown period until the peat had wasted away, the exact length of time during which these soils have developed is difficult to estimate.

Most Devensian deposits in south-east England (loess, solifluction deposits and river terrace gravels) date from the Late Devensian (26 000–10 000 BP), and as temperate soil forming processes were almost completely subdued during this period, the soils on these deposits can be attributed largely to Flandrian pedogenesis. However, the Devensian deposits differed from the Flandrian alluvium in containing less water and organic matter, so that 'ripening' processes such as oxidation of primary organic matter, leaching of water-soluble salts and desiccation would have been less important in the early stages of soil development. The Devensian deposits are also more variable in lithology than the alluvium, and occur in more diverse geomorphological situations, so that the range of resulting soil types is wider. Initial drainage conditions were usually better and mottling is less common, usually resulting from localised reduction of originally brown (oxidized) sediments where humus has been incorporated. Weathering of sheet silicates (biotite, chlorite, illite, glauconite), pyroxenes, carbonate, sulphide and phosphate minerals is widespread, and decalcified B horizons commonly contain up to 10% additional clay, illuviated from higher horizons. Unaltered parent material usually occurs within 1.0 m of the surface.

Comparison of unburied soils on Devensian deposits with those covered in the mid-Flandrian by burial mounds or colluvium (resulting from Neolithic or later deforestation and soil erosion) suggests that most of the clay illuviation occurred early in the Flandrian. Subsequently, climatic change or soil degradation by early farming communities resulted in increasing acidity, weathering and leaching, with widespread podzolisation (mobilization of iron and aluminium oxides) on sandy deposits.

Devensian pedogenesis

Buried soils radiocarbon dated to Devensian interstadials, mainly the Late Devensian Windemere Interstadial (Coope & Pennington 1977), show only shallow incorporation of humus, as in rendzinas on the Chalk (Kerney 1963, 1965; Evans 1967), or weak podzolisation (Straw 1980), but little or no evidence of decalcification, mineral weathering or clay illuviation. However, features characteristic of arctic structure soils (ice-wedge casts, involutions) are widespread in buried and unburied soils on Devensian and older deposits. In Britain, it was cold enough for these to form during at least five separate periods in the Devensian, but most of those in south-east England are probably contemporaneous with the Late Devensian glacial maximum (15 000–18 000 BP), as they affect or immediately underlie the Late Devensian loess and solifluction deposits.

As the Late Devensian arctic structure soils on major interfluves and plateaux, such as the Chalk dipslope, are usually continuous and overlain by <1 m of Late Devensian aeolian deposits, the land surface configuration in these areas has undergone little modification over at least the last 20 000 years. In contrast, valley floors and footslopes have been altered much more; they have greater thicknesses of solifluction, colluvial and alluvial deposits, and the arctic structure soils, Devensian interstadial soils and even early Flandrian soils are often deeply buried. Even in the lowland catchment of the Gipping valley, Suffolk (Rose *et al*. 1980), discontinuous gullies >9 m deep and gravel, sand and peat deposits totalling 12 m resulted from changes in fluvial processes caused by climatic fluctuations in the past 13 000 years.

Soils on pre-Devensian deposits

Although many soils in south-east England are formed wholly or partly in pre-Devensian deposits, most show no greater degree of development than those in Devensian and Flandrian deposits. These include the soils in pre-Devensian deposits, which are shown on recent maps of the Soil Survey of England and Wales as Lithomorphic Soils, Pelosols, Brown calcareous earths, Brown calcareous sands, Brown earths, Brown sands, Argillic brown earths, Ferric podzols, Humus podzols, Humoferric podzols, Gley-podzols, Ground-water Gley Soils, and Typical stagnogley, Pelostagnogley, Cambic stagnogley and Sandy stagnogley soils (Avery 1980). In most of them unaltered parent material occurs within 1.0 m of the surface, amounts of illuvial clay in B horizons are <10%, and mineral weathering is of similar severity to that in soils on Devensian deposits. The upper horizons of many of these soils contain materials of different particle size distribution and mineralogy from the unaltered C horizons beneath. The differences cannot reasonably be attributed to weathering, and reflect either incorporation of Devensian loess (Catt 1978) or deposition of thin, loamy and often stony superficial deposits over large areas of almost flat Mesozoic clay outcrops, probably by solifluction. Areas occupied by these soils have suffered only localized erosion since the Late Devensian, but must have been almost completely stripped of pre-existing soils after the Ipswichian Interglacial yet before the last main periglacial phase of the Late Devensian.

Interglacial pedogenesis

The soils on pre-Devensian deposits capping many level plateaux and older river terraces in south-east England show features indicating longer and/or more intense development than that typical of the Flandrian Stage. These features include deeper decalcification and oxidative weathering, unaltered C horizons often occurring several metres below the surface, brighter brown or reddish colours not inherited from red pre-Quaternary rocks, and more pronounced orientation of clay as determined from birefringence patterns in thin section. The increased clay orientation results partly from stresses caused by repeated shrink-swell cycles (suggesting more pronounced wet and dry seasons than Britain experiences today), and partly from more extensive clay illuviation, the bodies of illuvial clay (argillans) forming as much as 30% of some horizons. Soils possessing these features are separated as various 'palaeoargillic' subgroups (Typical palaeo-argillic brown earths, Stagnogleyic palaeo-argillic brown earths, Gleyic palaeo-argillic brown earths, Palaeo-argillic podzols and Palaeoargillic stagnogley soils) in the current Soil Survey classification (Avery 1980), and are approximately equivalent to the *Pale*-great groups of

Alfisols and Ultisols in the American system of soil classification (Soil Survey Staff 1975).

The characteristics of colour and microstructure, which are defined more precisely by Avery (1980, pp. 30–1), are interpreted as resulting from pedogenesis during one or more pre-Devensian warm periods. Horizons showing them often underlie Late Devensian aeolian or solifluction deposits, and are commonly affected by features of Devensian (or possibly older) arctic structure soils. Also, a similar reddening (rubefication) occurs in buried interglacial soils in older tills and loesses in North America and Europe, and has been attributed to segregation of iron as hematite, a process requiring warmer and drier summers than Britain experiences today (Schwertmann & Taylor 1977).

As palaeo-argillic soils occur sporadically on the Anglican Lowestoft Till in Essex (Sturdy *et al*. 1979) and other parts of East Anglia, at least some of them formed during the Hoxnian and/or Ipswichian interglacials. The same features characterize the Valley Farm Rubified *Sol Lessivé*, which underlies the Lowestoft Till and is formed on pre-Anglian braided river deposits (the Kesgrave Sands and Gravels) of the proto-Thames (Rose & Allen 1977). The gravels are Pleistocene, because they contain intraformational ice-wedge casts and large erratics of Welsh volcanic rocks (Hey & Brenchley 1977), and Hey (1980) suggested that older (high-level) parts are equivalent to his Westland Green Gravels and to the Pre-Pastonian (i.e. post-Baventian) marine conglomerate on Beeston Regis foreshore (West 1980). The Valley Farm and possibly some other palaeo-argillic soils are therefore Cromerian or Pastonian, but not older. Deposits of the Baventian and earlier Quaternary stages are known mainly from marine deposits explored in boreholes in East Anglia, and no soil horizons have yet been identified in them, so we do not know whether pedogenesis in the earliest Quaternary warm stages produced palaeo-argillic horizons or not. However, palaeobotanical and other comparisons with later interglacials suggest that palaeo-argillic horizons would have formed at these times.

The Valley Farm soil has played an important role in clarifying the Middle Pleistocene stratigraphy of East Anglia, because it marks a land surface that can be traced at numerous sites. The regional height variation of this surface suggests it was a series of low-relief river terraces, the highest in central parts of East Anglia and the lower ones to the south-east, each sloping north-eastwards towards the North Sea.

This drainage direction continues the line of the pre-Anglian proto-Thames through the Vale of St Albans (Gibbard 1977), though considerable widening is indicated where the bedrock changes from Chalk to softer Tertiary sediments, and is perpendicular to the present drainage system in the area. The latter was initiated on the Anglican glacial deposits that partly buried the proto-Thames terraces and diverted the river into its present course.

Soils on successive terraces of other rivers in south-east England provide more detailed evidence of the age of palaeo-argillic horizons. For example, soils on the lowest (Devensian) terrace of the Kennet contain undisturbed argillans that are orange or yellowish-brown in thin section. Those on the next older (Wolstonian terrace) have egg-yellow argillans, which were disrupted by periglacial soil disturbance before or during deposition of loess in which yellowish-brown argillans subsequently developed. Soils on the two highest terraces (Anglian or older) contain evidence for three phases of clay illuviation, which resulted in formation of: (a) red ferriargillans, which were disrupted before formation of (b) egg-yellow argillans, which were in turn also disrupted before deposition of loess in which (c) yellowish-brown argillans subsequently developed. Chartres (1980) attributed the orange and yellowish-brown argillans to Flandrian pedogenesis and the egg-yellow and red argillans to Ipswichian and Hoxnian pedogenesis respectively. In terms of field colour and micromorphology, the soils on all three older terraces qualify as palaeo-argillic, but the distinction between slightly reddened soils with evidence of only two illuvial phases and those with much redder colours and more complex microfabrics may have wider significance. For example, Bullock & Murphy (1979) described a palaeo-argillic soil of the redder, more complex type from a high-level interfluve in Oxfordshire; the mode of formation of the parent material was uncertain, but similar reddish, weathered high-level deposits in the area (the Plateau or Northern Drift) are regarded by Shotton *et al*. (1980) as considerably older than Cromerian, an age that accords with the complex history of clay illuviation and periglacial disturbance deduced from thin sections of the soil.

Palaeo-argillic soils also occur on and within older solifluction deposits, such as the flinty Coombe Deposits overlying the Goodwood raised beach (Hoxnian) in Sussex (Hodgson 1967). However, these do not necessarily provide information about pedogenesis during a particular interglacial (in this case the Ipswichian), as they have been partly derived by mass movement from older soils upslope (Dalrymple 1957). The differentiation of interglacial soils, truncated and disturbed *in situ* by later cryoturbation, from soliflucted soil materials representing periods of pedogenesis much older than their stratigraphic position suggests, is a major problem in dating soil-forming processes.

Palaeo-argillic soils on pre-Quaternary deposits

Other Palaeo-argillic soils in south-east England occur on pre-Quaternary deposits, such as the Hythe Beds in west Kent (Fordham & Green 1980) and the Chalk, especially on the North Downs and Chilterns. These materials were previously mapped by the Institute of Geological Sciences as drift deposits, the Angular Chert Drift (Worssam 1963) and Clay-with-flints (Jukes-Browne 1906) respectively, but petrographic studies have shown that they are the weathered residues of pre-Quaternary formations with only minor additions of Quaternary sediment. The Clay-with-flints, for example, is formed mainly from a remnant veneer of Reading Beds left on the Chalk dipslope during subaerial exhumation of parts of the sub-Tertiary surface (Catt & Hodgson 1976). Incorporation of flints into the clayey veneer involved: (a) denudation of the Reading Beds until the cover was thin enough to be permeable, (b) dissolution of chalk beneath, and accumulation of flints and illuvial clay in the voids, thus forming a layer of Clay-with-flints *sensu stricto* (Loveday 1962), (c) deep disturbance by slumping and frost-heaving to mix this layer with the overlying veneer of weathered Reading Beds. Where the Clay-with-flints cover is several metres thick, palaeo-argillic horizons with red mottles and complex microfabrics are restricted to the upper layers, suggesting that the denudation of Reading Beds and conversion of the remnant veneer to Clay-with-flints predates at least the Hoxnian Stage.

As Jones (1980) showed, the sub-Tertiary erosion surface is not a simple plane resulting from a single marine transgression, but is a multi-facetted polygenetic surface, locally overlain by Thanet Beds, Blackheath Beds, London Clay and Bagshot Beds as well as by the Woolwich and Reading Beds. Textural variations with the Clay-with-flints and other forms of Plateau Drift on the Chalk dipslope can often be attributed to original lithological differences between the deposits thought to have overlain separate facets (e.g. on parts of the North Downs, John 1980).

Woolridge & Linton (1955) divided the Chalk dipslope plateau throughout south-east England into three main surfaces: (a) a narrow ridge along the crest of the escarpment, the dissected remnants of a mid-Tertiary peneplain, on which the soils have developed since the Miocene or early Pliocene to form 'true residual Clay-with-flints' from the Chalk, (b) a Pliocene or early Pleistocene marine bench, separated from the mid-Tertiary peneplain by a degraded cliff at approximately 200 m OD, and on which 'Clay-with-flints is absent or subordinate', the marine deposits remaining 'in sufficient bulk to modify soil conditions', and (c) at lower levels the sub-Tertiary surface, exhumed by later Pleistocene subaerial erosion, and on which 'Eocene outliers in all stages of demolition survive and the soil characteristics are often modified by the distributed residuum of these beds'.

Subsequent soil mapping on the Chilterns (Avery 1964), South Downs (Hodgson 1967) and North Downs (Green & Fordham 1973; John 1980) has not confirmed the postulated soil differences between these three facets. Also, much of the geomorphological and sedimentological evidence assembled by Wooldridge & Linton to support the tripartite subdivision of the dipslope can be explained in other ways (Jones 1974, 1980; Catt & Hodgson 1976; John 1980). In particular, the Plio-Pleistocene marine bench and cliff often reflect monoclinal flexures in the Chalk, which are mid-Tertiary and consequently deformed the originally flat facets of the sub-Tertiary surface. Also, the lower margin of the marine bench often coincides with the intersection between two sub-Tertiary facets. Most of the Chalk dipslope plateau is therefore best interpreted as the warped, polygenetic sub-Tertiary surface, which has been exhumed and then irregularly lowered by dissolution of Chalk beneath the remnant veneer of basal Tertiary sediment.

Although the Plio-Pleistocene seas did not bevel the Chalk dipslope to the extent Wooldridge & Linton envisaged, they did transgress much of what is now the Chalk outcrop, leaving fossiliferous deposits, isolated patches of which survive at Netley Heath, Rothamsted and Lenham. Some of the unfossiliferous sands and gravels Wooldridge & Linton correlated with the shelly deposits (e.g. at Little Heath, Gilbert 1920) may also be marine, but most are likely to be early Quaternary river gravels. Locally on the North Downs the marine deposits were laid directly on Chalk (John 1980), but other occurrences suggest deposition on Lower Tertiary sediments, which in some places were protected from later weathering and in others were subsequently transformed into Clay-with-flints. At Little Heath the marine gravels rest on a thin layer of unaltered Reading Beds, whereas at Rothamsted the 'Red Crag' fossils were from isolated, disturbed blocks of ferruginous sandstone found within Clay-with-flints (Dines & Chatwin 1930). So, although much denudation of Lower Tertiary deposits had probably occurred before the Plio-Pleistocene transgressions, and some may have been accomplished by them, the transformation of basal layers into Clay-with-flints seems to have occurred subsequently (i.e. during the Pleistocene). John (1980) also attributed this transformation to Pleistocene events, because the extent of rubefication in the Plateau Drift of the North Downs is no greater than in the Pleistocene marine sands of Netley Heath.

The conclusion that the characteristics of Clay-with-flints and other chalkland Plateau Drifts result from Quaternary rather than Tertiary pedogenesis conflicts with most earlier opinions. However, these were based partly on the defective geomorphological model of Wooldridge & Linton, and partly on some other features thought to indicate tropical soil development, which could not have occurred in Britain during even the most temperate parts of the Quaternary. But all such features are open to interpretations other than tropical weathering on erosion surfaces that truncate, and were formed long after deposition of, the Lower Tertiary sediments. For example, the sarsens, puddingstones and other forms of apparently secondary silica in Plateau Drifts need not be surface silcretes formed by later weathering of Reading Beds or other Lower Tertiary deposits. The secondary chalcedony reported by Loveday (1962) is in fact derived from silicified fossils in the Chalk (Brown *et al.* 1969), and sarsens showing root casts (Small 1980) could have formed by late diagenetic silicification at depth of estuarine or other non-marine Lower Tertiary sediments containing contemporary plant remains. Summerfield & Goudie (1980) quote the small titanium contents and grain-supported microfabrics of most sarsens and puddingstones as evidence for silica cementation below the weathering zone, but attribute a weathering profile origin to rare examples with greater titanium contents and grains that 'float' in a cryptocrystalline or opaline cement. Even if the latter are pedogenic silcretes, the possibility remains that they formed in the Palaeocene during periods of emergence between successive depositional phases of the non-marine Reading Beds. The occurrence of sarsen pebbles in the Blackheath Beds (Sherlock 1947) confirms that at least some of the sarsens had formed as early as this, and it is now recognized that silicification is more likely to have occurred in the uniformly warm, humid climate of the early Cenozoic than in the more variable and often quite arid conditions of the mid- and late-Cenozoic (Summerfield & Goudie 1980).

Mineralogical evidence for Tertiary weathering of Clay-with-flints (e.g. the presence of kaolinite) is open to similar reinterpretation. The Reading Beds, from which the Clay-with-flints was largely derived, probably contain some material from kaolinised granites in south-west England. Also, as they were deposited mainly in non-marine conditions, the Reading Beds probably contain much detritus derived without significant modification from Palaeocene soils, in which kaolinite may well have been formed by tropical weathering. It is most unlikely that any evidence of Tertiary tropical weathering can be safely used to support the existence of relict Tertiary land surfaces as significant geomorphological features anywhere in south-east England, though such surfaces may exist elsewhere in Britain or other parts of north-west Europe.

Conclusions

Both buried and unburied soils in south-east England can be divided into those originating principally in the Flandrian and those with evidence of more advanced interglacial development (palaeo-argillic soils). The Flandrian soils are developed on Flandrian and Devensian sediments, and also on many older deposits in areas that must have suffered extensive periglacial erosion during the earlier Devensian. The palaeo-argillic soils are restricted to pre-Devensian deposits occurring mainly on level plateaux and terraces, which survived most of the Devensian erosion but were commonly covered by thin layers of Late Devensian loess. Locally it is possible to distinguish Ipswichian palaeo-argillic soils from others that have redder colours and more complex microfabrics, and began development in the Cromerian or Hoxnian Stages.

In Chalk plateau soils, which have previously been attributed to Tertiary pedogenesis, the extent of profile development is no greater than in the more complex interglacial soils, and their stratigraphic relationships with the few undoubted remnants of Plio-Pleistocene marine deposits suggest formation in the Quaternary after the sea had receded. Until more direct evidence is available of pre-Quaternary pedogenesis in south-east England (e.g. from buried soils within the Tertiary sedimentary successions), it is reasonable to attribute all features of unburied soils in the area to Quaternary events. This implies that no part of the land surface, even on upland areas south of all

glacial limits, is inherited without modification form the Tertiary. However, the erosion needed to remove evidence of pre-Quaternary soil development was not necessarily very great, so the absence of Tertiary pedogenic features does not conflict with the conclusion of Jones (1980) that 'Pleistocene denudation was mainly concerned with accentuating pre-existing landform patterns'.

References

AVERY, B. W. 1964. The soils and land use of the district around Aylesbury and Hemel Hempstead. *Mem. Soil Surv. GB* HMSO, London.

—— 1980. Soil classification for England and Wales (Higher Categories). *Soil Surv. Tech. Monogr.* 14.

BROWN, G., CATT, J. A., HOLLYER, S. E. & OLLIER, C. D. 1969. Partial silicification of chalk fossils from the Chilterns. *Geol. Mag.* 106, 583–6.

BULLOCK, P. & MURPHY, C. P. 1979. Evolution of a Paleo-argillic brown earth (Paleudalf) from Oxfordshire, England. *Geoderma*, 22, 225–52.

CATT, J. A. 1978. The contribution of loess to soils in lowland Britain. *In:* LIMBREY, S. & EVANS, J. G. (eds) *The Effect of Man on the Landscape: the lowland zone. Council Brit. Archaeol. Res. Rep.* 21, 12–20.

—— & HODGSON, J. M. 1976. Soils and geomorphology of the Chalk in south-east England. *Earth Surf. Processes*, 1, 181–93.

CHARTRES, C. J. 1980. A Quaternary soil sequence in the Kennet valley, central southern England. *Geoderma*, 23, 125–46.

COOPE, G. R. & PENNINGTON, W. 1977. The Windermere Interstadial of the Late Devensian. *Phil. Trans. R. Soc. B*, 280, 337–9.

DALRYMPLE, J. B. 1957. The Pleistocene deposits of Penfold's Pit, Slindon, Sussex, and their chronology. *Proc. geol. Ass. London*, 68, 294–303.

DINES, H. G. & CHATWIN, C. P. 1930. Pliocene sandstone from Rothamsted (Hertfordshire). *Summ. Prog. geol. Surv. GB 1929*, 1–7.

EVANS, J. G. 1967. Late-glacial and post-glacial subaerial deposits at Pitstone, Buckinghamshire. *Proc. geol. Ass. London*, 77, 347–64.

FORDHAM, S. J. & GREEN, R. D. 1980. Soils of Kent. *Bull. Soil Surv.* 9.

GIBBARD, P. L. 1977. Pleistocene history of the Vale of St. Albans. *Phil. Trans. R. Soc. B*, 280, 445–83.

GILBERT, C. J. 1920. On the occurrence of extensive deposits of high-level sands and gravels resting upon the Chalk at Little Heath, near Berkhamsted. *Q. Jl geol. Soc. London*, 75, 32–43.

GREEN, R. D. 1968. Soils of Romney Marsh. *Bull. Soil Surv. GB* 4.

—— & FORDHAM, S. J. 1973. Soils in Kent I Sheet TRO4 (Ashford). *Soil Surv. Record 14*.

HEY, R. W. 1980. Equivalents of the Westland Green Gravels in Essex and East Anglia. *Proc. geol. Ass. London*, 91, 279–90.

—— & BRENCHLEY, P. J. 1977. Volcanic pebbles from Pleistocene gravels in Norfolk and Essex. *Geol. Mag.* 114, 219–25.

HODGSON, J. M. 1967. Soils of the West Sussex Coastal Plain. *Bull. Soil Surv. GB* 3.

JOHN, D. T. 1980. The soils and superficial deposits on the North Downs of Surrey. *In:* JONES, D. K. C. (ed.) *The Shaping of Southern England. Spec. Publ. Inst. Brt. Geogr.* 11,. 101–30.

JONES, D. K. C. 1974. The influence of the Calabrian transgression on the drainage evolution of south-east England. *In:* BROWN, E. H. & WATERS, R. S. (eds) *Progress in Geomorphology. Spec. Publ. Inst. Brt. Geogr.* 7, 139–57.

—— 1980. The Tertiary evolution of south-east England with particular reference to the Weald. *In:* JONES, D. K. C. (ed.) *The Shaping of Southern England. Spec. Publ. Inst. Brt. Geogr.* 11, 13–47.

JUKES-BROWNE, A. J. 1906. The clay-with-flints; its origin and distribution. *Q. Jl geol. Soc. London*, 62, 132–64.

KERNEY, M. P. 1963. Late-glacial deposits on the chalk of south-east England. *Phil. Trans. R. Soc. B*, 246, 203–54.

—— 1965. Weichselian deposits in the Isle of Thanet, east Kent. *Proc. geol. Ass. London*, 76, 269–74.

LOVEDAY, J. 1962. Plateau deposits of the southern Chiltern Hills. *Proc. geol. Ass. London*, 73, 83–102.

ROSE, J. & ALLEN, P. 1977. Middle Pleistocene stratigraphy in south-east Suffolk. *J. geol. Soc. Lond.* 133, 83–102.

——, TURNER, C., COOPE, G. R. & BRYAN, M. D. 1980. Channel changes in a lowland river catchment over the last 13000 years. *In:* CULLINGFORD, R. A., DAVIDSON, D. A. & LEWIN, J. (eds). *Timescales in Geomorphology*, 159–75. Wiley, London.

SCHWERTMANN, U. & TAYLOR, R. M. 1977. Iron oxides. *In:* DIXON, J. B. & WEED, S. B. (eds) *Minerals in Soil Environments*, 145–80. Soil Sci. Soc. Am., Madison, U.S.A.

SEALE, R. S. 1975. Soils of the Ely district. *Mem. Soil Surv. G.B.* Harpenden.

SHERLOCK, R. L. 1947. *British Regional Geology. London and Thames Valley.* HMSO, London.

SHOTTON, F. W., GOUDIE, A. S., BRIGGS, D. J. & OSMASTON, H. A. 1980. Cromerian interglacial deposits at Sugworth, near Oxford, England, and their relation to the Plateau Drift of the Cotswolds and the terrace sequence of the Upper and

Middle Thames. *Phil. Trans. R. Soc. B*, **289**, 55–86.

SMALL, R. J. 1980. The Tertiary geomorphological evolution of south-east England: an alternative interpretation. *In:* JONES, D. K. C. (ed.) *The Shaping of Southern England. Spec. Publ. Inst. Brt. Geogr.* **11**, 49–70.

SOIL SURVEY STAFF, 1975. *Soil Taxonomy: a basic system of soil classification for making and interpreting soil surveys.* Agricultural Handbook 436, U.S. Department of Agriculture, Government Printing Office, Washington, DC.

STRAW, A. 1980. The age and geomorphological context of a Norfolk palaeosol. *In:* CULLINGFORD, R. A., DAVIDSON, D. A. & LEWIN, J. (eds) *Timescales in Geomorphology*, 305–15. Wiley, London.

STURDY, R. G., ALLEN, R. H., BULLOCK, P., CATT, J.

A. & GREENFIELD, S. 1979. Paleosols developed on Chalky Boulder Clay in Essex. *J. Soil Sci.* **30**, 117–37.

SUMMERFIELD, M. A. & GOUDIE, A. S. 1980. The sarsens of southern England: their palaeoenvironmental interpretation with reference to other silcretes. *In:* JONES, D. K. C. (ed.). *The Shaping of Southern England. Spec. Publ. Inst. Brt. Geogr.* **11**, 71–100.

WEST, R. G. 1980. *The Pre-glacial Pleistocene of the Norfolk and Suffolk Coasts.* Cambridge University Press.

WOOLDRIDGE, S. W. & LINTON, D. L. 1955. *Structure, Surface and Drainage in South-east England.* Philip, London.

WORSSAM, B. C. 1963. Geology of the country around Maidstone. *Mem. geol. Surv. GB.* HMSO, London.

JOHN A. CATT, Soils and Plant Nutrition Department, Rothamsted Experimental Station, Harpenden, Herts AL5 2JQ.